Indian Institute of Metals Series

About the Book Series:

The study of metallurgy and materials science is vital for developing advanced materials for diverse applications. In the last decade, the progress in this field has been rapid and extensive, giving us a new array of materials, with a wide range of applications, and a variety of possibilities for design of new materials, processing and characterizing the materials. In order to make this growing volume of knowledge available, an initiative to publish a series of books in Metallurgy and Materials Science was taken during the Diamond Jubilee year of the Indian Institute of Metals (IIM) in the year 2006, and has been published in partnership with Springer since 2016.

This book series publishes different categories of publications: textbooks to satisfy the requirements of students and beginners in the field, monographs on select topics by experts in the field, professional books to cater to the needs of practising engineers, and proceedings of select international conferences organized by IIM after mandatory peer review. The series publishes across all areas of materials sciences and metallurgy. An panel of eminent international and national experts serves as the advisory body in overseeing the selection of topics, important areas to be covered, and the selection of contributing authors.

R. Divakar · S. V. S. Narayana Murty · S. Srikanth ·
Amol A. Gokhale
Editors

Indian Metallurgy

The Platinum Years

IIM
Metallurgy
Materials Engineering

Springer

Editors
R. Divakar
Metallurgy and Materials Group
Indira Gandhi Centre for Atomic Research
Kalpakkam, Tamil Nadu, India

S. V. S. Narayana Murty
Materials and Manufacturing Entity
Liquid Propulsion Systems Centre
Trivandrum, India

S. Srikanth
National Metallurgical Laboratory CSIR
Chennai, Tamil Nadu, India

Amol A. Gokhale
Department of Mechanical Engineering
Indian Institute of Technology Bombay
Mumbai, Maharashtra, India

ISSN 2509-6400 ISSN 2509-6419 (electronic)
Indian Institute of Metals Series
ISBN 978-981-99-5059-1 ISBN 978-981-99-5060-7 (eBook)
https://doi.org/10.1007/978-981-99-5060-7

This Springer imprint is published by the registered company Springer Nature Singapore Pte Ltd.
The registered company address is: 152 Beach Road, #21-01/04 Gateway East, Singapore 189721, Singapore

Paper in this product is recyclable.

Series Editor's Preface

The *Indian Institute of Metals Series* is an institutional partnership series focusing on metallurgy and materials science and engineering.

About the Indian Institute of Metals

The Indian Institute of Metals (IIM) is a premier professional body (since 1947) representing an eminent and dynamic group of metallurgists and materials scientists and engineers from R&D institutions, academia, and industry, mostly from India. It is a registered professional institute with the primary objective of promoting and advancing the study and practice of the science and technology of metals, alloys, and novel materials. The institute is actively engaged in promoting academia–research and institute–industry interactions.

Genesis and History of the Series

The study of metallurgy and materials science and engineering is vital for developing advanced materials for diverse applications. In the last decade, progress in this field has been rapid and extensive, giving us a new array of materials, with a wide range of applications and a variety of possibilities for processing and characterizing the materials. In order to make this growing volume of knowledge available, an initiative to publish a series of books in metallurgy and materials science and engineering was taken during the Diamond Jubilee year of the Indian Institute of Metals (IIM) in the year 2006. IIM entered into a partnership with Universities Press, Hyderabad, in 2006, and from 2016 the book series is under MoU with M/s Springer Nature, and as part of the IIM book series, a total of 22 books were published till 2022. The books were authored by eminent professionals in academia, industry, and R&D with outstanding background in their respective domains, thus generating unique resources of validated expertise of interest in metallurgy. The international character of the authors' and editors has enabled the books to command national and global readership. This book series includes different categories of publications: textbooks to satisfy the requirements of undergraduates and beginners in the field, monographs on select topics by experts in the field, and proceedings of select international conferences organized by IIM, after mandatory peer review. An eminent panel of international and national experts constitutes the advisory body in overseeing the selection of topics, important areas to be covered, in the books and the selection of contributing authors.

About "Indian Metallurgy: The Platinum Years"

The Book on "Indian Metallurgy: The Platinum Years" with R. Divakar, S. V. S. Narayana Murty, S. Srikanth, and Amol A. Gokhale as editors lucidly narrates the country's growth story in mining and metallurgy since independence. The book discusses about the developments made by the industries, the innovations and inventions made by the R&D Laboratories and the human resources developed by the academic institutes over the last 75 years. The selected thirty-two articles from the industry, R&D laboratories and academic institutes, along with the perspectives of some of the eminent metallurgists, highlight the achievements made in the country over the years. These articles have been classified into four categories, viz, Industry, R&D Organization, Institute and Individuals.

All the thirty-two articles have been well presented by well-known experts in the field from India, and finely edited and delivered by the editors, R. Divakar, S. V. S. Narayana Murty, S. Srikanth, and Amol A. Gokhale. The editors have made excellent efforts to coordinate with professionals with rich experience in the respective filed to provide the articles of high relevance. This book will be a rich treasure for those who are interested in learning the historical perspectives of Indian metallurgy and materials development, in addition to the role of institutes and organisations involved in various metallurgical achievements in the last 75 years. The IIM-Springer Series gratefully national and global readership. This book series includes different categories of publications: textbooks to satisfy the requirements of undergraduates and beginners in the field, monographs on select topics by experts in the field, and proceedings of select international conferences organized by IIM, after mandatory peer review. An eminent panel of international and national experts constitutes the advisory body in overseeing the selection of topics, important areas to be covered, in the books and the selection of contributing authors.

Dr. U. Kamachi Mudali
Editor-in-Chief, Series in
Metallurgy and Materials Engineering,
Vice Chancellor, DAE
Homi Bhaba National Institute
Mumbai, India

Preface

It is the 75th year of Indian Independence as the country celebrates the Azaadi ka Amrit Mahotsav. The country has come a long way since then, from being an impoverished nation with a GDP of less than Rs. 3 Lakh Crores to being the fifth largest economy in the world with a GDP of close to Rs. 37 Lakh Crores (~2.85 trillion dollars) and contributing to about 3.28% of the world's GDP. The phenomenal growth of the country has been propelled by services, manufacturing industry, agriculture and consumption. The Mining & Metals sector has played a significant role in shaping this growth story accounting for about 2.5% of the GDP and is expected to increase to more than 8% by the end of this decade. India today produces 95 minerals including 3 atomic minerals. Only four steel companies (Bengal Iron & Steel Company, Tata Iron & Steel Company, Indian Iron & Steel Company & Mysore State Iron and Steel Works later renamed VISL) and Indian Aluminium Co. existed in India at the time of independence. The country that was producing only meagre amounts of pig iron (17 Lakh tonnes per annum), aluminium (25,000 tonnes per annum) and minor amounts of gold today is the 2nd largest producer of crude steel (144 million tonnes per annum), 4th largest producer of aluminium (4,032,000 tpa), lead (191,000 tpa), zinc (776,000 tpa), 6th largest producer of titanium (500 tpa) and 11th in uranium production (400 tpa). The country also produces significant amounts of copper (525,000 tpa), silver and a range of ferroalloys. From being a non-existent player in renewable energy, space technologies and atomic energy, today we rank 3rd in renewable energy (4th in solar energy with more than 50 GWe installed capacity), 5th among international space agencies and 12th in atomic energy (22 reactors with close to 6780 MW of installed capacity). The strategic sector in the country has created weapon systems including intercontinental ballistic missiles such as Agni, anti-satellite and anti-missile missiles, fighter aircrafts and aircraft carriers, satellites for communications, remote sensing global positioning systems, satellite launch vehicles (PSLV, GSLV and LMV3), had an Indian in space, orbited the Moon and Mars, and are a technological force to reckon with internationally. In all of these national missions, development of strategic and critical materials has played a key role in the ultimate success of the mission. A country that had no R&D & technology development base at the time of independence can boast of a string of materials-related R&D laboratories in defence (DRDO established in 1958), space (ISRO established in 1969), atomic energy (DAE established in 1954), Council of Scientific & Industrial Research labs (established in 1942), Department of Science & Technology labs (established in 1971), R&D Centres for Iron & Steel, Non-ferrous research centres and so on.

On the academic sphere, before independence, the country had only two colleges with graduation in metallurgical engineering (Banaras Hindu University and Bengal Engineering College) producing less than 30 graduates annually. There was no post-graduate or Ph.D. programme in Metallurgical Engineering in the country. Today there are around 60 colleges offering graduate or diploma programme in metallurgical engineering and about 26 of them have Ph.D. programs in metallurgy. As many as 4000–5000 students graduate in metallurgical engineering and materials science annually and more than 50 candidates secure their doctorate in metallurgy and materials science every year. In Materials Science and Technology Research, India ranks sixth with more than 12,500 SCI publications after China, US, Japan, Germany and South Korea.

However, for a country whose civilization is at least 10,000 years old, it does not do justice to mention only the contemporary progress. India has a rich tradition and legacy of metallurgy. There is archaeological evidence to suggest that iron-making in India could date back to 1800 BC. It is reported that Indian and the Persian army used arrows tipped with iron and ancient Romans used armour and cutlery made of Indian iron. Crucible steel (named wootz after the Tamil word Urukku and also the Kannada and Telugu word Ukku) was first developed in India around 300 BC. It is recorded that Indian Wootz steel was exported to the Middle East where it was used around 1000 CE to produce the legendary Damascus steel. The iron pillar at Delhi, believed to belong to the fifth century AD, stands testimony to the old iron-making heritage in India. Zinc was extracted in India as early as in the fourth to third century BCE and is the earliest known civilization to have produced zinc on a large scale. Copper (bronze) swords discovered from the Harappan site date back to 2300 BCE. Archaeological evidence of the Harappan site suggests that lead was also used in ancient India. Lead was used for coinage in South India during the second or third century BC (Sangam period). Lead oxide was also used for glazing pottery. Investment casting or the Lost Wax Process that today is used to cast complex aero-engine components was invented in India. There were ancient silver mines in northwest India. The Coinage of India dates back to first to the sixth century BCE and consisted mainly of copper and silver coins. It is reported that gold coins were first issued by Indo-Greeks in India at around 270 BC. India's heritage in Knowledge and Learning of metallurgy is also rich and goes back to the Vedic age. The Rigveda as well as Yajur Veda uses the Sanskrit term *Ayas* for metal. The Atharva Veda mentions the procedure of production of lead shots or granules. The Arthashastra describes a hierarchic structure for mining and metal extraction and the roles of the Director of Metals and the Director of Mining. Several ancient Sanskrit texts from ninth to twelfth century AD (*Brihad-vimana-shastram, Rasendra-sara-sangrahah, Shilpa-ratnam, and Rasa-ratna-samucchaya*) compile procedures for metal work, such as the building of furnaces, making of bellows, producing metal powders or binders or glue. There are as many as 44 ancient texts that describe the process of Indian metallurgy, the most popular one being the *Rasaratna Samucchaya*, which describes several aspects of metals extraction including the furnace (*kosthi yantra*), vessels for containing chemicals (*tiryak patana yantra*), the distillation pot (*dheki yantram*) etc. Takshashila University existed as early as the fifth century BC and Nalanda University established in the fifth century AD was internationally acclaimed.

The year 2021 also marks the 75th year of the establishment of the Indian Institute of Metals, which was inaugurated by Dr. Shyama Prasad Mukherjee, the then Minister for Industry and Supply on 29 December 1947 at the Royal Asiatic Society Hall, Calcutta with Sir Jahangir J. Gandhy as the first President and Dr. Dara P. Antia as the first Honorary Secretary of the Institute. The Indian Institute of Metals has also grown from operating in a small room with 42 members to having an elegant Metal House at Salt Lake, Kolkata with more than 12000 members today.

It was felt that on this occasion of the Azaadi ka Amrit Mahotsav and the Platinum Jubilee of the Indian Institute of Metals, the country's growth story in mining and metallurgy since independence must be chronicled. It must include the developments made by the industries, the innovations and inventions made by the Research & Development Laboratories and the human resources developed by the academic institutes which helped to realize these achievements. This book, consisting of a compilation of articles from the industry, R&D laboratories and academic institutes, is an outcome of this. The perspectives of some of the senior metallurgists have also been included.

We invited more than 60 industries (manufacturing, mining, ferrous, non-ferrous and metal working industries), R&D labs (Defence, Atomic Energy, CSIR, industrial R&D labs) and academic institutes (IISc, IIT's, NITs, College of Engineering Pune, PSG College of Technology, IIEST—formerly Bengal Engineering College and Jadavpur University) to contribute to this volume. Thirty-two articles (9 from industry, 10 from R&D labs, 8 from academic institutes and 5 from prominent individuals) were finally received and have been included in this volume. The articles not only highlight the achievements made in these 75 years but also the challenges faced and a vision for the future. We hope that the information compiled in this volume will have a long-lasting archival value.

Kalpakkam, India — R. Divakar
Trivandrum, India — S. V. S. Narayana Murty
Chennai, India — S. Srikanth
Mumbai, India — Amol A. Gokhale

Growth of IIM and Metallurgy in India During the Last 75 Years

History of growth of Indian Institute of Metals (IIM) and metallurgical industries as well as metallurgical education and materials research started soon after independence of India in 1947 followed by the 5-Year Plans undertaken by the Government of India. The first 5-Year Plan was dedicated to agriculture and irrigation. The second 5-Year Plan focussed on industrial development when metallurgical education and research, setting up of Indian Institutes of Technology (IITs), creation of Council of Scientific and Industrial Research (CSIR) laboratories such as National Metallurgical Laboratory (NML), Defence Metallurgical Research Laboratory (DMRL), Bhabha Atomic Research Centre (BARC), Indian Space Research Organization (ISRO) etc. were the major initiatives.

Hindustan Steel Limited was created to set up Durgapur, Rourkela, Bhilai, Bokaro and later Vizag steel plants. Hindustan Copper Limited and Hindustan Zinc Limited were nationalized. Meanwhile, privative companies like Tata Steel modernized their facilities and expanded production capacity at Jamshedpur and recently at Kalinganagar also. Indian Aluminium Company, a subsidiary of ALCAN Aluminium of Canada expanded capacities through brownfield and greenfield expansions. Government of India set up BALCO in Madhya Pradesh and NALCO in Odisha. Birla Group set up HINDALCO for production of Aluminium and downstream products in Renukoot, UP. Later ESSAR, now taken over by Arcelor Mittal and JSW/JSPL also added substantial steel-making capacities. Currently, the production capacity of steel in India is about 120 million tonnes, production capacity of aluminium is around 4 million tonnes, production capacity of copper at 1 million tonnes and production capacity of zinc is 0.8 million tonnes and production capacity of primary lead is 250,000 tonnes while recycling capacity of secondary lead has grown to over 0.8 million tonnes.

Unfortunately, Indian metals industry remained focused largely on domestic market with very little interest in international markets. Economic liberalization brought in by Government of India in 1991 made a sea change in the outlook of Indian entrepreneurs and private sector. Brown field modernization/expansion, greenfield expansion and takeover of some overseas companies and inefficient medium and small-scale industries became the norm for augmenting national capacity. Meanwhile, China outperformed India in all areas of metal production as they focused on global markets with least cost of production in the world. India continued to remain both as a technology buyer as well as equipment buyer from overseas countries like South Korea, China, Germany and France. Even the engineering components for macro and micro industries are being imported from China. India, once a major producer of engineering goods gradually opted for computer and software sectors ignoring hardware. Heavy engineering, automobiles and electronic industry continued to be assembly lines with foreign supply of mechanical, auto components as well as electronic components. There was of course no focus on manufacture of silicon wafers for solar panels.

In the field of metallurgical education and research, BARC, DMRL and ISRO have achieved excellent progress but our academic institutions and CSIR laboratories have not made any significant path–breaking R&D achievement. Indian research on nanotechnology has done well but this knowledge is mostly being used by European and American companies for the global market. India has been largely providing scientific and engineering manpower to many developed economies and importing nanoproducts and components from overseas countries.

During the same period, IIM grew from an organization housed in a hired accommodation in 1945 to its own premises in Kolkata. Membership grew to over 12,000 in the late 90s and the number of chapters also grew to around 60. Amongst the major chapters, Delhi Chapter has made significant contribution to IIM both technically and financially. Later this year, we shall organize Platinum Jubilee Celebrations of IIM in Delhi under the Presidentship of Mr. T. V. Narendran. During the last year, in spite of Covid-19, IIM could organize many programme in online mode including its flagship event NMD-ATM2020. We sincerely hope that 2021 will be a turnaround year for the world including India. ATMANIRBHAR BHARAT, an initiative taken by the Government of India will hopefully bring in a new kind of indigenous developments, focused on global markets. The world is eagerly looking towards India as a reliable alternative to China.

India being a mineral-rich Country, our focus, for the next 50 years, should be on sound metallurgical education, research and a very large number of metallurgical industries, both large as well as MSME units.

I also feel very strongly that the importance of IIM in India will continue to be relevant as it is in many other mineral-rich countries like China, Russia, South Korea, Brazil, South Africa and Australia. IIM should also become the largest family of the metallurgical fraternity spanning industry, research and teaching.

R. N. Parbat
Former President of Indian Institute
of Metals

Current Series Information

To increase the readership and to ensure wide dissemination among global readers, this new chapter of the series has been initiated with Springer in the year 2016. The goal is to continue publishing high-value content on metallurgy and materials science and engineering, focusing on current trends and applications. So far, eleven important books on state of the art in metallurgy and materials science and engineering have been published and three books were released during IIM-ATM 2022 at Hyderabad. Readers who are interested in writing books for the Series may contact the Series Editor-in-Chief, Dr. U. Kamachi Mudali, Former President of IIM and Vice Chancellor of Homi Bhabha National Institute (HBNI), DAE, Mumbai at ukmudali1@gmail.com, vicechancellor@hbni.ac.in or the Springer Editorial Director, Ms. Swati Meherishi at swati.meherishi@springer.com

Contents

Editors and Contributors

About the Editors

Dr. R. Divakar joined the Indira Gandhi Centre for Atomic Research (IGCAR) at Kalpakkam in 1989 with a Bachelor's degree in Metallurgy and subsequently earned Ph.D. degree in 2001, both from Indian Institute of Science. He specialized in the applications of electron microscopy to physical metallurgical studies and established electron optical techniques for post-irradiation examination at IGCAR. During 2007–09, he worked as a post-doctoral associate at the University of Minnesota, USA. Dr. Divakar presently leads metallurgy and materials science activities at IGCAR on projects for the development of new materials and related fabrication and inspection technologies in support of fast reactor technology and Indian Advanced Supercritical power plants, and accelerator-based materials science studies. He is a Fellow of Electron Microscope Society of India (2016) and is a recipient of Metallurgist of the Year Award in Metal Science Category from the Ministry of Steel, Government of India (2018). Dr. Divakar also contributes to academic programmes of Homi Bhabha National Institute and is a Life Member of Indian Institute of Metals.

S. V. S. Narayana Murty has received his B.E. from Andhra University, Visakhapatnam, M.E. from IISc Bangalore, and Ph.D. from IIT Bombay (2003), specializing in metallurgical engineering. He joined the Vikram Sarabhai Space Centre, Trivandrum, in 1993. He was a post-doctoral fellow at the National Institute for Materials Science, Japan (2003–2006). Presently, he works as General Manager of Material Development and Production Group at Liquid Propulsion Systems Centre, Trivandrum. He is a recipient of Young Metallurgist Award (2001) and Metallurgist of the Year Award (2018), both from Ministry of Steel, Government of India. He was a member of ISRO Team Excellence Award for 'Indigenization of Copper alloy Cu-Cr-Ti-Zr thrust chamber of cryogenic and semi-cryogenic engines' (2012). He received the AIDA Technology Innovation Award by Japan Society for Technology of Plasticity (2009) for development of ultrafine grained steels. He was awarded the G.D. Birla Gold Medal (2022) by the Indian Institute of Metals.

Dr. S. Srikanth is a highly accomplished scientist and engineer with a distinguished career in the field of metallurgy. Dr. Srikanth obtained his B.Sc. (Engg.) in Metallurgy from Regional Engineering College, Rourkela, India, in 1982 and continued his education at the prestigious IISc Bangalore, where he earned his M.E. in Metallurgy in 1984 and completed his Ph.D. in Metallurgy in 1990, on the thermodynamics of alloys. Subsequently, he embarked on a post-doctoral fellowship at McMaster University in Ontario, Canada, where he explored the field of material science and engineering and further worked as Visiting Scientist at the University of Leeds, UK, and the Institute of Physics, Czech Academy of Sciences in Prague. In 1990, Dr. Srikanth joined the National Metallurgical Laboratory (NML) in Jamshedpur, India, as Scientist C, rising to the position of Director, CSIR-NML, by 2010. Over the years, he made significant contributions to the field of metallurgy, specializing in thermodynamics and kinetic modelling of metallurgical processes.

Dr. Amol A. Gokhale completed his B.Tech. in Metallurgical Engineering from IIT Bombay in 1978 and M.S. and Ph.D. in Metallurgical Engineering from University of Pittsburgh, USA, in 1980 and 1985, respectively. He served in the Defence Metallurgical Research Laboratory (DMRL), Hyderabad (DRDO), from 1985 to 2015 in various capacities, such as Head of the Light Alloy Casting Group, Solidification Technology Division, Extraction Technology Division and finally as Distinguished Scientist and Director of the Laboratory. At DMRL, he led research on aluminium lithium alloys for aerospace applications, aluminium alloy components for torpedoes, crash-resistant aluminium foams, additive manufacturing of Ni-based superalloys and very high-temperature materials for hypersonic vehicles. He steered a few major technology transfer initiatives in critical materials on behalf of the Laboratory. He was a director on the board of the Mishra Dhatu Nigam Ltd., Hyderabad.

Contributors

C. Andrew Electroplating and Metal Finishing Division, CSIR Central Electrochemical Research Institute, Karaikudi, Tamil Nadu, India

P. K. Banerjee Hindalco Industries Ltd., Worli, Mumbai, Maharashtra, India

Satadipa Banerjee Department of Metallurgical and Materials Engineering, National Institute of Technology, Durgapur 713209, India

A. Basu National Institute of Technology Rourkela, Department of Metallurgical and Materials Engineering, Rourkela, Odisha, India

Debashish Bhattacharjee Tata Steel Limited, Kolkata, India

Sajal Kumar Bhattacherjee MECON Limited, Ranchi, Jharkhand, India

Uday Chakkingal Department of Metallurgical & Materials Engineering, Indian Institute of Technology-Madras (IIT Madras), Chennai, India

Naveen Chandrasekaran Electroplating and Metal Finishing Division, CSIR Central Electrochemical Research Institute, Karaikudi, Tamil Nadu, India

K. Chattopadhyay Department of Materials Engineering, Indian Institute of Science, Bangalore, India

I. Chattoraj CSIR-National Metallurgical Laboratory, Jamshedpur, India

Paritosh Chaudhari Institute for Plasma Research, Near Bhat, Gandhinagar, Gujarat, India;
Homi Bhabha National Institute, Training School Complex, Anushakti Nagar, Mumbai, India

Roy M. Cherian Vikram Sarabhai Space Centre, Trivandrum, India

P. P. Deshpande Department of Metallurgy and Materials Engineering, COEP Technological University (COEP Tech), (A Unitary Public University of Government of Maharashtra), Formerly College of Engineering Pune, Pune, India

James Ebenezer Electroplating and Metal Finishing Division, CSIR Central Electrochemical Research Institute, Karaikudi, Tamil Nadu, India

B. P. Gautham Gerald Tennyson, Sreedhar Reddy and Pradip, TCS Research, Tata Consultancy Services, Pune, India

Vinayak P. Gawade Walchandnagar Industries Limited, Walchandnagar, Indapur, Pune, India

Amol S. Gengaje Walchandnagar Industries Limited, Walchandnagar, Indapur, Pune, India

K. S. Ghosh Department of Metallurgical and Materials Engineering, National Institute of Technology, Durgapur 713209, India

M. M. Ghosh Department of Metallurgical and Materials Engineering, National Institute of Technology, Durgapur 713209, India

R. Gopalan International Advanced Research Center for Powder Metallurgy and New Materials (ARCI), Hyderabad, India

Sumit S. Gorde Walchandnagar Industries Limited, Walchandnagar, Indapur, Pune, India

Amit Gupta Aditya Birla Science and Technology Company Pvt Ltd., MIDC Taloja, Panvel, Maharashtra, India

Nirav Jamnapara Institute for Plasma Research, Near Bhat, Gandhinagar, Gujarat, India;
Homi Bhabha National Institute, Training School Complex, Anushakti Nagar, Mumbai, India

M. Jayakumar Electroplating and Metal Finishing Division, CSIR Central Electrochemical Research Institute, Karaikudi, Tamil Nadu, India

Bijesh Jha Hindalco Industries Ltd., Worli, Mumbai, Maharashtra, India

L. John Berchmans Electroplating and Metal Finishing Division, CSIR Central Electrochemical Research Institute, Karaikudi, Tamil Nadu, India

Roy Johnson International Advanced Research Center for Powder Metallurgy and New Materials (ARCI), Hyderabad, India

Alphonsa Joseph Institute for Plasma Research, Near Bhat, Gandhinagar, Gujarat, India;
Homi Bhabha National Institute, Training School Complex, Anushakti Nagar, Mumbai, India

Vivekanand Kain Bhabha Atomic Research Centre, Trombay, Mumbai, India

Samir Khirwadkar Institute for Plasma Research, Near Bhat, Gandhinagar, Gujarat, India;
Homi Bhabha National Institute, Training School Complex, Anushakti Nagar, Mumbai, India

J. Krishnamoorthi Department of Metallurgical Engineering, PSG College of Technology, Coimbatore, TN, India

Hrishikesh Kumar Department of Metallurgical and Materials Engineering, National Institute of Technology, Durgapur 713209, India

P. Kumar Department of Materials Engineering, Indian Institute of Science, Bengaluru, India

Ravi Kumar Department of Metallurgical & Materials Engineering, Indian Institute of Technology-Madras (IIT Madras), Chennai, India

G. Madhusudhan Reddy Former Director, Defence Metallurgical Research Laboratory, Hyderabad, India

Sanjib Majumdar Bhabha Atomic Research Centre, Trombay, Mumbai, India

Archana Mallik National Institute of Technology Rourkela, Department of Metallurgical and Materials Engineering, Rourkela, Odisha, India

Anuttam Mishra Indian Rare Earths Limited, IREL (India) Limited, ECIL Building, Mumbai, India

Sanak Mishra Formerly with Research & Development Centre for Iron and Steel, Steel Authority of India Limited, Ranchi, Jharkhand, India

M. Mohan Vikram Sarabhai Space Centre, Trivandrum, India

Sunil Mohan Department of Metallurgical Engineering, Indian Institute of Technology (Banaras Hindu University), Varanasi, India

N. Mohanapriya Electroplating and Metal Finishing Division, CSIR Central Electrochemical Research Institute, Karaikudi, Tamil Nadu, India

Ajit Kumar Mohanty Bhabha Atomic Research Centre, Trombay, Mumbai, India

Debasis Mukherjee Formerly with Research & Development Centre for Iron and Steel, Steel Authority of India Limited, Ranchi, Jharkhand, India

N. K. Mukhopadhyay Department of Metallurgical Engineering, Indian Institute of Technology (Banaras Hindu University), Varanasi, India

Sunil Nair Mukand Ltd., Thane Belapur Road, Dighe, Kalwe, Thane, India

K. Narayan Prabhu Department of Metallurgical and Materials Engineering, National Institute of Technology Karnataka, Mangalore, India

Sudhir K. Nema Institute for Plasma Research, Near Bhat, Gandhinagar, Gujarat, India;
Homi Bhabha National Institute, Training School Complex, Anushakti Nagar, Mumbai, India

Sagar Pandit Hindalco Industries Ltd., Worli, Mumbai, Maharashtra, India

Abhijit Pati Bharat Aluminium Company Limited, New Delhi, India

Upendra Prasad Institute for Plasma Research, Near Bhat, Gandhinagar, Gujarat, India;
Homi Bhabha National Institute, Training School Complex, Anushakti Nagar, Mumbai, India

R. S. Prasannakumar Electroplating and Metal Finishing Division, CSIR Central Electrochemical Research Institute, Karaikudi, Tamil Nadu, India

L. Pugazhenthy Executive Director, India Lead Zinc Development Association, New Delhi, India;
Past President, The Indian Institute of Metals (2008-09), New Delhi, India

Kumud Ranjan MECON Limited, Ranchi, Jharkhand, India

R. Ranjan Department of Materials Engineering, Indian Institute of Science, Bengaluru, India

D. Srinivasa Rao International Advanced Research Center for Powder Metallurgy and New Materials (ARCI), Hyderabad, India

Tata Narasinga Rao International Advanced Research Center for Powder Metallurgy and New Materials (ARCI), Hyderabad, India

Vijay D. Raskar Walchandnagar Industries Limited, Walchandnagar, Indapur, Pune, India

Kalyan Kumar Ray Department of Metallurgical and Materials Engineering, Indian Institute of Technology Kharagpur, Kharagpur, India

Ranjit Kumar Ray Formerly, Professor of Materials & Metallurgical Engineering IIT, Kanpur, India

S. Sankaran Department of Metallurgical & Materials Engineering, Indian Institute of Technology-Madras (IIT Madras), Chennai, India

B. Nageswara Sarma Department of Metallurgical Engineering, Indian Institute of Technology (Banaras Hindu University), Varanasi, India

Aniruddha S. Sawargaonkar Walchandnagar Industries Limited, Walchandnagar, Indapur, Pune, India

Lokendra Raj Singh Vijayanagar Works, JSW Steel Ltd., Dist - Ballari, Karnataka, India

S. C. Sharma Vikram Sarabhai Space Centre, Trivandrum, India

Ankit Singhania Tata Steel Limited, Kolkata, India

Kaushal Kumar Sinha M. N. Dastur and Company (P) Ltd, Kolkata, India

Amit Sircar Institute for Plasma Research, Near Bhat, Gandhinagar, Gujarat, India; Homi Bhabha National Institute, Training School Complex, Anushakti Nagar, Mumbai, India

S. Somanath Vikram Sarabhai Space Centre, Trivandrum, India

Komal Kapoor Nuclear Fuel Complex, Hyderabad, India

Vivek Srivastava Hindalco Industries Ltd., Worli, Mumbai, Maharashtra, India

B. Subramanian Electroplating and Metal Finishing Division, CSIR Central Electrochemical Research Institute, Karaikudi, Tamil Nadu, India

S. Subramanian Department of Materials Engineering, Indian Institute of Science, Bengaluru, India

S. Suwas Department of Materials Engineering, Indian Institute of Science, Bengaluru, India

Mita Tarafder CSIR-National Metallurgical Laboratory, Jamshedpur, India

Raghvendra Tewari Bhabha Atomic Research Centre, Trombay, Mumbai, India

R. Vijay International Advanced Research Center for Powder Metallurgy and New Materials (ARCI), Hyderabad, India

U. K. Vishwakarma MECON Limited, Ranchi, Jharkhand, India

Part I
Industry

Chapter 1
Research and Development in Tata steel—The Past, the Present and the Future

Debashish Bhattacharjee and Ankit Singhania

Abstract Tata Steel had always been the beacon of technology advancement in the country. In a country where investment in industrial research has not been readily available, Tata Steel has been a remarkable exception. The path towards research and development in Tata Steel can roughly be broken down into three parts. In its early years, i.e. till the Second World War, Tata Steel was involved in the development of rail steel grades and steels for armour application and bridge construction. From then till the end of 1990s, the Jamshedpur R&D centre firmly upheld the business in improving yield at different phases of operations and in advancing new products for the Indian market to suit the production process. At the turn of the millennium, with rapidly changing internal and external environment, Tata Steel R&D outfitted to meet the strategic objectives of the organization for sustained differentiation at market place and value creation for stakeholders. This paper outlines Tata Steel's glorious and gritty research and development saga.

Keywords Tata Steel · Research and development · Rail steel grade · Process improvement · Product development · Intellectual property

1.1 Introduction

Industrial Revolution is one of the greatest landfall occasions of the human civilization. Among other things, it fundamentally changed the position of steel in the world. From an uncommon fascinating material used in weaponry, it turned into a generally accessible material utilized in a wide assortment of applications. This was rendered possible due to advancement in steel-making processes imagined during the Industrial Revolution. The twentieth century got known as the Century of Steel and the material got integral to monetary thriving, political force and worldwide renown.

D. Bhattacharjee (✉) · A. Singhania
Tata Steel Limited, Kolkata 700071, India
e-mail: debashish.bhattacharjee@tatasteel.com

R. Divakar et al. (eds.), *Indian Metallurgy*, Indian Institute of Metals Series,
https://doi.org/10.1007/978-981-99-5060-7_1

The Industrial Revolution came to colonial India in the latter half of the nineteenth Century. This corresponded with the beginning of the movement for self-rule that became the antecedent of the Independence Movement. Indian businessmen who had started as traders now turned industrialists. Among these industrialists was Jamsetji Nusserwanji Tata, a cotton textile head honcho from Bombay. A solid patriot and a broadly voyaged individual, Jamsetji, was convinced that steel would assume a focal part in making India financially independent. This propelled him to set up a modern steelwork in India which after a great deal of adversities at long last began operations in 1907.

The Industrial Revolution brought in changing paradigms of manufacturing, globalization and dismantling of trade barriers. It thus became imperative for enterprises to be globally competitive. One of the principal means to remain ahead is to attain leadership in technology by promoting industrial research. Tata Steel incorporated the spirit of research and innovation right from the beginning, when the nation's organized industry was still in its early years. How Tata Steel has introduced, handled and polished all the facets of industrial research makes a fascinating tale. This paper intends to tell that story.

1.2 The Inception

"Do you mean to say that Tatas propose to make steel rails to British specifications? Why, I will undertake to eat every pound of steel rail they succeed in making."

That was the contemptuous comment that was apparently made by Sir Frederick Upcott, the then Chief Commissioner for the Indian Railways. Repudiating the distrust, the newborn Tata Steel succeeded commendably in making steel rails that were satisfactory to the Indian Railways. It likewise exported 1500 tonnes of the material to Mesopotamia to help the Allied lobby during World War I. Tata Steel tapped out its first rail-heat in February 1912. Around then, just one open hearth was running. Second one was included in September 1912. The first batch of rails produced by the organization was of 41 ¼ pounds section on a request from the Baroda State Railway (Fig. 1.1).

The standing of Tata Steel to produce sound rail steel in India had effectively been set up when the first world war broke out in August 1914. By then, the Company had finished a Bar Mill, increased the number of furnaces and raised a Physical Test House for tensile and other tests (Fig. 1.2). This could be viewed as the begetter of in-house testing and research.

At the point when the Great War broke out the organization assumed great significance as it offered the lone potential supply of iron and steel east of the Suez Canal. It was furnished with gigantic orders for shell steel bar of various sizes and rails for the construction of military tracks. The Company had just open-hearth furnaces at that point and shell steel had never been made in such furnaces. Notwithstanding, Tata Steel succeeded in making the shells under the hounded and skilful order of its first

Fig. 1.1 Surface inspection of rails at the new rail finishing mill

Fig. 1.2 100-tonne horizontal multiple lever Bucton Tensile testing machine at the Physical Laboratory

General Manager Temple W. Tutwiler. The organization didn't have the methods at that point to press the steel into shells. Tutwiler stripped each rail route and ship-building workshop that had a lathe to drill five-inch adjusts into shells. They worked and made all the difference in Mesopotamia for British and Indian soldiers. Not a solitary tonne of Tata Steel's shell steel was rejected. Special grades of steel for combat helmets and jerry cans were likewise evolved and provided.

The conflict took nearly 80% of the organization's production. When the First World War was finished, Tata Steel had effectively provided 2400 km of special Steel Rails and 3,00,000 tonnes of special grade steel for the manufacture of shells, combat helmets, jerry cans and other materials.

In its earliest stages, the frameworks and techniques which were set down, especially in inspection and testing, established the foundation for research and development work in Tata Steel.

1.3 The Formative Years

The manufacturing organizations that arose not long after the Industrial Revolution understood that continuing innovation was the key to remain competitive. Nonetheless, the expanding intricacy of new advances implied that development required devoted groups of researchers and engineers working together in a coordinated way. This paved the way for Industrial Research and Development (R&D).

One of the world's first such modern R&D centres was created by Philips in Netherlands, which in 1914 was set up as an exploration research facility to study physical and chemical phenomena and to invigorate product innovation. In India, Tata Steel became the first organization to set up a composite and complete corporate centre for R&D. The research centre was officially opened by Sir Nowroji Saklatwala, then Chairman of the Tata Group on 14 September 1937 (Fig. 1.3).

The new Control and Research Laboratory was intended to manage matters relating to research work notwithstanding routine work under the accompanying heads:

i. Control of raw materials, including analytical and chemical problems, for purposes of choice or examination.

Fig. 1.3 Mr A R Dalal (Director of Tata Sons) delivering his speech during the laying of the foundation stone for the R&D building

ii. The investigation, perception and oversight of all metallurgical activities carried out inside the steel plant.
iii. The properties of special iron and steel.
iv. Refractory materials.
v. Corrosion issues.
vi. Development of new steel and new products of all kinds.
vii. Fuel research facility.

Metallurgy, Chemistry and Refractories were the three main sections in the research centre. The Metallurgical section was furnished with rooms and hardware for mechanical testing, heat treatment, physical and corrosion testing, dark rooms for metallography and so forth. The Chemical area had facilities for wet analysis, gas analysis, colorimetry, electro-chemistry and specimen preparation rooms. The Refractory area, spread over 6000 square feet, had dust-proof pulverizing rooms, heaters for spalling, fusion, load and slag tests, a brick making room, electric furnaces and a special microscope room.

The building was intended to meet the research requirements of the steel organization for at least the following 20 years. That it still obliges the enhanced R&D requirements of the company, is a declaration of the far-sightedness of the designers and the developers.

1.4 From Howrah Bridge to 'Tatanagars'

The shiny steel grid of the Howrah Bridge overshadows the Hooghly River in Kolkata. It is perceived the world over as the image of this Indian city. Relatively few individuals realize that in the mid-1940s Tata Steel provided the majority of its 26,500 tonnes of steel. Delivering steel for the Howrah Bridge was a significant test for Tata Steel, since it had not yet wandered into low alloy structural steels. The specifications of steel to be utilized required a tensile strength of 37 to 43 tonnes for every square inch. Developing this grade became the major focus of its new in-house research centre that was opened only a few years prior.

At the time, an Electric Arc Furnace and a 0.5 tonne High Frequency Induction Furnace were just installed in the steel melting shop and were used to produce the alloy steel. The researchers and engineers of Tata Steel faced several challenges including difficulty in rolling of sections, difficulty in removal of a tightly adhering oxide scale and problems related to surface cracking. However, they could overcome all these challenges and the new steel product they created was christened 'Tiscrom' (Fig. 1.4).

Tiscrom and other early sorts of high-tensile steel were quarternary steels and were not amenable to welding. Tata Steel's researchers therefore also started developing a high-strength structural steel that was suitable for welding. They concocted 'Tiscor', a quinary steel containing Chromium, Copper, Silicon and Phosphorus. Tiscor's high yield strength enabled its use in thinner sections, while its weldability advanced its

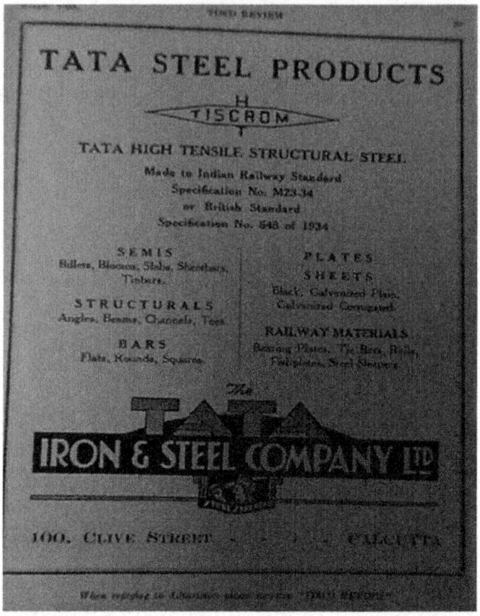

Fig. 1.4 The first branded product of Tata Steel—TISCROM

use in freight cars, ships, trams and various other vehicles (Fig. 1.5). Tiscor was also found to be more corrosion resistant than plain carbon structural steel.

While Tiscrom and Tiscor took care of the steel necessities of the civilian sector, one of Tata Steel's best hours in innovativeness came during the Second World War. During five years of the conflict, researchers at Tata Steel managed to create upwards of 110 distinct varieties of steel. The most remarkable accomplishment was

Fig. 1.5 A wagon made from Tiscor Steel plates

Fig. 1.6 The Tatanagar, a light armoured vehicle built at Jamshedpur by Tata Steel (1940)

the development of a bullet-proof armour plate. This new bullet-proof aromour plate was first utilized for reinforced vehicles called 'Tatanagars' (Fig. 1.6). They were widely utilized by the British Army deployed on the North African front.

The Second World War set off an enormous spray of imagination at Tata Steel's fledgling Control and Research Laboratory. This empowered the organization to make a rich variety of alloy steels for shear blades, machine tools, parachute harness and even razor blades. Tata Steel also succeeded in making magnetic steel bars and high-speed tool steels. A fascinating side line was the production of mint die steel. This could have been the first attempt at the manufacture of quality tool steel in India.

1.5 Overcoming Challenges in Iron Making

The global steel industry has become highly energy and resource efficient because of sustained technological improvements over many decades. Tata Steel has led the way in India. Its endeavours in R&D and its scientific approaches to operational excellence have created world-class iron-making operations, despite deficiencies in the locally accessible raw materials: a high ash content in the coking coal and a high alumina content in the iron ore (Fig. 1.7).

Coke is arguably the main factor in a steady blast furnace operation. There are numerous approaches to describe coke quality. However, the coke strength after reaction (CSR) and the ash content in the coke are amongst the decisive factors. A high ash content and a poor CSR were viewed as the key reasons behind low blast furnace productivity.

Fig. 1.7 The first blast furnace (1911)

The ash content of coal from Tata Steel's West Bokaro mine varies from 30 to 35%. This ash content is very high. However, a washing process is now able to reduce the ash content to around 15 to 18% and yet give a clean coal yield of 40 to 50%. Years of research and improvement in beneficiation processes—dense medium cyclone and flotation process—have helped improve yield with better control of ash. Recently, Tata Steel R&D has developed a novel dense media cyclone christened as Tata JK dense media cyclone and a flotation column with external sparger. The target is to improve clean coal yield by a further 5%. In addition to new process developments, Tata Steel has also developed seam-specific reagents for better selectivity of captive coal. These initiatives will not only help improve the fuel efficiency at blast furnaces but would also help extend the mine life.

In the early 1980s, Tata steel embraced another significant coal technology—Stamp Charging. This technology had initially been developed in Germany and was examined, adjusted and progressed in Tata Steel. Coking coal includes just 15% of India's 200 billion tonnes of coal holds, 80% of the coking coal reserves are of the medium coking type. During the last part of the 1970s and early 1980s, the R&D division of Tata Steel completed broad tests to augment utilization of domestic medium coking coal for coke making. They assessed different pre-carbonization technologies including stamp charging. It was plainly shown that stamp charging was not just the most fitting pre-carbonization technology for coke making at Tata Steel, but it was also likely to produce coke of much better quality than the top charging process. In view of the above findings, it was decided to progressively introduce stamp charging in the coke oven batteries at Tata Steel.

The coke making industry has seen significant technological developments over the last decade. Conventional coke ovens (slot type ovens) with raw gas recovery have reached dimensions of more than 8 m in height and 90 m3 in useful oven volume, with a single coke oven producing ~40 tonnes.

In the last decade, the topic of environmental sustainability has gained increased momentum. Coke oven batteries with non-recovery technology have found increased acceptance amongst coke makers. Coke plants with non-recovery technology are in operation in USA, Australia and India. Tata Steel also has a non-recovery coke plant which helps produce power from waste gas. Introduction of refined coke quenching technologies in Tata Steel such as Coke Dry Quenching (CDQ) is also helping coke makers to further reduce the CO_2 footprint.

Parallel to these changes, Tata Steel R&D is also leveraging computational methods to design coal blend and predict coke quality. Successful demonstration of advanced coke making technology SCOPE 21 shows that coke making has a huge potential to grow and evolve.

When iron ore is mined, it normally produces a combination of lumpy iron ore rock and iron ore fines. Lumpy ore can be used directly in the blast furnace, while iron ore fines can be agglomerated either in sinter plants, which fuse the fines into a solitary permeable mass, or in pellet plants, which form small iron ore balls that are subsequently baked.

Tata Steel's relationship with sinter making began in the year 1959 with the development of India's first Sinter Plant. The utilization of limestone to improve sinter quality began in 1968 and dolomite was additionally added in 1978. Several other developments during the mid-1980s significantly improved blast furnace productivity.

With sinter in limited supply, the extent of sinter in the blast furnace burden dropped to 30% and a need was felt to increase the percentage of sinter. However, the presence of significant levels of alumina in the iron ore fines (normally 5% alumina) represented an issue to keep up the quality of sinter. A vital breakthrough came in the form of 'blue dust'—very fine ore that is rich in iron (67%) and low in alumina (<2%). A nitty gritty investigation by R&D reasoned that up to 40% blue dust can be mixed with iron ore fines without any unfavourable impact on sinter productivity. With the utilization of blue dust and commissioning of the second sinter plant in 1989, the extent of sinter in the blast furnaces expanded to 50% in 1990 and 63% in 1992.

In the late 1990s, another drive named 'P40' was embraced to improve the gross sinter productivity by 50%, from 27 to 40 tonnes/m^2/day. As part of this project, R&D conducted several trials to optimize the ignition intensity, to improve the granulation of sinter mix and to minimize air leakages into the sinter machine. These initiatives ensured that this difficult objective was met.

To further the cause of environmental sustainability and increasing the life of iron ore reserves, use of binary and ternary ore blends of Noamundi, Joda and Khondbond ore fines were introduced in Sinter Plants. Use of micro-pelletized ore fines and solid waste was introduced to improve sinter Tumbler Index and productivity. Solid fuels are now being used in sinter plant to achieve energy efficiency and to keep emissions under control. Sinter plants at Tata Steel consume > 6% of reverts in sinter mix, thereby achieving 100% recycling of all iron and carbon bearing waste generated inside the steel plant. Commissioning of sinter plant at Tata Steel Kalinganar (TSK) has also increased the capability of the company to consume > 30% iron ore concentrates.

In the same period when blue dust was adopted as a measure to reduce the alumina in sinter, different investigations were likewise in progress to bring down the alumina content of the feed ore. Through a series of experiments in the last part of the 1980s and early 1990s, it was shown that the alumina content could be decreased by a combination of washing iron ore lumps and jigging iron ore fines. These initiatives helped decrease blast furnace slag volume from 330 to 280 kg/tonne of hot metal in 15 years.

Today, automated mining, sizing of iron ore and iron ore beneficiation are progressively being utilized in India. A side effect is that more iron ore fines and ultra-fines are being produced. The capacity to utilize such fines in sinter making is restricted. Pelletizing is globally the most used process to agglomerate iron ore fines that cannot be used in sintering. In 2012, Tata Steel commissioned a new 6 million tonnes per year pelletizing plant.

An interesting side line which evolved due to introduction of pellets in blast furnace burden was improvement in sinter quality and productivity. With introduction of pellets, iron ore fines with lower alumina content and lower super-fines were directed towards sinter making. This helped in increasing sinter productivity to > 44 tonnes/m^2/day. Also with a drop in sinter quantity in the blast furnace burden, a need was felt to increase the sinter basicity to more than 2.4 to minimize the addition of raw flux at blast furnaces. This new regime of sinter chemistry helped tackle the problem of poor sinter Reduction Degradation Index (RDI), a classic problem of sinter with higher alumina iron ore fines.

Other than managing issues identified with raw materials, Tata Steel's R&D and Scientific Services has contributed significantly to improvements in other aspects of iron making. For instance, its researchers have utilized different downsized models of blast furnaces to explore different avenues of burden distribution. A downsized model of D blast furnace was first utilized in 1981 to design moveable throat armour to streamline the outspread dispersion of the burden. Comparable models at a 1 to 10 scale were consequently constructed. Prior to commissioning of G blast furnace, R&D carried out detailed study to understand the burden distribution and burden movement. On 10th December 2010, a team from R&D in Jamshedpur and Tata Steel Europe also successfully deployed an advanced trajectory probe to measure the path of burden materials when being charged in the blast furnace (Fig. 1.8).

Modernization of ironmaking operations started in 1992 with commissioning of G blast furnace. Subsequently, H, I and TSK blast furnace I were commissioned in 2008, 2012 and 2016 respectively. All these modern blast furnaces are equipped with Pulverized Coal Injection system to replace expensive coke, enhanced agglomerate addition, Top Gas Pressure Recovery Turbines (TRT) and use of Dry Quenched Coke. This has made blast furnaces energy efficient with benchmark carbon footprint and productivity (>2.8 tonnes/m^3/day). Today, the average Pulverized coal injection and coke rate are at > 200 kg/tonne and < 350 kg/tonne of liquid iron produced. Indigenous simulation and model-based process insights, tuyere design, adoption of non-invasive technologies to monitor the burden distribution and Blast Furnace health monitoring are some of the in-house developments which has supported this transition. In the

Fig. 1.8 Probing the blast furnace using an advanced material trajectory probe (2010)

near future, the addition of TSK phase-2 ultra-modern blast furnace with state-of-the-art technologies will take the energy efficiency and low-carbon footprint operation to new benchmarks in global iron and steelmaking operations.

The quest for better iron making technologies is a continuous cycle. Future challenges for researchers lie in such areas as reducing the carbon footprint, smelting reduction, use of non-coking coal and full process automation. Tata Steel's R&D division is outfitting to meet these new objectives.

1.6 Perfecting Steelmaking

Tata Steel began its steel making venture with four 40-tonne stationary open-hearth furnaces served by a 300-tonne hot metal mixer. At the point when the first world war broke out, there was a sudden increase in demand for steel and promptly plans were made to increase annual production to half a million tonnes of ingot steel. Tata Steel embraced the Duplex process in the early 1920s (Fig. 1.10). The Duplex process combined blowing hot metal in a Bessemer converter and afterward refining the blown metal in an open-hearth furnace (Fig. 1.9).

During the early 1940s, Tata Steel's attention was drawn to the potential outcomes of the Perrin Process that was worked effectively by Ugine in France. It included pouring partially refined steel into a highly oxidizing basic slag and enabled rapid steel production at lower cost. Around the same time, the Second World War broke out in Europe. It constrained imports of steel and Tata Steel was encouraged to deliver steels for wheels, tyres and axles to meet the necessities of Indian railways. The Perrin process was therefore modified. The modified process was called a Triplex Process and involved a combination of an acid Bessemer process, a basic open-hearth process and acid open hearth process. This endeavour ended up being an altogether

Fig. 1.9 First ingot casting steel from ladle into ingot mould (2 December 1911)

Fig. 1.10 A group of three bessemer converters at duplex plant (1928)

accomplishment at that point. Notwithstanding, later it was demonstrated that the basic open-hearth process was also able to produce steels suitable for railway wheels and axles. The Triplex process was abandoned in 1950s.

During the time, Tata Steel's administration confronted a significant issue of whether to invest in the new Linz-Donawitz (LD) oxygen steel making process, which began from Austria. After actual trials at Linz in Austria, Tata Steel concluded that the process would be uneconomical in India. A key explanation was that the locally accessible refractory materials were not expected to last long enough under the severe process conditions. Another obstruction was the specifics of Indian raw materials that result in a high phosphorus content in the liquid iron and require explicit operating practices during steelmaking. At last, in its third expansion phase, Tata Steel did switch to oxygen steel making. Its first LD oxygen steel making plant (LD1) was commissioned in 1983 and had two LD converters of 130-tonne capacity (Fig. 1.11). LD2 was commissioned ten years later with the installation of two 130-tonne converters. One more converter was added later.

Various enhancements were done in oxygen steel making process over the years. It included substitution of the old three-hole lance tip by a new six-hole lance and the asymmetric location of eight tuyeres at the bottom of Basic Oxygen Furnace (BOF) vessel at LD1. These drives helped reduce the turn down phosphorus from an initial of 0.035 to 0.015% by early 2000s. It likewise expanded the life of the converter lining in LD1 dramatically, from 160 heats in 1983 to 1211 in 1997 and more than 2000 heats in 2002. This was a world-beating accomplishment for a lining made with tar bonded dolomite bricks. By 2005, the life of the dolomite lining was raised further to more than 2300 heats because of introducing other strategies. Today the lining survives more than 7000 heats following the changeover to a different type of lining using Mag-carbon bricks in 2006.

In 2004, R&D launched a new strategic development project to bring down the phosphorus in steel. Its aim was to exhibit the production of steel with a phosphorus content of below 0.015% from hot metal with phosphorus content of up to 0.22%. The new drive focused on a non-uniform flow from eight injection points located in a symmetric pattern in the bottom of converters at LD2 and use of Calcium Ferrite in the LD vessel.

Fig. 1.11 The first LD Oxygen steel making plant in Jamshedpur (1983)

Slag splashing studies were carried out using water model which helped increasing the campaign life of LD vessels. Advanced data analytics combined with fundamental models are being used to predict the end-point temperature of the vessel. New technologies, like audiometer-based smart lance, have been adopted to minimize sloping in vessel and to move towards auto vessel operation mode.

Today liquid steel is continuously cast, cooled and solidified into different shapes, such as slabs or billets. Tata Steel's first six strand continuous billet caster was commissioned at LD1 in 1983. Two more continuous casters were added later. The first single strand slab caster at LD2 came in 1992 and two more strands were added later.

R&D at Tata Steel has completed extensive studies to comprehend and improve the fundamental processes of continuous casting. A lot of work has been carried out to optimize porous plug location through various numerical and water modelling studies. This helped in improving the mixing time and enhanced the floatation of non-metallic inclusions in the ladle furnace. Several studies were carried out in RH process (secondary refining system) to improve decarbonization and currently, Tata Steel can achieve < 15 parts per million (PPM) of Carbon in most of the heats. Work also has been done with respect to inclusion characterization and improving the cleanliness of steel thereby, reducing several inclusion-related defects in the rolled product and reducing submerged entry nozzle clogging during continuous casting practices.

This story on steel making and steel casting research shows that the impact of R&D is multi-faceted. Frequently its job is to help the adoption and indigenization of a novel technology that was first piloted outside India. Even then, the technical challenges are huge. Throughout the years, the steel making and casting research group of the R&D division has grown very strong modelling capabilities, both in physical water models and virtual models in the form of computational fluid dynamics. With the development of progressively advanced computational models and the adoption of more precise measuring procedures, we can expect new breakthroughs and see researchers flourish in this intricate and exciting field of steel making.

1.7 Quest for New Products

Right from its inception, Tata Steel's essential objective was to make products that meet the prerequisites of customers in the Indian market. Back in 1912, the very first products to roll out of the new steel plant were rails for the nation's quickly extending rail routes. The First and Second World Wars also triggered a spurt in innovation in Tata Steel.

From the mid-1950s, India saw fast and pervasive advancement of its infrastructure. However, the steel prices were directed by the Joint Plant Committee (JPC). Unfortunately, the costs fixed by the JPC were low to such an extent that Tata Steel could scarcely create benefits. Luckily, special steels and alloy steels were kept external to the domain of JPC and their prices were determined simply by market patterns. Management therefore initiated the development of new products to increase Tata Steel's profitability. The new developments included rimming steels for the manufacture of wire rods and cold rolled steel strip and tubes for boiler applications. Tata Steel also developed LPG sheets, weldable steel plates, high strength billets and gothic section bars to produce seamless tubes.

The twenty years from 1970 to 1990 saw a wide range of product introductions. During the early 1970s, Tata Steel chose to diversify into special steel products for the engineering industry. A few in-plant activities were changed considering the necessities of the engineering industry. Low-Alloy High-Strength Structural Steels were likewise evolved and delivered by quench and temper techniques utilizing natively accessible ferroalloys. During the mid-1970's micro-alloyed plates, sheets, strips and structural products were produced by refining the grain size of low-carbon ferrite-pearlite steels with the utilization of modest quantities of niobium and/or vanadium.

During the mid-1980s further adjustments were made in the steelmaking operations to reduce the cost of micro-alloyed steels. Extra deep drawing quality sheets were developed to improve formability and strain ageing characteristics. Also, cold drawn seamless tubes were developed for defence purposes. Tata Steel additionally developed wear-resistant plates and closed die forged blanks. A new product called 'Tiscral' was sold in the form of plates and was the first wear-resistant steel produced in India. In addition, Tata Steel developed a wide variety of plain carbon and low alloy steel strip with a high degree of cleanliness which were commercialized for a wide scope of uses, such as hack swas, razor blades, springs and chains.

With new facilities in Tata Steel, a variety of new products was added to the bushel of existing products. Spring steel flats for automobiles, rolled-forged rings for the bearing industry, deep drawing quality steels in the form of strips for the cold rolling industry and the automobile industry, low alloy clean forging quality steels for crank shafts, gears and forged products for the engineering industry are some examples. Through intensive research, creep-resistant steel for use in boilers was developed and commercialized. Tata Steel likewise created and provided for 70% of the low hydrogen steel for axles made in the Wheel and Axle Plant set up in Bangalore by the Railways. It also provided more than 5000 tonnes of seamless bars to clients like Bharat Heavy Electricals to produce seamless tubes.

During these years, the product offerings from Tata Steel grew richer as an ever-increasing number of segments were identified. Nonetheless, this widespread broadening of Tata Steel's product portfolio ended unexpectedly in 1991 when the Indian economy opened-up. Although the economic liberalization brought about a welcome relief in operational terms, it also meant that Tata Steel had to stand up and compete with imports from top international steel makers. Thus, Tata Steel had to deploy new competitive strategies including a rationalization of its wide product portfolio and an increased focus on core competencies in selected markets.

Changing rules for foreign direct investment resulted in the advent of global manufacturing goliaths into India, particularly in automobiles. One such newcomer was the Suzuki Motor Corporation, a major Japanese carmaker. During the 1980s it set up a cutting-edge vehicle fabricating plant close to Delhi in a joint endeavour undertaking with Maruti Udyog Ltd. and the Government of India. At first, Maruti needed to source steel from Japan as the Indian steel industry couldn't meet the quality necessities. Tata Steel responded to the call to address Maruti's issues and could do as such with intense research and development of special steels.

This drive was likewise extended to the Indian bike and three-wheeler areas. For instance, Indian bike producers used to make petroleum tanks with imported materials such as Terne coated and electro-galvanized steel sheets. In 2003, Tata Steel developed a corrosion resistance chromate coating on galvannealed steel sheet and supplanted the above items with superior performance. In another such case, low-carbon low alloy steel strip has been produced for Bajaj Auto. Tata Steel also developed a cold rolled batch annealed bake hardening steel. It permits 20% abatement in thickness of car body panels, making cars lighter and eco-friendly.

It isn't only the automobile business that R&D focuses for its new product developments. For instance, low-carbon steel using boron were developed for making ERW tubes for town water transport, supplanting the expensive FM tubes. New coatings are being developed for steel which utilize polymers and nanoparticles of silica, titania and alumina to supplant costly zinc and harmful hexavalent chromium. Thin organic coating formulations are developed for galvanized products to double their corrosion resistance. Several advanced coatings are likewise being worked on at R&D.

The latest addition in the flat product complex at Jamshedpur is the 'Thin Slab Caster and Rolling Mill (TSCR)'. Commissioned in February 2012, with a capacity of 2.40 mtpa, the mill is equipped with state-of-the-art technologies to cater to the various customer requirements. This is an integrated shop where the steel made in steel melting shop is continuously cast and rolled.

Since inception of this plant, R&D has been involved in process optimization and new product development through mathematical modelling, Gleeble simulation, lab/pilot scale studies and in-depth microstructure and mechanical property characterization. Research in the direction of dynamic soft reduction during casting, modelling of the reheating furnace heat treatment cycle are some of the examples of the work carried out by R&D in the TSCR complex. Some of the most prominent grades that were churned out of TSCR include Dual Phase 600, SPFH 540 and SPFH 590.

Over the years Tata Steel has migrated from ingot rolling to Hot Strip Mill (HSM) processing. In this context, R&D accepted a major challenge to develop new chemistries and rolling schedules for different grades that are processed through HSM route. Over the years, Thermo-mechanically controlled processing (TMCP) rolling concept with microalloying additions has been successfully introduced to produce various Advanced High-Strength Steel (AHSS) grades through HSM. Advanced thermomechanical simulation using Gleeble simulator (Fig. 1.12) and advanced characterization facilities such as Scanning Electron Microscope (SEM), Electron Back Scattered Diffraction (EBSD) and Transmission Electron Microscope (TEM) have been deployed to develop these products (Fig. 1.13).

Tata Steel has commissioned a new Hot Strip Mill at Kalinganagar, Odisha, in the year 2015. Apart from reheating furnaces, it has a couple of roughing mill stands, a seven-stand finishing mill and a Run-out-Table (RoT) facility capable of intense

Fig. 1.12 Gleeble simulator for advanced thermomechanical simulation

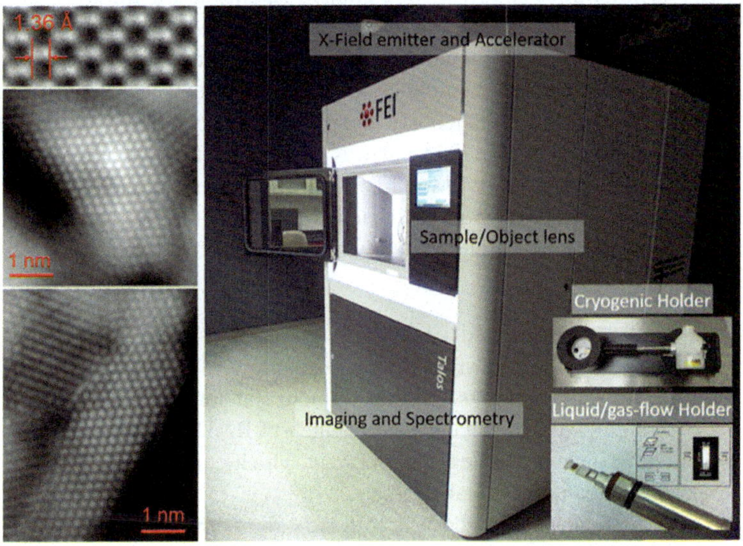

Fig. 1.13 Transmission Electron Microscope (TEM) for advanced characterization

cooling. Next generation high strength steels such as ultra-high strength DP steel and Transformation Induced Plasticity (TRIP) steel were developed from these mills for different automotive parts. Lifting & Excavation grade steels and line pipe steels (up to X80) have also been developed from this mill. Moreover, a special Quenched & Partitioned HR grade steel has also been developed using fast cooling facilities.

As India progresses forward, it will draw in an increasing number of innovative enterprises with varied and evolving needs of steel. Tata Steel anticipates addressing the issues of these new clients. It will be a contest of innovation and competition. Knowing its history, we can infer that Tata Steel with its R&D and Scientific Services Division is capable to confront this test.

1.8 Pursuing Intellectual Property

We are living in a knowledge-driven economy and the economic growth in twenty-first century will be largely driven by intellectual capital creation. Tata Steel recognizes this fact and thus, consciously creates Intellectual Property (IP) which are of strategic interest to the company. Today, Tata Steel (India) is recognized as one of the leading Innovators in the Indian Steel Sector. It has a patent portfolio of more than 1300 patent families and over 1000 granted patents across the world.

In 1938, Tata Steel documented its first patent (Fig. 1.14). It was granted in 1939 and was the principal patent of the Indian steel industry. Be that as it may, before the millennium's over, the complete number of granted patents was just 8, with another 16 in application stage. Simultaneously, the primary indications of solidification in the global steel industry were arising. It was anticipated that, when the worldwide innovation pioneers arise, they will solidify their positions on the exclusive ownership of technology. Tata Steel likewise expected to improve the protection of its own intellectual property. Three significant advances were taken during 2001–2008 for the equivalent:

i. A patent cell was constituted to help execute numerous initiatives pointed toward spreading attention to the significance of intellectual property, the approaches to secure new developments and the pertinent standards and limitations.
ii. A second significant advance was the launch of R&D Thrust Areas—an initiative aimed to develop strategic technology which can help Tata Steel create a competitive advantage at market place.
iii. The third significant advance was to fortify the administration instrument on Intellectual Property by formation of review committees.

Recently there has been a paradigm change in the way a company innovates. Open innovation has become a 'new favourite' of the industry wherein, the roles of academia and research institutes become important. Tata Steel has tapped into this new trend and has forged strong tie-ups with research institutes across the world. Tata Steel has put the best practices in place to collaborate with institutes of eminence

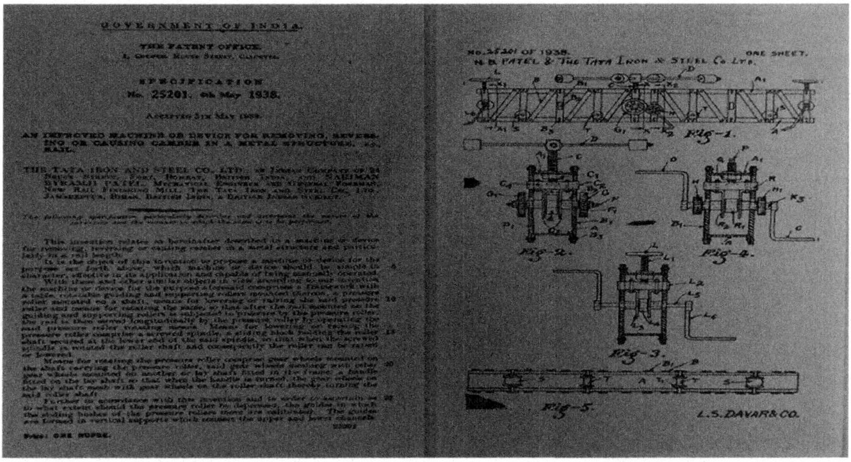

Fig. 1.14 First patent for Tata Steel

creating value for all the stakeholders. It has on an average more than 50 research projects running with the collaborators at any point in time.

Considering its patent filings, Tata Steel India is presently perceived as a benchmark in the Indian business arena. It has won many prestigious and coveted awards such as the Government of India Annual IP Award 2011 for 'the highest number of granted patents during the past 5 years amongst all Indian private companies. The company also won the Thomson Reuters Innovation Award in 2011 and 2015 that recognized Tata Steel as 'the most innovative private company amongst high-tech corporate India'. It won the CII National Intellectual Property Award in 2011 & 2017 and the Asia IP Elite award at IPBC Asia Conference in 2019. Recently in 2020, Tata Steel was awarded the CII Industrial Intellectual Property Award for the 'Best Patent Portfolio in the Large Enterprises (manufacturing/engineering) category'.

Tata Steel's IP strategy has been largely revolving around protecting its own turf to ensure that it has freedom-to-operate and can minimize IP licensing from foreign companies. These efforts have paid off in recent times and Tata Steel has been developing new products either by itself or through collaboration with the Indian research institutes. Having said that, Tata Steel has also been actively exploring licensing and commercialization activities—both in-licensing and out-licensing. Tata Steel has successfully licensed a few patents in the past and has also acquired IP to strengthen its portfolio. Tata Steel is actively evaluating opportunities to own IP in the areas of its strategic interest.

1.9 A Leap into the Future

R&D and innovation are essential for progress in all industries. The steel industry is no different. The launch of Tata Steel's Control & Research Laboratory in 1937 was the very first start of corporate R&D in all of India. Clearly, Tata Steel has always felt an urgency to pioneer.

Till the end of the twentieth Century, R&D at Tata Steel was mostly driven by the need to improve existing products or to overcome process constraints that were specific to the local conditions. However, the onset of the twenty-first Century saw significant changes in Tata Steel's internal and external environment. As such, R&D at Tata Steel saw a paradigm shift in its programmes with focus on strategic imperatives. A company-wide reorganization also happened in 2001 to help the company adapt to these changes. Responsibility for the process and quality control was taken out of the purview of Metallurgical Services and was placed with the manufacturing units. In addition, the R&D and Scientific Services (SS) departments merged and the Refractory Technology Group was added to form the R&D and SS Division. This centralized all laboratories into a single division for the effective management of these facilities and their associated know-how.

It was in 2017, during the formulation of the strategic roadmap for Tata Steel for 2030, that Technology Leadership was identified as one of the Strategic Enablers. Tata Steel aspires to be 'Top 5 in technology in steel industry globally'. This was based on the consideration that in-house technology is becoming critical as ease of technology access is reducing. The objectives of this enabler are to create new market segments (to provide "new to world" products, services and solutions), to maintain sustainable leadership (to reduce environmental impact) and to make Tata Steel an innovation, capability & process benchmark (to have world-leading infrastructure and people for driving innovation and achieving benchmarks in current process and product).

In pursuit of the aspiration of technology leadership "Top 5" goal, Tata Steel embarked on the Technology Roadmap journey in 2018. Road-mapping as a tool helped identify the applications that Tata Steel needed to pursue and the capabilities required driving the identified applications. The Applications were termed as Technology Leadership Areas (TLAs) and the Capabilities were clubbed together under Technology & Innovation Management System (ITM).

Till now, six TLAs have been identified (Fig. 1.15) which include coating solutions, leveraging low-quality raw materials, use of Hydrogen in steel value chain, Carbon capture and use, materials for future mobility and water efficiency. Each of the TLAs has multiple projects which is being driven under the lens of the Technology & Innovation Management System (ITM).

The Technology and Innovation Management System in Tata Steel has been institutionalized around four pillars (Fig. 1.16).

i. Organizational Structure and Governance—This pillar helped unify the technology vertical and establish a robust governance and review mechanism for TLAs & Technology Management System.

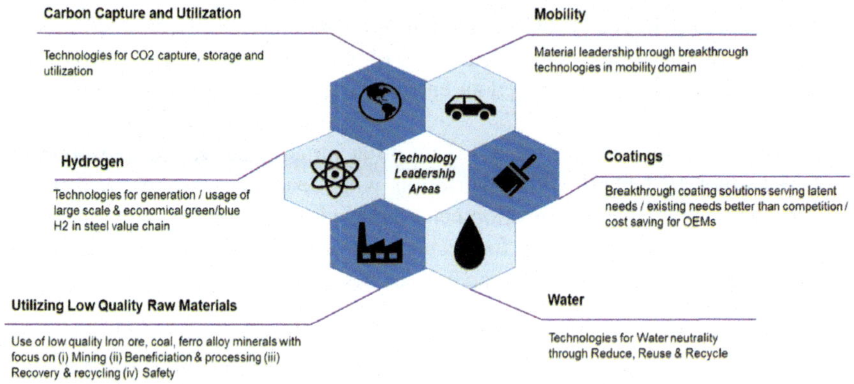

Fig. 1.15 Technology Leadership Areas (TLAs) identified by Tata Steel

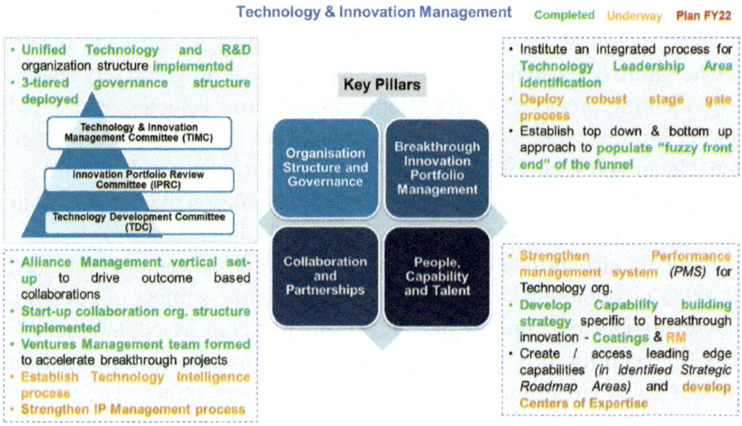

Fig. 1.16 The four pillars of technology and innovation management system at Tata Steel

ii. Breakthrough Innovation Portfolio Management—This pillar aims to institutionalize a strong mechanism for management of health and speed of TLA projects' funnel. A Project Management Office (PMO) has been set up to drive the TLA projects which generally cut across the various functions in the organization.

iii. Collaborations and Partnerships—This pillar aims to establish systems to leverage external ecosystem for strategic partnerships and processes to monetize in-house technologies and know-how. Under this pillar a team has been formed which looks at forging collaborations with Academia, Industry and Government Institutions to drive projects of mutual interest. A second team has been constituted to help Tata Steel plug meaningfully in the start-up ecosystem. A separate business unit has been established with the sole responsibility of monetizing in-house IPs. Recently, the incubation of an in-house IP has resulted in steel-making slag finding a commercial route of use as a soil conditioner. Branded as Dhurvi Gold, it has witnessed an encouraging customer response during pilot sales.

iv. People, capabilities and talent—This pillar aims to deliver people processes contextual to technology organization with respect to hiring, performance management, capability development, etc. Under this pillar a revamped Performance Management System for Technology vertical and creation of 'Centres of Expertise' with leading edge capabilities has been envisaged.

Both the TLAs and the ITM are dynamic in nature which gets updated as new challenge areas emerge. For example, a seventh TLA has been identified recently as TLA-Digital. The ITMs have also added two new pillars 'Value enhancement through commercialization' and 'Democratization of Technology and Innovation'. Identifying and successfully driving the Technology Leadership Projects will be a key differentiator for Tata Steel moving forward, making the company future ready.

One such area in which TLAs are making Tata Steel future ready is in Sustainability. The future is Carbon-free. This means the use of progressively lesser Carbon in the Ironmaking process, greater adoption of electric vehicles, etc. Tata Steel is working on multiple routes for Carbon capture and utilization and for generation of green, cheap Hydrogen. The first pilot plant at 5 tonne per day scale is already working in Jamshedpur. A hydrogen-rich syngas generation plant using biomass and municipal solid waste is being set up at Tata Steel Kalinganagar.

Tata Steel has also initiated new line of business and has ventured into developing breakthrough technological interventions to address the subject of circularity. Tata Steel has set up steel recycling business to prepare for scrap-based steelmaking, which has low CO_2 footprint. It is also working on novel extractive metallurgy technology for getting Ni and Co from overburden of chromite ore mines. The amount of Ni and Co extracted from this source can provide around 40% of national need of these elements by 2030. Tata Steel is also developing business in recycling of e-waste and battery waste.

In its ambition to put India in the world map of new materials by being an India-based global supplier of high-quality materials and solutions, Tata Steel has set up a new business vertical on non-steel, new materials. The vertical which was set up in FY18 started with Composites and Graphene. Recently, in FY21, Medical Materials was added to the list.

In composites three large market sectors are being addressed—Railways, Infrastructure and Industrial products. The products use multiple manufacturing techniques such as filament winding, compression moulding and pultrusion. The composites business has made the first composite footbridge for public use in India.

Tata Steel also has one of the world's largest Graphene manufacturing plants in Jamshedpur with an annual capacity of 100 tonne. The plant has been commissioned in FY21. Graphene is applied to conveyor belts, idler rolls, HDPE pipes and is also used as coating on steel structures for corrosion resistance.

Tata Steel has also taken first steps in medical materials through manufacture of world-class quality Hydroxyapatite and Collagen. The intention is to get into implants.

1.10 Conclusion

Tata Steel has always been at the forefront of technology evolution in India serving as a benchmark not just for the steel industry but for other industries as well. This has been made possible by the strong R&D team that Tata Steel has. Questioning convention, meticulous research and the excitement of discovery have consistently been and will consistently keep on fascinating its specialists and researchers. As the world becomes increasingly VUCA (Volatile, Uncertain, Complex and Ambiguous), it is now important more than ever to challenge the status quo and stretch beyond the known. It is the joint aspirations of the management and researchers that will ensure that they live up to the legacy of Tata Steel; continually endeavouring to make an enduring and positive imprint anyplace it works.

References

Amit Chatterjee, Ramesh Singh, Tata Tech., Tata Steel, pp. 1–8, (March 1993) Glimpses of R&D at Tata Steel

Mukherjee, T.: Tata Tech, Tata Steel, pp. 1–10, (December 1987) Research in the Steel Industry

Sachitanand, N.N., Mark, B., Denys, P.K., Banerjee.: Pathfinders of excellence: 75 years of research at Tata Steel, (2012)

Chapter 2
WIL's Contribution in Development of Welding for Exotic Materials

Amol S. Gengaje, Aniruddha S. Sawargaonkar, Vinayak P. Gawade, Sumit S. Gorde, and Vijay D. Raskar

Abstract Walchandnagar Industries Ltd. (WIL) was founded in 1908 and has been a significant Indian engineering company throughout the platinum years. Founded as a sugarcane farming firm and later diversified into making sugar refined spirits, plastic goods, cement plants, paper and pulp plants, and water tubes has risen to the challenge of projects in nuclear power, aerospace, missile defense, oil and gas and steam generation plants, among others, on the basis of its strong engineering, project management and manufacturing infrastructure. The company has several firsts to its name, including manufacturing critical components for India's nuclear reactor and submarine missile launchers, among others, and has expertise in fabrication and machining of ferrous and non-ferrous materials, and welding development in highly critical and exotic materials in defense, aerospace.

2.1 Walchandnagar Industries Ltd: Feather in Our Nation's Cap with a Benchmark for Engineering Excellence

Walchand Hirachand Doshi (1882–1953)—The visionary who gave India its first aircraft factory, shipyard and car factory was known for patriotism and innovative spirit. His entrepreneurship in shipping, aviation and automobiles earned him the title 'Father of transportation in India' (Fig. 2.1).

Pioneer in Aircraft manufacturing: He founded 'Hindustan Aircraft Limited' which manufactured planes like Harlow Trainer, Glider, Hawk-P36, Harlow-PC5, etc. Later it was nationalised by Government of India.

Pioneer in Maritime Shipping: He founded Scindia Steam Navigation Company. On 5 April 1919 the ship 'SS Loyalty' sailed on its maiden voyage, a crucial step when sea routes were controlled by the British. 5 April is now celebrated as 'National Maritime Day' in India.

A. S. Gengaje (✉) · A. S. Sawargaonkar · V. P. Gawade · S. S. Gorde · V. D. Raskar
Walchandnagar Industries Limited, Walchandnagar, Indapur, Pune 413114, India
e-mail: amol.gengaje@walchand.com

R. Divakar et al. (eds.), *Indian Metallurgy*, Indian Institute of Metals Series,
https://doi.org/10.1007/978-981-99-5060-7_2

Fig. 2.1 Shri Walchand Hirachand Doshi (1882 - 1953)

Pioneer in Shipbuilding Industry: He founded Hindustan Shipyard, Vizag. He recognised the need of country's infrastructure for ship building and established a shipyard in Vishakhapatnam in 1948. The company's 1st ship 'Jal-Usha' was commissioned by India's first Prime Minister Pandit Jawaharlal Nehru. Later on the shipyard was nationalised by Government of India.

2.2 Birth of the Organisation

2.2.1 Early Years

After spending a few years in his father's family business of banking and cotton trades, Seth Walchand realised he was not interested in the family business. He became a railway contractor for constructions in partnership with a former railway clerk. The partnership later became Phatak-Walchand private limited. It was in the construction business, first as a railway contractor, and then as a contractor to other departments of government & the partnership lasted till 1915. He proved to be a successful railway contractor but he was open to other business ideas as well. Projects undertaken by him were grand in design. Most of his projects were highly leveraged. With nationalisation, the government took over some of the projects like the shipyard and the aircraft factory. The government also needed these industries as they directly supported war efforts.

2.3 Walchandnagar Industries Ltd

Seth Walchand successfully transformed the barren, rock-strewn, practically uncultivated land near Kalamb village into lush green sugarcane fields and organised sugar cane farming and founded the Walchand Industries Ltd (WIL) in 1908, which was started as a large-scale sugar farming firm. Later it diversified into making sugar refined spirits, sugar machinery, plastic goods, cement plant, paper and pulp plant and water tubes.

Walchandnagar Industries Limited (WIL) is an ISO 9001: 2015 certified Indian company with global presence and diversified business portfolio in Projects, Products and High-tech Manufacturing. Carrying a legacy of more than 110 years of Engineering Excellence, WIL has established its name as one of the best in its operational areas. WIL is known for pioneering achievements in Indian engineering industry and for its contribution to nation building activities.

WIL has a strong engineering, project management and manufacturing infrastructure to undertake projects and supply of machinery and equipment, in the fields of Nuclear Power, Aerospace, Missile, Defence, Oil & Gas, Steam generation plants, Independent power projects, Turnkey Cement and Sugar plants. WIL has a large proven reference list of satisfied customers across the world.

2.4 Achievements of WIL at a Glance

2.4.1 First Ever

- Indigenously built Sugar project.
- Company to export Sugar machinery and complete Turnkey project from India.
- Export of complete Cement plant with largest ever Kiln manufactured by any Indian company.
- Independent Power project based on municipal solid waste in India.

2.4.2 First to Manufacture

- Critical components for Nuclear Reactor.
- Main propulsion gear boxes for Indian built Navy Frigates.
- Components for Satellite launch vehicle for ISRO.
- Critical components for 235 MW & 540 MWe Nuclear Power projects.
- Road mobile launchers for DRDO.
- Missile firing containers in Nuclear Submarines.
- One of the largest Optical Telescope in Asia.
- Critical components for India's first Moon mission 'CHANDRAYAAN-I'.

- Major critical components for India's Intercontinental ballistic missile (ICBM) programme 'Agni V'.
- Critical components for India's first Mars mission 'MANGALYAAN'.

The trend has been continued and the organisation has developed hardware for India's most ambitious project—mission Gaganyaan and its proof pressure testing facility in India.

WIL has a much diversified portfolio & expertise in fabrication & machining of ferrous and non-ferrous materials like 80HELS, MDN 250, Titanium alloys, Aluminium, 15CDV6, AL Bronze, S690QL, Borated steel, Inconel, Monel and so on. A lot of experiments have been carried out on the heat treatment cycles, welding techniques in order to evolve the most complicated and prestigious projects by utilising these materials.

Although it is not possible to mention about WIL's contribution in evolving welding, heat treatment, etc. about all of those highly critical & exotic materials in detail, a few of the cases have been flashed which have a major contribution to Defence, Aerospace and Nuclear sectors.

2.5 Welding Development in ESR Modified 15CDV6 Material [2-3]

Welding is a major step in the fabrication of most of the pressure vessels, structures and equipment. Steels with carbon equivalent in excess of 0.40 wt% show a tendency to form martensitic on welding, and therefore are considered difficult to weld. This ESR modified 15CDV6 material has a carbon equivalent value of nearly 1.0 that classified it as a very difficult to weld steel.

In this ESR modified 15CDV6 material welding was carried out successfully by Auto GTAW as well as Manual GTAW, without preheating. This was carried out by modification of welding parameters, cleaning process, tacking sequence, etc. Mechanical Properties of the Weld, met the requirements. WIL carried out experiments by using gas tungsten arc welding (GTAW) process with 8CD12 filler wire to join the 20 mm thickness annealed 15CDV6 material respectively.

ESR Modified 15CDV6 Welding by Auto GTAW process can be possible without preheating by proper parameter setting and precautions. Oil Quenching Treatment to the Weld Coupon gives success rate of 100% than the Water Quenched Weld Coupons & Improvement in mechanical properties as compared to Water Quenching treatment. Oil Quenched Weld Coupon gives 12.4% more UTS & 14% (maximum) more 0.2% proof stress value than Water Quenched Weld Coupon. Not much improvement in %elongation observed either by Water Quenching or by Oil quenching, however both methods meet minimum specified elongation values.

2.5.1 Optimization of HT Procedure for 15CDV6 Material

The welded test coupon was subjected to different hardening, tempering heat treatment cycles. The experiment results indicated that by optimising the hardening, tempering temperature and their soaking time, a wide range of mechanical properties can be obtained.

Ultimately, with lots of trials, heat treatment cycle was finalised. By following the evolved cycle, desired mechanical properties were achieved.

2.6 Metallurgical Study of Aluminium Bronze (Non-ferrous Overlay) on Low Alloy Steel

Welding procedure development of Aluminium bronze metal cladding on steel AB2PKM required proper selection of welding process and careful execution of set of trials to establish sound welding. This is particularly important for a component having close geometrical, dimension tolerances and stringent NDE requirements. Aluminium bronze cladding was carried out on a low alloy steel component. Here we had carried out various welding trials conducted by using SMAW and GMAW processes for carrying out the cladding. The processes are compared with respect to weldability in dissimilar metal, consistency in welding quality and welding deposit efficiency. Mechanical testing and metallurgical analysis of weld interface is also provided for the understanding of metallurgical behaviour at weld interface of dissimilar metal cladding joint (Fig. 2.2).

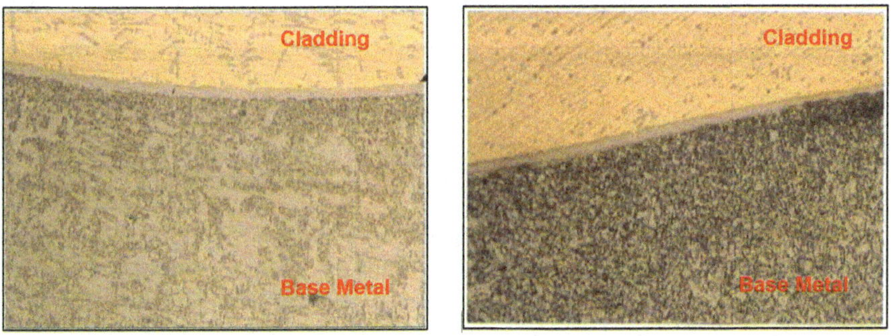

Fig. 2.2 Microstructure analysis, Etchant: Vilellas, magnification used: 200X (Left) SMAW process, (Right) GMAW process

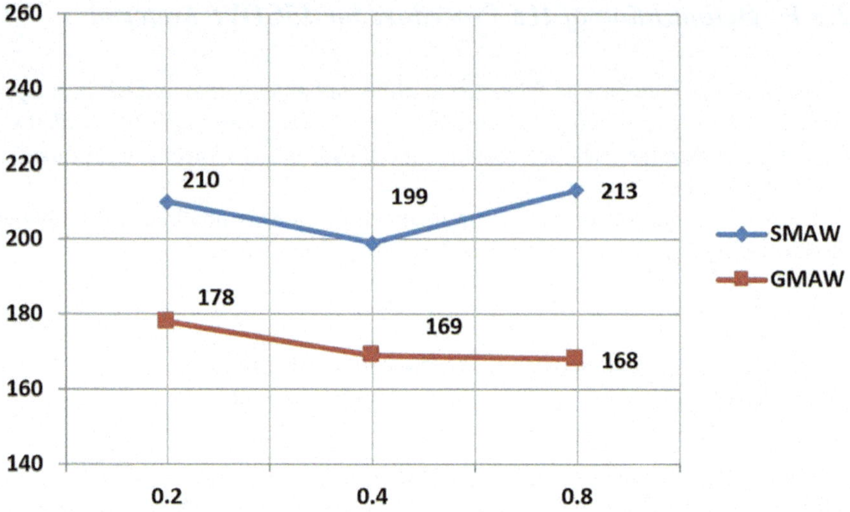

Fig. 2.3 Hardness points intervals in mm from weld interface towards cladding side

2.6.1 Metallurgical Analysis

Microstructure in cladding layer shows equiaxed dual phase (Alpha + Beta). Aluminium bronze in SMAW as well as GMAW process. Microstructure in base metal shows bainite and tempered martensite. Diffusion bonding at cladding free from defects.

The microstructure at base metal side shows coarser grains of bainite in tempered martensite for SMAW process as compared to GMAW process. This is due to higher heat input in SMAW process.

2.6.2 Hardness Traverse Analysis

The hardness readings in cladding and base metal are compared and shown graphically for GMAW & SMAW. It is observed that the hardness is on lower side for GMAW process indicating lesser heat input than SMAW (Fig. 2.3).

2.6.3 Conclusion

In the aforementioned work, it is evident that GMAW process yielded better results as compared to SMAW process for cladding layer of Aluminium Bronze. The problems such as porosity, cracks and lack of fusion encountered in SMAW process were successfully eliminated by adoption of GMAW process. The hardness readings show lesser heat input for GMAW process as compared to SMAW process.

The welding quality was found consistent in all layers.

Savings in weld consumable cost up to 50% were possible with GMAW because of continuous filler wire and better process efficiency of GMAW over SMAW.

2.7 Welding Technology for Stainless Steel Containing 1.2% Boron Content [1]

WIL bagged the order for the major portion, for the fabrication of PFBR components for 500 MWe project. Out of the various equipment manufactured by WIL for PFBR, the Intermediate Heat Exchanger (IHX) involves fabrication of A887 type 304B4 Gr.B—Borated stainless steel.

The basic purpose of using Borated steel in Nuclear fabrication is to reduce the intensity of Neutron bombardment & hence, to control the chain reaction in heart of the Reactor core.

A887 type 304B4 Gr.B is a 1.2% Boron containing, typical stainless steel type of material, not classified in P Number under ASME Sec. IX for welding procedure Qualification. Also, such types of steels do not find any place in routine fabrication. In absence of any Technical input for welding of borated steel, WIL had to initiate for development work.

Lots of trials were conducted on the following modes:

(a) Deploying welding consumables on Borated steel, as per ASME Sec. II C.
   - Consumables used for trials: E-316–15 (Mod.), E Ni Cr Fe3, E-308 L-15, E-316 L-15, E-309 L-15, E-312–16.

(b) Deploying welding consumables on Borated steel, which are not in line with any classification.
   - WIL obtained imported sample electrodes dia.3.2 mm containing Boron. These electrodes were deployed for welding Borated steel plate of approx.150 mm length. No defects were seen while welding the entire thickness of test coupon. Both visual & radiography were found satisfactory.
   - Considering the job schedule & lead time to purchase imported electrodes, it was decided to develop indigenous welding consumable. Consumable manufacturers for developing suitable electrodes were approached. One of the consumable manufacturers came forward, took this challenge & manufactured E-308–16(Nearest) electrode containing Boron.
   - The trial test coupons were welded by using E-308–16(Nearest) electrode. Visual examination, radiography test & mechanical test [tensile & hardness] found satisfactory after conducting various trials and modify weld edge preparation.

- Observations: E-308–16(Nearest) welding consumable with modified weld edge preparation met stipulated radiographic requirements. The repeatability was also established.

Acknowledgements It gives us great satisfaction in presenting the Article on 'WIL's Contribution in development of welding for exotic materials'. We would like to add a few heartfelt with deep sense of gratitude, the encouragement received from Shri V P Shukla (Sr. Vice President Manufacturing) & Shri N M Nadaph (Advisor Nuclear Business), it would have been otherwise difficult to complete this article work without his enthusiastic and timely advice. We take this opportunity to express sincere thanks to Shri Chirag Doshi (MD & CEO of Walchandnagar Industries Limited) for giving the opportunity to present the article.

References

ASME Sec II A, ASME Sec II C & ASME Sec IX

Suresh, M.R., Sinha, P.P., Sarma, D.S., Ballal, N.B., Krishna Rao, P.: Study of Welding characteristics of 0.3C-CrMoV (ESR) ultrahigh strength steel (2007)

Bandyopadhyay, T.R., Rao, P.K., Prabhu, N.: Improvement in mechanical Properties of Standard 15CDV6 Steel by increasing carbon and chromium content and inoculation with Titanium during ESR (2012)

Sreekumar, K., Murthy, M.S.P., Natarajan, A., Sinha, P.P., Nagarajan, K.V.: Identification and morphology of structures in 15CDV6 steel. Trans. Indian Inst. Met. **35**(4), 349–355 (1982)

Chapter 3
Mukand—Quality Leadership in Stainless Steel, and Special and Alloy Steel Long Products

Sunil Nair

Abstract The article traces the growth of Mukand as a specialist in the production of special, alloy and stainless steel long products, starting with re-rolling mills and a foundry in Mumbai and Lahore in 1937. Over 400 grades of steel including stainless steel long products, alloy and special steel long products for the automotive industry are manufactured today. The company was a pioneer in adopting and making a success of continuous billet casting technology, Vacuum Oxygen Decarburisation technology and Oxygen Top and Bottom blown Converter technology for the manufacture of stainless steel. Today, Mukand is a leading premium producer and exporter of stainless steel long products of austenitic, ferritic, martensitic, precipitation hardening and duplex varieties.

3.1 Introduction

Born in the pre-independence era, Mukand Ltd. was incorporated in the year 1937. The company commenced operations with re-rolling mills and a foundry in Mumbai and Lahore respectively. After India gained independence from the British and the subsequent Partition, the Lahore operations shifted to Mumbai.

Today the company is a large multi division, multi-location, engineering company specialising in the manufacture of Special, Alloy and Stainless Steel Long products. The company also designs, manufactures and commissions Heavy Industrial Machinery including execution of turnkey engineering projects. Mukand's growth story has traversed several milestones (Fig. 3.1).

S. Nair (✉)
Mukand Ltd., Thane Belapur Road, Dighe, Kalwe, Thane 400605, India
e-mail: sunilnair@mukandsumi.com

R. Divakar et al. (eds.), *Indian Metallurgy*, Indian Institute of Metals Series,
https://doi.org/10.1007/978-981-99-5060-7_3

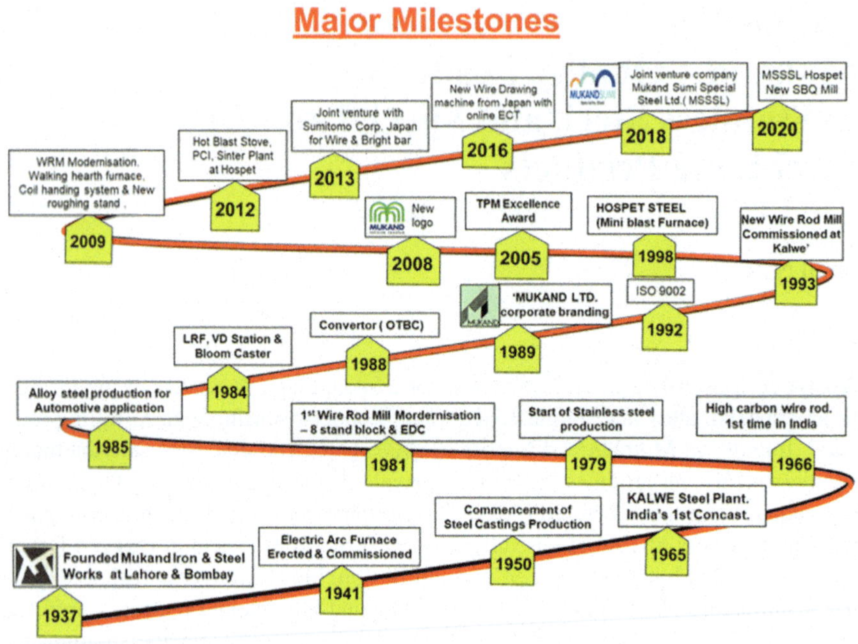

Fig. 3.1 Major milestones of Mukand's growth story

3.2 The Foundry

The Steel Foundry of Mukand used to be the largest in the non-government sector in India until it shut down in the year 2000. The foundry produced castings ranging from tiny components weighing a few kilo grammes to large castings weighing 100 tonnes a piece to customer specifications and drawings. It was one of the first in the country to develop and manufacture high speed cast steel bogies for freight cars. Mukand's foundry was the largest supplier of bogies and couplers to the Indian Railways.

Mukand exported bogies and couplers to Germany, Korea, Taiwan, United Kingdom, USSR (erstwhile) and the USA apart from many countries in Asia and Africa.

Mukand developed the CASNUB bogies for the Indian Railways and became one of its major suppliers (Fig. 3.2).

Fig. 3.2 CASNUB bogie developed for the Indian Railways and alloys steel castings of Kurla foundry

3.3 The Steel Plant

It is rightly said that Mukand is steel as the company today manufactures over 400 different grades of Stainless Steel and Alloy Steels. It is a leading exporter of stainless steel long products, and one of the country's leading suppliers of alloy and special steel long products to the automotive industry.

3.4 Pioneering Spirit

Mukand erected its first Electric Arc Furnace in 1941 and in the year 1965, was the first in the country to adopt and make a success of the continuous billet casting technology using the 'S' type continuous casting machine when it was still nascent in the world. The company was also the first in India to adopt the Vacuum Oxygen Decarburisation (VOD) technology and the first to instal an Oxygen Top and Bottom blown Converter (OTBC) for the manufacture of stainless steel when there were only three other plants that used the technology in the world.

In 1987, Mukand installed an Ultra High Powered Furnace which was then one of the most modern anywhere in the world, equipped with computerised process controls, scrap pre-heating arrangements operating on a waste recovery system and with eccentric bottom tapping and oxyfuel burners as part of its configuration. A ladle refining furnace was added as part of the secondary refining station.

Mukand installed a fully automated wire rod cum bar mill with its walking-beam type cooling bed and an eight stand, no twist block mill. Mukand was the fourth plant in the world to introduce the Easy Draw Continuous cooling system (EDC) for wire rods.

At the time of installation in India, it was the first EDC in India for modifying the microstructure of high carbon wire rods using a hot water quenching technique. The resultant fine pearlitic structure allowed for direct drawing of rolled wire rods to wire with a very high reduction, eliminating the need for patenting. High tensile wire so produced found applications in areas such as the Prestressed Concrete (PC) wire.

3.5 Development of Stainless Steel at Mukand

Mukand commenced production of stainless steel in the year 1979 using the conventional dilution and remelting technique in an Electric Arc Furnace. This process was not economical as it involved using expensive low carbon ferro-alloys, especially low carbon ferro-chrome which was imported. This was a time when India had to depend on imports for all its stainless steel requirements. Mukand continued with the conventional process till 1987, when the introduction of new technologies revolutionised the stainless steel production. The UHPF-OTBC-VOD process flow was unique to Mukand and was christened the TRIPLEX process (Fig. 3.3).

One of the challenges faced in the manufacture of stainless steel in the initial period was how to maintain carbon below 0.08% which was the limit for the popular austenitic 18–8 grade. This was resolved by maintaining the slag basicity at <1.3 to prevent Carbon pickup from slag.

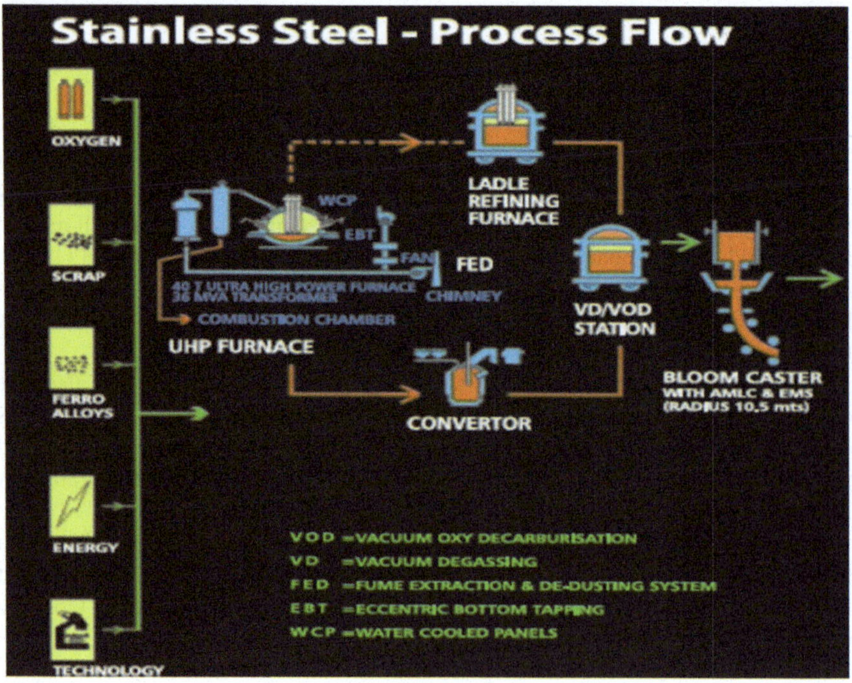

Fig. 3.3 Stainless steel TRIPLEX process route

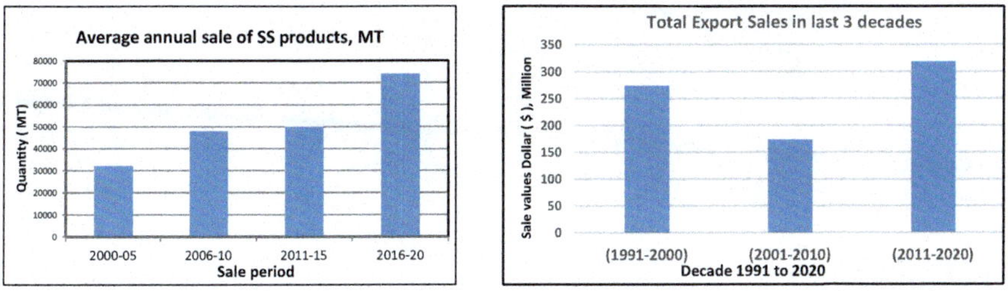

Fig. 3.4 **a** Average annual sale of SS long products, **b** Total SS export sale value for last 3 decades

The other challenge was to prevent excessive oxidation of Chromium. This was done by maintaining a controlled Silicon level in the steel between 0.5 and 0.6%. This process continued for around 5 years after which the TRIPLEX process route and closed stream continuous casting revolutionised stainless steel making. Improved quality and productivity levels could now be easily achieved.

Mukand today is a leading premium producer and exporter of stainless steel long products of austenitic, ferritic, martensitic, precipitation hardening and duplex varieties.

The Quality leadership and wide product mix has helped Mukand achieve significant growth in stainless steel sales, both domestic and export in the last decade (Fig. 3.4a and b).

3.6 Specialised Alloy Steel Development at Mukand

In 1985, Mukand decided to produce alloy steels for the automotive industry. Japanese specifications, however, demanded cleaner steel with Oxygen not higher than 15 ppm. With improvisations in the secondary refining process at the Ladle Refining Furnace (LRF), improved quality of steel with lower Oxygen levels was achieved. The closed stream continuous casting process also prevented the re-oxidation of steel.

Mukand's steel experts were able to achieve the required results and in 1986–87 began supplies to respective component makers. Initial production was high carbon wire rods for cycle spoke and PC wire application, and boron steels for cold heading fastener application.

In the early stages, customers were apprehensive of steel cast through the continuous caster. However, this was overcome with the effective use of Mould EMS and Secondary EMS thereby improving the internal structure of the steel. With extensive testing and trials, customers were convinced that even with 1:6 reduction ratio for Continuous Cast material, the product quality was equivalent to steel produced through ingot casting.

A rigorous process engineering was done to ensure and control Oxygen below 15 ppm. It was achieved by controlling the alumina inclusions and ensuring low calcium level in the steel. This process standardisation also the paved the way for

Fig. 3.5 Rolled steel products and auto component parts produced from Mukand material

ball bearing steel development with oxygen < 10 ppm, which was required for better fatigue life in the bearings.

Mukand installed the Short Time Cycle Furnace (STC) in 1992 for spheroidised annealing. The controlled atmosphere furnace imported from the USA and made with patented Japanese technology, ensured minimum surface decarburisation and a uniform spheroidised structure to meet the requirements of the cold heading process.

All this has helped Mukand supply a variety of steel products for a variety of applications to the Indian steel market (Fig. 3.5).

3.7 Increase in Production as Markets Mature

As the automobile market in India grew, the Original Equipment Manufacturers (OEMs) increased their sourcing of raw materials from within the country. In order to be globally competitive and also become a leading supplier to the auto component industry, Mukand decided to set up a green field steel plant based on the blast furnace route. This plant was commissioned in the year 1998 as a joint venture company in Hospet, near Bellary region, where the high grade iron ore reserves are situated. Best in class technology was adopted early on to ensure high productivity and to meet the stringent quality requirement of the multinational automotive companies who had set up base in India.

The Mini Blast Furnace (MBF)—Energy Optimising Furnace (EOF)—LRF—Vacuum Degassing route for Alloy Steel production entered the growth phase from 2001 onwards (Fig. 3.6). The Hospet Steel Plant now produces almost 0.7 Million MT steel per annum.

Today, Mukand is at the forefront of 'clean steel' production in the alloy steel long product sector by adopting superior process design. All the major passenger vehicle and two-wheeler manufacturers in the country use Mukand's steel for the manufacture of critical components. Several European and American auto makers too have approved Mukand steel for the manufacture of components that are then exported to various countries across the globe.

Annual sales of Special and Alloy Steel have doubled in the last 20 years (Fig. 3.7).

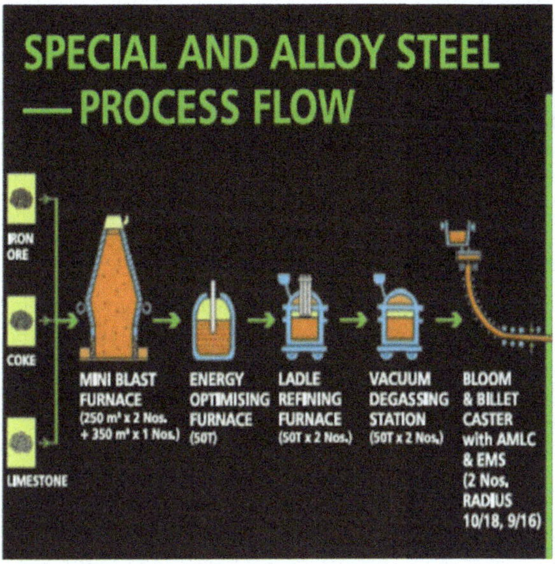

Fig. 3.6 Process flow for special and alloy steel making

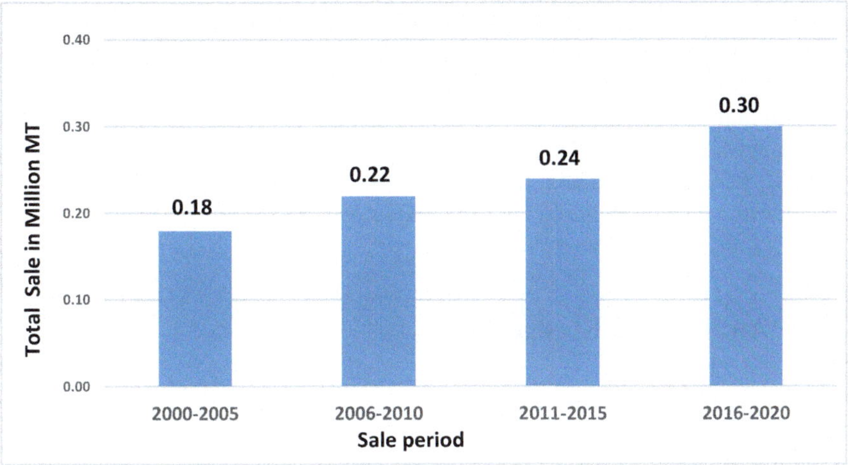

Fig. 3.7 Average annual sales of special and alloy steel

3.8 Research and Development (R&D)

Research and Development has been at the forefront of Mukand's growth in the steel sector. The Company established the R&D department in 1985. 'Super steel—60' dual phase steel bars for the construction industry, controlled surface layer (CSL)

process for colouring stainless steel parts for decorative purposes were some of the achievements in the initial years of this department. The company invested in equipment such as the Scanning Electron Microscope (SEM) and automatic computer controlled polishing machines way back in the 1990s. The R&D unit designed and installed a pilot plant for descaling of stainless steel wire rods through a 'Rapid Electrolytic Descaling' process. R&D activities were carried out in heat treatment and development of import substitution steel, viz., ball bearing, cold heading quality steel and high-speed machining steels. These achievements were well recognised in the metallurgical field and the R&D division of Mukand won 'Outstanding in-house R&D' award of 1993 in the Secondary steel Sector from Department of Scientific and Industrial Research (DSIR), Government of India. Recent focus of R&D has been on the development of new verities of Duplex stainless steels, Microalloyed steels and process improvements for clean steel production. Several technical papers have been published in national and international journals.

3.9 Total Quality Management

Mukand was one of the first companies in implementing Total Quality Management (TQM) systems in the steel sector. TQM was implemented by the company in the late 80s, starting with Quality Circles, and Juran Quality Improvement (JQI) programmes. The ISO 9000 standards were also adopted, as a Quality Management System. Mukand was the first steel company in the country to be certified to this standard in the early 90s.

In the year 1999, Total Productive Maintenance (TPM) was initiated with the help of JIPM, Japan. Mukand won the TPM Excellence Award from JIPM, in the year 2003 for its Machine Building Division and in 2005 for its Steel Division.

3.10 Awards and Recognitions

Mukand's leadership in consistent high-quality products has been well recognised. The company won the National Quality Award four times during the period 1994 to 2000 in the Mini steel sector category from the Indian Institute of Metals (IIM). Mukand won the Gold Certificate from the Ministry of Steel for the year 2016–17 for contribution to the secondary steel sector. Mukand has been regularly winning Supplier Quality Awards from its customers. Prominent among these have been the Global Quality award for the year 2012 and the Global best supplier award for the year 2019 from SKF, Sweden.

3.11 Gearing Up for the Millennium

Having forged a joint venture with Sumitomo Corporation of Japan, Mukand recently commissioned yet another green field project of a state-of-the-art modern Bar and Wire Rod rolling mill in Hospet, Karnataka with an investment of Rs. 6,500 Million. This modern bar and rod mill with PSM technology of SMS Germany has a capacity of 0.6 million MT.

This fully automated facility aims to embrace the Industry 4.0 norms. Becoming carbon neutral and embracing clean energy technologies to reduce the carbon footprint is the next giant step in sustainable development that the company hopes to achieve.

References

Amit Ganguly, Srinivasan, K.R., Sunil Nair: Specialty stainless steel from Mukand, mineral & metals review, **XXXII**(9) p. 53 (Sept.-06)

Mukand company annual reports

Niranjan More, Sunil Nair, Srinivasan, K.R., Athalye, A.S.: Microalloyed steel for long products for autocomponent applications. In: 20th International Forging Congrees 2011, Conference Proceedings, p. 325

Srinivasan, K.R., Amit Ganguly.: Mukand's Contribution towards Stainless Steel Development, Steel Scenario, **7**(Q-2) p. 69 (Oct-Dec'97)

Chapter 4
History of Indian Rare Earths Limited

Anuttam Mishra

Abstract In the pre-independence era, a chance discovery of monazite in Kerala led to its exploitation for the benefit of Europe. Recognizing the importance of rare earth minerals for the Indian nuclear program, Indian Rare Earths Limited (IREL), a central government undertaking was incorporated in 1950 to process minerals for the production of rare earths and extraction of thorium. This article describes the genesis and achievements of IREL over the past seven decades, highlighting IREL's contribution to the development of nuclear power industry in India. Having established production facilities for the present requirements of Indian nuclear program, IREL is implementing projects to meet requirements of other strategic applications such as rare earth magnets, and to establish plants to bring new products to the value chain.

4.1 Genesis

Initiation of beach sand mineral (BSM) industry in India dates back to 1908, when a German chemist, Herr Schomberg identified monazite in coir exported from Kerala. Those days were the days of gas mantle and thorium oxide was in huge demand. Encouraged by this supply chain, Herr Schomberg established the first plant at Manavalakurichi (MK) in 1910 and subsequently another plant in Chavara for separation of Monazite. Both the plants were closed down during the first World War, when Herr Schomberg was arrested on charges of being a German Spy. Thereafter, the London Cosmopolitan Mineral Company took over these plants. Hopkins & Williams, another Company based out of London, took over the operations and first export of ilmenite from Chavara started in the year 1922. Indian ilmenite maintained monopoly in the global market as a feed source for making white titanium pigment.

Post-independence, when the Government of India set up Atomic Energy Commission under the chairmanship of Dr. Homi J. Bhabha, one of the first steps

A. Mishra (✉)
Indian Rare Earths Limited, IREL (India) Limited, ECIL Building, Veer Savarkar Marg, Prabhadevi, Mumbai 400 028, India
e-mail: anuttammishra@irel.co.in

R. Divakar et al. (eds.), *Indian Metallurgy*, Indian Institute of Metals Series,
https://doi.org/10.1007/978-981-99-5060-7_4

taken by the commission was a stop the export of monazite and evaluate the possibility of setting up a facility of processing the minerals for production of rare earth and extraction of thorium on a commercial scale. In August 1950, Indian Rare Earth Limited, now IREL (India) Limited, was incorporated at Bombay as a private Limited Company jointly owned by the Government of India and Travancore, Cochin. The construction work of the Rare Earth plant at Aluva in Kerala commenced in April 1951 in collaboration with the French firm, *Société des Produits Chimiques des Terres Rares*, now Solvey, by providing the necessary process details and technical assessment (Fig. 4.1). In the same year, IREL became a full-fledged Central Government Undertaking under the Department of Atomic Energy.

On December 24, 1952, the late Prime Minister Pt. Jawaharlal Nehru dedicated the Rare Earth Plant to the nation which had an installed capacity of processing 1,500 tons per annum (tpa) monazite to produce Rare Earths and Thorium compounds.

During the decade, IREL was entrusted by the Atomic Energy Commission to setup a Thorium plant at Trombay to process crude thorium hydroxide generated at the rare earth plant to extract uranium and produce thorium nitrate and thorium oxide for use in manufacture of incandescent gas mantle and for nuclear power program respectively. The thorium plant, commissioned in the year 1955, was one of the largest in the world meeting the requirement of the then vast gas mental industry in India as well as abroad.

Fig. 4.1 Rare earth division, Kerala

In the early stages of development of BARC, then the Atomic Energy Establishment, Trombay, IREL played an important role in construction of Canada India Reactor (CIRUS) as well as construction and commissioning of a Uranium metal plant at Trombay. The uranium produced from IREL's Thorium plant in Trombay was first put in the Cirus Reactor for furthering the nuclear research in India. Considering the success in project implementation by IREL, the company was entrusted to setup the Uranium Mill Project at Jaduguda. The project was completed in record time and handed over to Uranium Corporation of India in the year 1966.

After becoming a full-fledged central government undertaking in 1963, under the administrative control of Department of Atomic Energy (DAE), IREL took over a number of private companies (Hopkins & Williams, Travancore Minerals Limited) engaged in mining and separation of beach and minerals in the southern part of the country and established two mineral divisions one at Chavara in Kerala (Fig. 4.2) and other at Manavalakuruchi in Tamil Nadu (Fig. 4.4), and thus started the mineral operations of the Company.

The year 1977 witnessed IREL commencing its work on its flagship division, Orissa Sand Complex (OSCOM) at Chatrapur, Odisha (Fig. 4.4) where commercial operation started in the year 1986. A new Thorium Plant was commissioned in OSCOM Unit of IREL in 1998 and production of thorium nitrate in Trombay plant was discontinued.

Fig. 4.2 Chavara plant, Kerala

Fig. 4.3 Manavalakuruchi plant, Tamil Nadu

Beginning of the new century witnessed closure of the monazite processing operation in RED, Aluva on the recommendations of statutory body, Atomic Energy Radiological Board (AERB). By that time, the plant, which was designed with initial capacity of 1,500 tonnes per annum, (tpa) had been upscaled to 4,200 tpa. The operations of the plant were retrofitted and IREL took up Thorium Retrieval Uranium Recovery Restorage (THRUST) Operation to process the crude thorium hydroxide stored in silos to recover the uranium values and store the thorium values in purer form as thorium oxalate.

The monazite processing operations were established in a higher scale in OSCOM unit of Odisha. The Rare Earth Processing Plant (REEP) was commissioned with a capacity to process 10,000 tpa of monazite. To add value to the mixed rare earths chloride produced from monazite processing operations, the operations of IREL in RED, Aluva were modified based on in-house R&D to set up a plant to produce separated high pure rare earths.

4.2 Operation

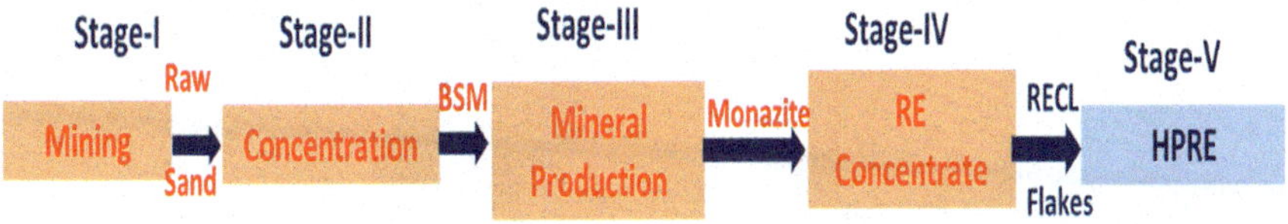

Stage-1

Mining of ore (beach sand) containing 10%–25% Atomic/ Heavy minerals.

Mining Areas: OSCOM Deposit, Odisha Block II, IEE, IV & IV EE in Chavara, Kerala, Vettmadai-Kottumangalam-Pelliyarcoil deposit, Kanyakumari, Tamil Nadu.

Stage-2

Up-gradation of the mined-out sand to concentrate of about 7 Lakh tonnes containing about 95–97% of Atomic/heavy Minerals.

Stage-3

Mineral processing/separation based on physical properties to produce individual Atomic Minerals which interalia include monazite, zircon, ilmenite, rutile, leucoxene, sillimanite, and garnet.

Stages 2 & 3 operations are carried out in Mineral Separation Plants at OSCOM, Odisha; Chavara, Kerala and Manavalakurichi, Tamil Nadu.

Stage-4

Cracking of Monazite for production of RE concentrate, Thorium, NGADU, and Tri-Sodium Phosphate. The operations are carried out in OSCOM Plant in Odisha.

Stage-5

Processing of RE concentrate to produce separated high pure RE in the form of oxides/ compounds of samarium, neodymium, praseodymium, lanthanum, and cerium. The operations are carried out in RED, Aluva, Kerala.

The product profile of IREL involves multi-mineral and chemical products, totalling around 30 products:

Minerals: Ilmenite (3 grades), Rutile (3 grades), Zircon (3 grades), Monazite (3 grades), Sillimanite (2 grades), and Garnet (5 grades).

NB: Grades are denoted by Q-Chavara, MK-Manavalakurichi, OR-OSCOM.

Chemicals: MRECL, Nuclear Grade Ammonium Di-Uranate (NGADU), oxides/ compounds of Samarium, Neodymium, Praseodymium, gadolinium, dysprosium, etc.

Fig. 4.4 Orissa sand complex (OSCOM)

By-products: Tri-Sodium Phosphate (TSP), thorium nitrate/oxalate, oxides/compounds of Lanthanum, Cerium.

4.3 Market Outreach

Initially established with the objective to supply strategic compounds, IREL gradually widened its product profile to cater to the strategic as well as commercial sectors.

While rare earths such as dysprosium and gadolinium are supplied to the Department, thorium values are stored in engineered trenches for use in the 3rd stage nuclear power program of the Country. Zircon is supplied to Nuclear Fuel Complex (NFC) to manufacture structural material to house nuclear fuel. Further, a number of thermal barrier coating materials based on zircon and rare earths are produced based on in-house R&D and supplied to Defence sector for test works.

The other minerals are supplied to number of MSEs, who have set up operations in the value chain. The mineral production and revenue from operations are continuously increasing by the day. For the first time in the history of the company, the revenue from operations crossed the Rs. 1,000 crore mark, last year (Fig. 4.5).

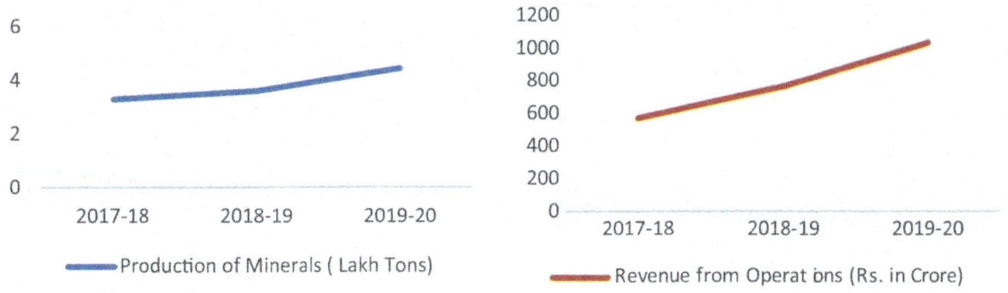

Fig. 4.5 **a** Mineral production, **b** revenue from operations

4.4 Current Projects

With the objective to broaden its strategic alliance and commercial footprint, IREL has embarked on the following projects:

4.4.1 Projects in the Value Chain of Rare Earths and Minerals

4.4.1.1 Rare Earth Permanent Magnetic Plant

The plant is being set up in BARC complex at Vizag (Fig. 4.6) for producing 3 tpa of samarium cobalt permanent magnet for use by DAE, Defence, and Space sector. Construction activities of the plant have started post receipt of all statutory clearances from such as environment clearance, consent to establish, and BARC safety committee. The plant will be operational in FY 2022–23.

4.4.1.2 Rare Earth and Titanium Theme Park

Intermediate and final value chain of rare earth is absent in the country, on account of which India is dependent on import for these items. In order to enhance the consumption of rare earth within the country, IREL is setting up a Rare Earth and Titanium Theme Park in Bhopal (Fig. 4.7) which involves the following:

- Upscaling of the scientific interventions developed by BARC in laboratory scale to pilot scale.
- Demonstration of these pilot technologies to entrepreneurs so as to encourage them to set up commercial plants.
- Provide training to develop skilled manpower for the future.

The theme park will be operational in FY 2023–24.

Fig. 4.6 Rare earth permanent magnet plant

Fig. 4.7 Rare Earth and Titanium Theme Park, Bhopal

4.4.2 Pilot plant for Titanium Dioxide and Zirconium Oxy-Chloride

Pilot plant is being set up to produce high purity titanium dioxide and zirconium oxychloride from titanium slag and zircon of OSCOM grade. The objective of the plant is to standardize operating parameters and gain confidence to set up commercial plants. Environment clearance for the project is received and preparation of DPR is underway.

4.4.3 *Titanium Slag*

IREL has entered into a Memorandum of Understanding with UKTMP JSC, Kazakhstan for setting up of facility for producing titanium slag, which will be used as the feed material for production of titanium sponge. Trial for finding out the suitability of OSCOM grade ilmenite for conversion to titanium slag and sponge has been taken up and discussions are on with the Agency to translate the MoU into a Joint Venture.

4.5 Mineral Projects

4.5.1 *Capacity expansion of OSCOM*

Capacity expansion program envisages to increase the capacity of OSCOM Plant to about 6.3 Lakh tpa (increase by 3.0 Lakh tons) of total minerals. All statuary clearances, viz., Environment, CRZ, Forest stage I, and consent to establish have been obtained and the agency for execution has been appointed. The construction at site has commenced. The project is slated for completion in the last quarter of fiscal 2021–22.

4.5.2 *Hrushikulya-BajrakotBrahmapur Deposit (BSM), Odisha*

IREL has formed its first subsidiary by forming a Joint Venture Company with Industrial Development Corporation of India Limited, Odisha to exploit the mineral sands deposits in Odisha. IREL intended to obtain license for 2542 ha deposit and would be getting the license shortly as per the provisions under AMCR 2016. Project DPR is ready and pre-project activities such as environment/CRZ clearance and preparation of mining plan is in process. The project on completion will enhance the mineral production capacity of IREL by additional 2.5 Lakh tonnes pa of total minerals by 2024–25.

4.5.3 *Bramhagiri Mineral Sands Deposit (BSM), Odisha*

In line with the provisions of AMCR 2016, IREL has initiated the process of obtaining the license for deposit to the extent of 2499 ha from Government of Odisha. The project will be taken up in two phases and each phase will increase the mineral producing capacity by additional 2.5 lakh tons and total of 5.0 lakh tonnes pa.

4.5.4 *Inayam-Midalam Deposit (BSM), Tamil Nadu*

IREL is aspiring to obtain license for exploitation of the Inayam-Midalam deposit over an extent of 1174 ha which would be augumenting its current capacity at MK plant since existing mining lease of the area associated with plant is almost completed. Government of Tamil Nadu have forwarded the reservation proposal of IREL to the Ministry of Mines. This project is targeted at sustaining the present operations of MK Plant.

4.5.5 *Kudiraimuzhi Deposit (BSM) in Tamil Nadu*

IREL has the plan to harness the deposit to the extent of 6000 ha by forming a JV with the Government of Tamil Nadu PSU; Tamil Nadu Minerals Limited (TAMIN). The MoU has been approved by IREL Board and is awaiting approval in Chief Minister of TN Office. The capacity of IREL plant would be 2.5 Lakh tonnes pa.

4.5.6 *Donkuru-Boruva (BSM) Deposit in Andhra Pradesh*

IREL is expressed intent to exploit 1590 ha BSM deposit of Bonkuru Boruva in AP, which is in neighborhood of OSCOM, which would be capable of producing 2.5 Lakh tons of minerals per annum.

4.5.7 *BoruvaBendi (BSM) Deposit in Andhra Pradesh*

IREL is expressed intent to exploit 1590 ha BSM deposit of Bonkuru Boruva in AP, which is in neighborhood of OSCOM, which would be capable of producing 2.5 lakh tons of minerals per annum.

4.6 RE Source Augmentation

4.6.1 Ambadongar Deposit

A Carbonatite deposit bearing Rare Earth Elements (REE) is explored in Ambadongar village of Chhota Udepur District, Gujarat spread over an area of about 400 ha. The average grade of RE in the ore is 1.34%. Initially IREL is preparing to exploit about 73,400 tonnes of REO. Preparation of TEFR & DPR of the project to ascertain its technical feasibility and economic viability to take up pre-project activities. The initial proposal of IREL is to set up a plant to produce 2,500 tpa of RE concentrate containing about 1,000 tons of REO.

4.6.2 Bramhagiri RE Plant

A replica of existing Rare Earths Extraction plant shall be established along with exploitation of BSM at Brahmagiri. Thus, additional capacity of 11,000 tonnes of RE concentrate would be established based on monazite produced from aforementioned deposits of BSM.

4.7 Summary

In summary, the history of Indian Rare Earths Limited (IREL) dates back to the early twentieth century when monazite was identified in Kerala, leading to the establishment of the first plant for separation of monazite. Over the years, IREL has played a significant role in India's rare earths, providing zircon for extraction of elements crucial for nuclear reactor structural materials, and thorium for the upcoming third stage of Indian nuclear program. Looking ahead, India's demand for rare earths is expected to grow significantly in the coming years due to the increasing use of rare earths in various sectors such as electronics, energy, and defense. Hence, in recent years, IREL has focused on adding minerals to its product profile and to identify newer sources of rare earths.

Chapter 5
The DASTUR Story

Kaushal Kumar Sinha

Abstract The story of M. N. Dastur & Company (P) Ltd. is one of dreams and determination of a young man who set up and grew, over almost 50 years, India's first consulting engineering organization in the private sector, based entirely on indigenous talent and expertise. It is also a tale of challenges overcome and a firm's emergence over time as a globally acclaimed consulting leader. The organization has grown from a handful of professionals, each an expert in his own domain, into a large, vibrant multi-disciplinary team that is synonymous with excellence in engineering, business and technology consulting services.

5.1 The Making of a Visionary

In 1945, Minu Nariman Dastur (Fig. 5.1), a bright young engineer from a middle-class Parsi family who had graduated in electrical and mechanical engineering from Banaras Hindu University (BHU) in 1938, moved to the US to take up a Tata Endowment Scholarship at Massachusetts Institute of Technology (MIT). He had already worked for Tata Steel for three years, first as a Graduate Apprentice, then as Second Helper in the open-hearth furnaces at Jamshedpur. Subsequently he helped plan the plant's expansion as a Technical Assistant in the General Superintendent's Office.

At MIT, Dastur worked under the legendary Dr John Chipman, completing his doctorate on 'Principles of Steelmaking' in 1948 while taking a minor in business administration. Over the next six years, Dr Dastur gained experience working for two US consulting engineering firms, H. A. Brassert & Co. (where he also headed their Ore Research and Metallurgical Laboratory in Greenwich, CT), and Ramseyer & Miller Inc. During this time, he became an expert in pelletization, in direct reduction of iron ores, and in conceptualizing/planning steel plants in South America.

In 1954, Dr Dastur was sent to India by the US Agency for International Development, in the capacity of "American expert", to advise Mysore Iron and Steel Works on its plant expansion programme. During this mission, Dr Dastur met Prime Minister

K. K. Sinha (✉)
M. N. Dastur and Company (P) Ltd, Kolkata 700013, India
e-mail: Kaushal.S@dastur.com

R. Divakar et al. (eds.), *Indian Metallurgy*, Indian Institute of Metals Series,
https://doi.org/10.1007/978-981-99-5060-7_5

Fig. 5.1 Founder Chairman Dr M. N. Dastur

Nehru, who, realizing that the American expert happened to be Indian, invited him to return to India to set up a consulting engineering organization to support the government's planned greenfield steel plants. Following a further meeting with the Minister for Commerce and Industry, Dr Dastur decided to return to India to help develop his country's fledgeling national steel industry.

5.1.1 A Firm Believer in the Concept of 'Atmanirbhar Bharat'

Dr Dastur believed that India's industrial development strategy, especially for the iron and steel sector, should be primarily based on indigenous talent and expertise. As early as 1953, when India was finalizing plans with the German company Krupp and Demag to install a greenfield integrated steel plant at Rourkela, one of three plants to be set up with foreign technical and financial assistance, Dr Dastur expressed strong reservations about the unrestricted import of foreign technical and engineering skills. In a letter to India's Consul General in New York, Dr Dastur cautioned that it is "... absolutely necessary to exercise a check on the work of the foreign engineers on a project of such magnitude costing hundreds of millions of dollars", and that "... 70 to 80% of the engineering can be done locally, a good part of the future plants fabricated in India, and only the most essential items imported from abroad".

Dr Dastur was convinced that technical control and management of projects in the country should be "concentrated in a group of competent Indian engineers who are familiar with local conditions and future needs". On his return to India in 1955, he set up M. N. Dastur & Company (P) Ltd. (DASTUR) in Calcutta–India's first consulting engineering organization—primarily aimed at providing comprehensive consultancy, design and engineering services for the national iron and steel industry. The young visionary had returned to his roots.

5.2 Giving Shape to a Vision

Dr Dastur started operations from a single floor at P-17, Mission Row Extension, in the busy commercial area of Calcutta, with a small team of talented and experienced professionals with substantial expertise in the design and operation of iron and steel units. Initially, he faced considerable challenges in convincing potential clients of DASTUR's capability to handle projects effectively without recourse to off-shore expertise.

The first project was for Tata Steel to set up a ferromanganese plant at Joda, designed to produce 30,000 tons annually, with a provision for expansion to 100,000 tonnes per annum. The project was awarded based on global tenders, and DASTUR had to compete with bidders from the United States, Britain, Germany and Japan. The plant was to be set up on a turnkey basis, which included stringent penalty clauses for each month of delay. In spite of numerous local problems involving local tribes, and the Suez crisis, the project was completed eight months ahead of schedule, earning DASTUR a bonus (Fig. 5.2).

The Joda ferromanganese plant not only laid the foundations of India's ferroalloy industry, but it also established DASTUR's credibility as an Indian firm capable of putting a metallurgical plant into operation in only 24 months, starting with nothing but a bare site and under adverse conditions.

Fig. 5.2 Joda ferromanganese plant

The success of Joda gave shape to Dr Dastur's vision of being a partner in the industrialization efforts of India by providing comprehensive consulting engineering services "from concept to completion". A spate of assignments followed, including design of an electric arc furnace shop, continuous casting and wire rod mill for Mukand Iron & Steel Works, Bombay; a ferro-alloy plant utilizing low-grade ores for the Raja of Sandur; a uranium milling plant at Jadugoda for the Atomic Energy Commission; an industrial estate for manufacturing light engineering goods for Voltas Limited, Bombay; and civil engineering of the Rourkela Fertilizer Plant.

In the early 1960s, the government appointed DASTUR as consulting engineers to set up an alloy steels plant for Hindustan Steel Limited (now SAIL) at Durgapur. The plant, an excellent model of international collaboration, was designed to produce 80,000 ingot tonnes per annum of a complete range of alloy and special steels in bar and sheet form. While DASTUR had overall responsibility for the engineering and supervision of the construction of the plant, Atlas Steels of Canada provided specialized training and know-how for the alloy steel production. The equipment was purchased from leading international suppliers on the basis of global tenders and recommendations by DASTUR.

It was around this time that Dr Dastur was invited by the Institution of Engineers (India) to deliver the fourth Sir M. Visvesvaraya Lecture at the 41st Annual Convention in Bombay in February 1961. Dr Dastur's lecture, entitled 'Steel in India—Economic and Technological Possibilities' had all the hallmarks of a visionary par excellence. The ground-breaking ideas presented in the lecture included:

- the need for India to plan for a potential steel demand of 110 million ingot tonnes by the turn of the century, considering the pivotal role of steel in ensuring the growth of the country's economy;
- a phased programme of capacity build-up through an appropriate mix of large multi-million-tonne integrated plants and a number of smaller-capacity integrated/non-integrated plants;
- planning the layout of steel plants to provide for rational and unrestricted growth and diversification to cater to evolving markets;
- conservation of coking coal resources through the blending of inferior coking coals with prime varieties;
- introduction of more intense burden preparation techniques and the adoption of appropriate modern technologies/operating practices to suit indigenous raw materials.

5.3 Dreams Take Wings

As a result of its contribution to the alloy steels plant, DASTUR enjoyed the government's full confidence as a competent, reliable consulting engineering organization. In 1963, the Steel Minister appointed DASTUR as the overall consultants for the Bokaro project, although its role was reduced to the design and engineering of auxiliary units when the government accepted a package of financial and technical assistance from the Soviet Union. Despite this setback, DASTUR secured several prestigious assignments, many of them pioneering in nature, during the 1960s and 1970s. Notable amongst them were.

- first continuous caster at Mukand Kalwa
- uranium milling project for Indian Rare Earths at Jadugoda
- engineering support for nuclear research at Nuclear Fuels Complex
- first ESR unit for Firth Sterling
- first rotary kiln sponge iron unit at Kothagudem for Andhra Pradesh Industrial Development Corporation
- India's only super alloys plant for MIDHANI

Two significant projects initiated during this period were the modernization/capacity augmentation of Tata Steel's Jamshedpur works, and a large integrated steel plant based on natural gas-based direct reduction and electric arc furnace steelmaking at Misurata in Libya (Fig. 5.3). A unique feature of the Jamshedpur project was the phased implementation, with the surplus revenue generated after implementing each phase being utilized for part-funding the subsequent phases.

Fig. 5.3 DASTUR engineers at Misurata site

5.4 Distinguishing Features of DASTUR

The distinguishing feature of DASTUR has been its capability of providing integrated design and engineering services, from concept to commissioning, for a wide range of projects. With a large multi-disciplinary team of professionals and technical staff, DASTUR provides a unique blend of experience and talent. As client demand grew, the organization outgrew its headquarters in Kolkata, opening branch offices in Madras (Chennai), Bombay (Mumbai), Bangalore (Bengaluru), New Delhi, Bhubaneswar and Hyderabad. An international presence with offices in Düsseldorf, Tokyo, Abu Dhabi and the USA has allowed the organization to also cater to its global clientele.

DASTUR has earned global appreciation for its dedication and teamwork and is internationally recognized as one of the largest independent consulting engineering organizations in the world. DASTUR has advised on the development of India's steel sector as Retainer Consultant to the Steel Ministry, and as a Member of Steel Ministry Committees has been nominated to prepare a number of White Papers. UNIDO has also invited DASTUR to prepare master plans for various countries. The global recognition of its technical competence has enabled DASTUR to secure international assignments against stiff competition, notable amongst them being SIDOR in Venezuela, ARCO Steel in Egypt, ArcelorMittal Europe, Perwaja in Malaysia, STELCO in Canada, as well as GPH and Basundhara in Bangladesh, Qatar Steel, and Senaat.

The following figure depicts some of the important landmarks in the growth of the DASTUR organization (Fig. 5.4).

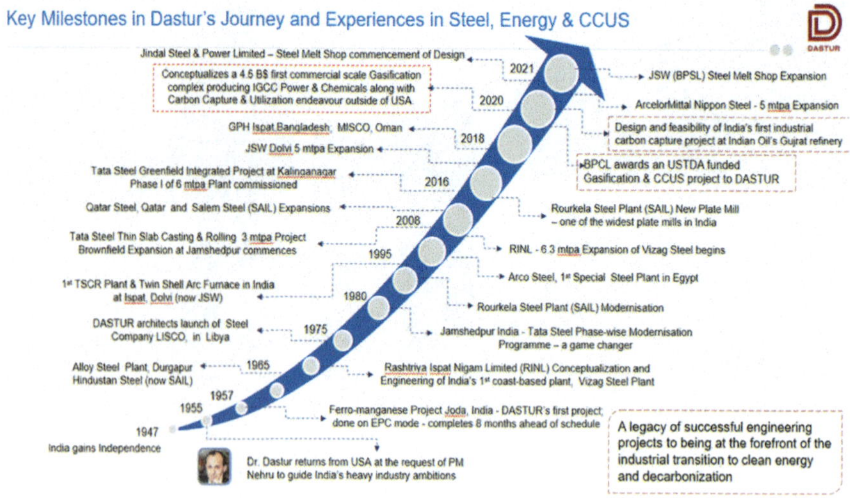

Fig. 5.4 Landmarks in the growth of the DASTUR organization

5.5 Professional Integrity and Conviction

As an independent organisation, DASTUR has always provided objective and unbiased advice in the best interests of its client and their project.

At the beginning of the Tata Steel Jamshedpur modernization project, intense deliberations took place between Tata and DASTUR's experts over the choice between the proven BOF steelmaking process favoured by DASTUR and the relatively recent and unproven bottom blown process. Ultimately Mr J. R. D. Tata, the final decision maker, chose the BOF process. Another of DASTUR's recommendations that was accepted was the future provision for bottom purging of inert gas.

Similarly, when the Malaysian government awarded work on the integrated Perwaja DR-EAF plant to a Japanese consortium on a turnkey basis under its "look East" policy, although the NSC-DR process being offered had not been proven in commercial operations, project consultants DASTUR insisted on the inclusion of stringent performance guarantee conditions with consequential liquidated damages in the contract to safeguard the client's interest. Ultimately the DR unit failed to perform as per the guaranteed conditions and the contractor had to suitably compensate the Malaysian client.

For the first integrated steel plant in Libya, DASTUR had recommended Misurata as the most suitable location for construction of a captive deep-water port to serve the plant, although Col. Gaddafi wanted the plant located at his home town of Sirte. Not to be intimidated, Dr Dastur reiterated to the Libyan authorities that Misurata was the most suitable location, based on a detailed location selection study. Ultimately Col. Gaddafi himself agreed to locate the plant at Misurata.

DASTUR's consistent record of professional excellence, its stringent business ethics, and its pursuit of the sole objective of satisfying customer's needs and expectations have led to numerous accolades from its global clientele. A few such examples are shown below (Fig. 5.5).

5.6 Catering to Evolving Market Requirements

In response to the needs of globalization and growth, DASTUR has added a new dimension to its extensive array of services in the form of Dastur Business & Technology Consulting (DBTC). DBTC leverages the wealth of DASTUR's existing in-house capabilities and the combined expertise of its veteran professionals and specialized services to cater to clients globally. DBTC focuses on the development of innovative technology-enabled solutions for clients and operates worldwide, continuously evolving to meet client requirements and market dynamics.

Dastur Innovation Labs (DIL) was founded in Toronto in 2017 and is the applied industrial research wing of DASTUR. DIL carries out cutting-edge technology-driven applied research in the areas of ironmaking and steelmaking, oil and gas,

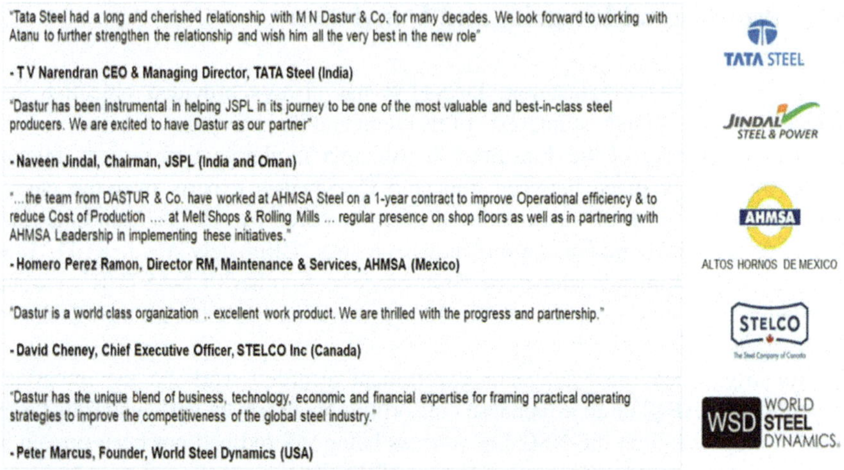

Fig. 5.5 Examples of accolades and appreciation from M. N. Dastur's global clientele

chemicals and petrochemicals, gasification and carbon capture utilization and storage (CCUS), and feedstock and fuel supply chains. DIL focuses on the development of industrial solutions that are innovative, sustainable and implementable. The DIL global team consists of multi-disciplinary experts in process modelling and simulation, materials/mechanical/chemical engineering, supply chains and logistics science, sensors and automation, computation, statistics, ML and AI, and works collaboratively with experts from both industry and academia. DIL's industrial solutions are based on the fundamental principles of engineering, multi-scale process modelling, fluid dynamics, mass and energy balance, high-temperature thermochemistry and rate kinetics, coupled with data analytics, process economics and operational expertise.

Dastur Energy was formed in 2020 and designs solutions for the rapidly evolving areas of clean energy systems and carbon abatement. It works with governments, enterprises and institutions on areas related to clean energy systems design, energy policy, energy engineering, and carbon dioxide removal in industrial sectors from petrochemicals, oil and chemicals to iron and steel.

5.7 Vision for the Future

As a consulting engineering firm, DASTUR's vision for the future is to help clients solve challenging and compelling engineering problems, be it charting appropriate growth strategies through organic and inorganic routes, increasing operational efficiency and reducing waste, or ensuring the sustainability of their business by managing their environmental impact and carbon footprint. To this end, DASTUR is increasingly looking at deployment of the latest technologies and know-how in a wide range of service offerings for clients: designing, engineering and assisting in the commissioning of capital projects in the shortest possible timeframes without compromising on quality, safety or environmental considerations; or providing objective opinions to clients to aid in their investment decisions; or providing innovative solutions to manage solid waste and gaseous emissions.

One of DASTUR's key focus areas going forward is clean energy and ensuring the sustainability of industrial systems, and with its expertise, know-how and IP, the organization anticipates being the first port of call for industrial clients in this area. With industry contributing to over 30% of global anthropogenic CO_2 emissions, industrial clients the world over face the key challenge of transitioning to clean energy systems. DASTUR is working with clients across North America, the Middle East and India to design economically viable and flexible industrial-scale clean energy and carbon capture systems enabling clients to transform their clean energy transition and sustainability challenge into a competitive advantage and stay ahead of the regulatory curve.

As a consulting firm, DASTUR's most important and perhaps only assets are its people. DASTUR strives to be a company where employees not only feel fully engaged and intellectually challenged in solving client problems but are also adequately rewarded, recognized and cared for. During the COVID-19 pandemic, DASTUR transformed its operational model from bricks-and-mortar office based to fully functional "work from home" within a matter of days. Working from home is currently the default mode of operations at DASTUR, ensuring employee safety and well-being as well as business continuity.

5.8 DASTUR's Association with The Indian Institute of Metals (IIM)

As a Patron Member, DASTUR has enjoyed a long and enduring association with IIM, and provides support for its various activities. Dr Dastur was intimately involved from the Institute's early formative days. He was its President from 1967 to 1970 and was instrumental in the formation of the Calcutta Chapter in 1966. Mr S. Das Gupta, former Chairman and Managing Director at DASTUR, served as President from 1997 to 1998. In February 1990, DASTUR, in association with the Iron & Steel Division of IIM, organized a two-day national seminar in Kolkata on issues related to the modernization of steel plants, at a time when SAIL was about to embark on a massive program of modernization and expansion of its existing units (Figs. 5.6 and 5.7).

Fig. 5.6 Dr Dastur with president V. V. Giri during NMD-ATM in Delhi, 1972

Fig. 5.7 Dr Dastur participating at an IIM function

5.9 DASTUR—The Consistent Pursuit of Excellence and Innovation

DASTUR is a knowledge-based organization continuously striving to solve the engineering problems of clients through the application of experience, expertise, people and IP. Its ongoing success hinges on being relevant to the ever-changing and evolving requirements of clients, and requires continuous learning and innovation. From its inception, DASTUR has continuously evolved and placed the highest importance on the quality and excellence of its work and the independence of its advice. As the world becomes more competitive, DASTUR's quest for excellence and its passion for providing quality service and products to clients, prioritizing their needs and requirements, and being the first port of call for their engineering and technology problems, are the basic tenets to steer the organization going forward. The management of DASTUR is fully committed to this quest for excellence and innovation, both in traditional areas of strength in the metals and mining sectors, and in the new vistas of clean energy and carbon engineering. As a global company but with its roots firmly grounded in India, DASTUR is committed to bringing cutting-edge technology and engineering from around the world and using it to solve India's most pressing industrial problems and challenges.

Chapter 6
Hindalco's Journey Over a Period of Six Decades: Making the World Greener-Stronger-Smarter

P. K. Banerjee, Sagar Pandit, Amit Gupta, Vivek Srivastava, and Bijesh Jha

Abstract Hindalco's growth trajectory started in 1958 with setting up of India's first integrated aluminium facility at Renukoot, in the eastern fringe of Uttar Pradesh, India. The plant was set up by the premier industrialist of India, Shri G.D Birla, in collaboration with Kaiser Aluminium Corporation Ltd. of USA, the 3rd largest producer of aluminium then. The growth continued with brownfield expansions at Renukoot, INDAL acquisition (2001), Novelis (2007), brownfield expansion of some of these plants and major greenfield projects at Utkal, Mahan, Aditya and Hirakud FRP (2014). Hindalco's acquisition of Aleris Corporation (April 2020), through Novelis Inc., has cemented the Company's position as the world's largest flat-rolled products player and recycler of aluminium. Today, Hindalco is an industry leader in aluminium and copper, with a consolidated turnover of US$18 billion. Hindalco ranks among the global aluminium majors as an integrated producer with footprints in 9 countries outside India running the gamut of operations from bauxite mining, alumina refining, coal mining, captive power plants and aluminium smelting, to downstream rolling, extrusions and foils. Their product basket serves many industries like Electrical, Mass Transportation, Automotive, Packaging, Cookware, Defence, Building, Construction and Architecture. With its DNA embedded in the Company's purpose statement that reads—"We manufacture materials that make the world Greener—Stronger—Smarter", Hindalco has been rated the world's most sustainable aluminium company in the Dow Jones Sustainability Indices (DJSI) 2020. Hindalco got a total score of 75 points as against the industry average of 51.

Keywords Bauxite · Alumina · Aluminium smelting · Aluminium downstream · Technology · Innovation

---

P. K. Banerjee (✉) · S. Pandit · V. Srivastava · B. Jha
Hindalco Industries Ltd., 7 Floor, Birla Centurion, Pandurang Budhkar Marg, Worli, Mumbai, Maharashtra 400030, India
e-mail: pradip.b@adityabirla.com

A. Gupta
Aditya Birla Science and Technology Company Pvt Ltd., MIDC Taloja, Panvel, Maharashtra 410208, India

R. Divakar et al. (eds.), *Indian Metallurgy*, Indian Institute of Metals Series,
https://doi.org/10.1007/978-981-99-5060-7_6

6.1 Introduction

Hindalco Industries Limited, the metals flagship company of the Aditya Birla Group, has a consolidated turnover of US$18 billion, making it an industry leader in aluminium and copper. The Company most recently expanded via acquisition of Aleris Corporation in April 2020, through its subsidiary Novelis Inc., cementing its position as the world's largest flat-rolled products player and recycler of aluminium. Hindalco's state-of-art copper facility comprises a world-class copper smelter which is one of Asia's largest at a single location, a fertilizer plant and a captive jetty (Hindalco Corporate, 2021).

Today, Hindalco has its presence in 9 countries, ranking among the top integrated aluminium producers, globally. The Birla Copper unit produces copper cathodes and continuous cast copper rods, and other by-products, including gold, silver, and DAP fertilizers. It is India's largest producer of gold. Its aluminium is accepted for delivery under the High-Grade Aluminium Contract on the London Metal Exchange (LME), while its copper quality is also registered on the LME with Grade A accreditation. Hindalco's products serve industries like Electrical, Mass Transportation, Automotive, Packaging, Cookware, Defence, Building and Construction and Architectural.

6.2 Hindalco's Journey

The following section describes the journey of Hindalco in technology and operations over the last six decades as shown in Fig. 6.1.

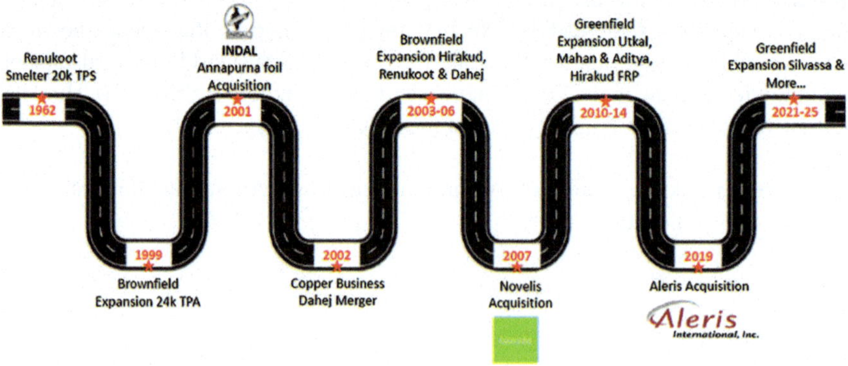

Fig. 6.1 Hindalco journey over last six decades

6.2.1 *The Beginnings*

Hindalco's story dates to the young Indian democracy of the 1950s. In 1958, Hindalco made a significant contribution to the vision of an industrial India by setting up its first integrated aluminium facility at Renukoot, Uttar Pradesh. The plant was set up by the premier industrialist of India, Shri G D Birla, in collaboration with Kaiser Aluminium Corporation Ltd. of USA, which at that time was the 3rd largest producer of aluminium. The intense desire of the visionary leader to industrialize India and provide employment to thousands of countrymen led to the birth of Hindustan Aluminium Corp Ltd. (presently, Hindalco Industries Ltd.) Renukoot.

India's first modern aluminium smelter was commissioned in 1962. The very concept of sustainability had been built into the design of the plant since the beginning. The energy requirement was met through green source, the Rihand Hydel Power Plant, situated near the smelter. The best available technology from Kaiser Aluminium corporation was deployed for aluminium smelting. To further strengthen the steady supply of electricity to the plant, Hindalco established a captive power plant (CPP) at Renusagar in 1967. This, along with a co-generation power units at Renukoot ensured continuous supply of power to the smelter and other operations (Hindalco Internal Reports). Concept of CPP was later replicated in various industries in India that helped in building the country's industrial base.

6.2.2 *Brownfield Expansion and Technology Upgradation*

Hindalco's integrated operations at Renukoot comprise Smelting Unit, Alumina Refinery, manufacturing facilities for downstream products and power generation plant supported by co-generation facility. The entire alumina product is transported to own smelting unit. Bauxite mining takes place in adjacent states at distances varying from 200 to 400 km.

Hindalco's Alumina Refinery at Renukoot created a landmark by enhancing its production capacity through radical modernization and upgradation schemes. The innovative approach involved introduction of Sweetening Process, Modernization of Digestion Units and Upgradation of Precipitation Circuit. The state-of-the-art Alusuisse Precipitation Technology was successfully incorporated to increase liquor productivity from 58 to 77 g/L (Shah et al. 2004).

The brownfield expansion of Hindalco Alumina Refinery was carried out with great success with respect to both process upgradation and efficient capital utilization. With fully stabilized and optimized operations, the operating cost was reduced. The challenges and constraints adequately addressed during the project implementation developed high level of confidence in the refinery team to understand that more such types of upgradation could be accomplished in self-reliant ways (Shah et al. 2004).

Fig. 6.2 Renukoot smelter plant before and after modernization-Covered pots with auto feeding

On the smelter front, Kaiser's smelter technology relied on manual alumina feeding, voltage adjustment, anode-effect termination and most other operations. Owing to high energy consumption and rising energy prices, during mid 80s serious efforts were made to modernize Renukoot smelting facilities by retrofitting new technologies, which can be depicted from Fig. 6.2. The guiding principle in updating the technology of production process was energy conservation, pollution control, improvements in norms of inputs, improved efficiency and recovery, leading to reduction in cost of production and improvement in metal quality (Tandon and Prasad 2003).

In the FRP business, the capacity and technology upgradation involved investments in the rolling plant to increase the size of rolling ingot to 2,200 kg during 1978–79. These investments led to production crossing 25 kt by 1985. Installation of continuous caster in 1990–91 and commissioning of the second cold mill in 1993–94 led to further capacity increase to 35 kt/year. In the extrusion business, increase in production capacity was coupled with product innovations leading to addition of several new products in the portfolio (Hindalco Internal Reports).

6.2.3 Greenfield Expansions and Acquisitions

The growth trajectory of Hindalco took a sharp upturn in the twenty-first century with a series of investments in greenfield capacity expansions and acquisitions. Acquisition of INDAL in 2001 was one key turning point. It significantly expanded the geographical footprint of the Company, brought in several upstream and downstream assets and led to diversification of the product portfolio. INDAL's mining operations in Lohardaga, refinery operations at Muri and Belgaum and interests in Utkal Alumina were significant additions to Hindalco's business.

Muri alumina refinery, commissioned in 1948, is one of India's first alumina refineries constructed by INDAL. The refinery was merged with Hindalco in 2005 and produces smelter grade of alumina. Hirakud smelter was earlier setup by INDAL

Fig. 6.3 Mahan and Aditya Smelters are doing better than their 720 kT nameplate capacity with our own technology enhancement

in partnership with Alcan, with HSS (horizontal stub Soderberg) technology, which had end-to-end pots operating at 55 kA.

INDAL had a sizable downstream business. Half of INDAL's turnover was coming from its sheet business; its plants at Belur (West Bengal) and Taloja (Maharashtra) produced import substitution products like lithographic sheet, foil stock, fin stock, closure stock and heavy gauge circles. Manufacturing plants for foils and packaging business were located at Kalwa and Kolur producing bare, laminated, lacquered and printed foils for pharma, cigarette and food industry. Apart from INDAL acquisition, major greenfield expansions were undertaken at Aditya, Utkal, Hirakud FRP (Odisha) and Mahan (Madhya Pradesh) plants. The state-of-the-art AP36 technology was used for greenfield smelters at Mahan and Aditya, shown in Fig. 6.3.

6.2.4 Securing Raw Materials

Hindalco operates four coal mines: Gare Palma IV/4 and IV/5 mines at Chhattisgarh and Kathautia and Dumri mines in Jharkhand. These mines are a combination of underground and open cast mining. The mines are focused on maintaining safe and sustainable operations through appropriate methodology. These captive sources, together with linkage coal sources from Coal India Ltd., provide coal security to meet Hindalco's energy needs.

Bauxite is the ore used in the production of alumina, which in turn is the basic raw material in the aluminium manufacturing process. Hindalco's bauxite mines are spread across four different states, namely, Jharkhand, Odisha, Chhattisgarh and Maharashtra. There are in total 27 leases of bauxite mines with Hindalco, some of which date back to1940s. These mines are spread over three geological zones—West coast region, Chotanagpur region and East coast region. The eco-friendly initiatives of Hindalco include Bagru ropeway and LDC (Long Distance Conveyor) installed at Baphlimali mines in Odisha. Bauxite mines have played a major role in Hindalco's sustainability drive. Water conservation through rainwater harvesting and other such initiatives have helped us in our journey towards a water positive company (Fig. 6.4).

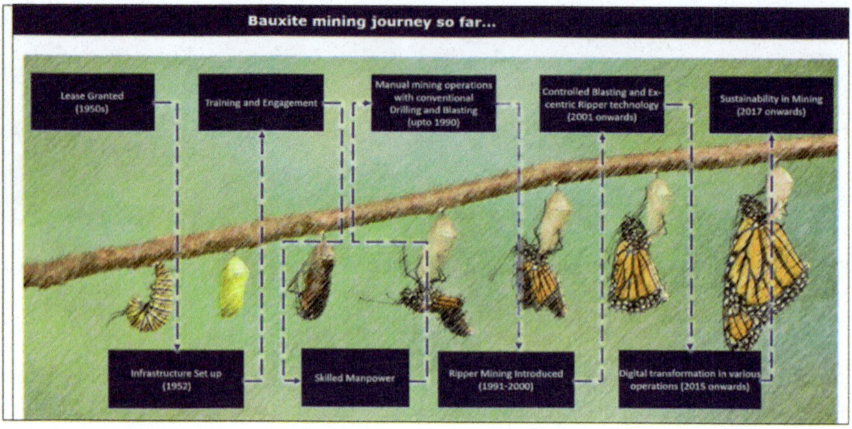

Fig. 6.4 Bauxite mining journey at Hindalco

Hindalco has believed in merging its development with that of societies. The Corporate Social Responsibility (CSR) activities are carried out to tackle the challenges in five social focus areas—namely health, education, infrastructure, sustainable livelihood and social causes. In one such example in Jharkhand, Hindalco is developing 23 villages and transforming them to model villages.

Hindalco Mining Group has won several awards in safety and environment. The mines have won the National Safety Award from the Hon'ble President of India in different categories in the years 2009, 2010, 2014, 2015 and 2018. FIMI has awarded the environmental award for 2018–19 to Bhusar bauxite mines (Hindalco Internal Reports).

Hindalco focuses on evolving mining into a prime sustainable developmental area of national interest. As a key sustainable drive, a Bio-Park (Dhumkuria and Birsa Upvan) was developed at Bagru mines in the mined-out area, as shown in Fig. 6.5. Also, projects of red mud disposal at the mined-out area at Baphlimali mines are being developed with various stakeholders that have potential of providing a clear sustainability path to global bauxite mines.

Several water-harvesting ponds have been created in each plateau along Hindalco's journey to water positivity. Also, 70% of the mined-out areas have been backfilled and subsequently handed over to the village landowners for agricultural purpose.

Fig. 6.5 Sustainable mining practices at Bauxite Mines of Hindalco

6.3 Technological Challenges and Innovations

Technology and Innovation have played an important role in the sustainable growth of Hindalco. The company operates two Hindalco Innovation Centres as shown in Fig. 6.6, one, HIC-Alumina at Belagavi working on R&D of bauxite, alumina and specialty alumina products, and another at Taloja, working around aluminium fabricated products and tribology. In addition, Hindalco engages the Aditya Birla Group's corporate research and development centre, Aditya Birla Science and Technology Company Private Limited (ABSTC), for conducting R&D in selected areas of work through chartered R&D projects. These are based on the domain expertise and R&D facilities available in ABSTC (ABSTCPL, 2021).

HIC-Alumina is recognized by the Department of Scientific & Industrial Research (DSIR), Government of India. With over 25 years of experience, the research & development team of expert scientists carry out research in the field of bauxite, bauxite residue utilization (Pandit and Jadhav 2017), processability studies of the Bayer process, product development, quality control and application research for enhanced understanding of the end-usage of specialty chemicals.

HIC-SemiFab started in 1979 as oil & lubrication laboratory, over the years, it has developed into an Innovation Centre for SemiFab and Tribology. It is located in

Fig. 6.6 Hindalco Innovation Centre—Alumina and SemiFab

Taloja FRP Plant and is a DSIR recognized R&D centre. In its global reach initiative, this facility works closely with Novelis plants.

6.3.1 Development of Specialty Alumina and Hydrates

The Belagavi Refinery was established in 1968, with a capacity of 75 ktpa alumina and 30 ktpa aluminium. It was expanded in phases to 260 ktpa alumina and subsequently the capacity was enhanced, through de-bottlenecking, to 380 ktpa. However, the smelter (aluminium) plant was de-energized in the 1990s, because of the non-availability of power from the grid. For the survival of the plant, extensive research was undertaken in converting the plant for producing specialty alumina and hydrate products. Now it is the largest producer of special grade alumina that goes for refractory to glass and cable manufacturing.

6.3.2 Bauxite Residue use Towards Circular Economy

Hindalco has become a world leader in the utilization of bauxite residue and has created a benchmark for the alumina industry across the world. Hindalco has worked closely with the cement manufactures across India to develop a circular economy model. Hindalco is the world's first company to achieve 100% red mud utilization across three of its refineries, namely, Belagavi, Muri and Renukoot (Pandit 2019; Surawar 2020) (Fig. 6.7).

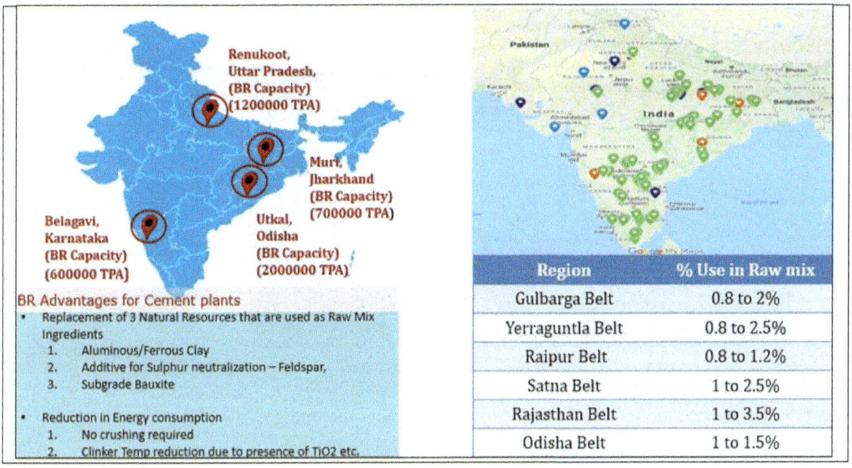

Fig. 6.7 Bauxite Residue generation and use in various Cement Plants

6.3.3 Development of Energy Efficient Aluminium Smelting Technology

Hindalco, in collaboration with corporate the R&D centre, ABSTC, have been working on reducing the energy consumption in their smelter. ABSTC has advanced computational modelling capability coupled with solid know-how of aluminium smelting technology, which were utilized while developing some of the key technologies like Cu-inserted collector bar, magnetically compensated busbar and PLC-based pot control system, shown in Fig. 6.8.

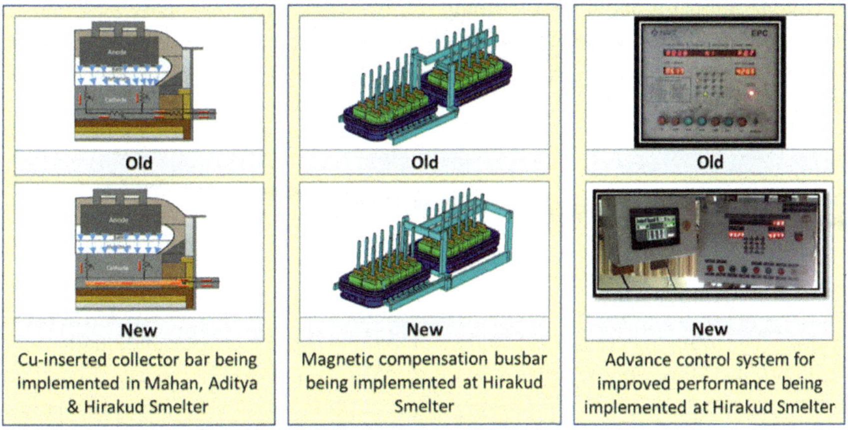

Fig. 6.8 Key enabler for energy efficient aluminium smelting technologies

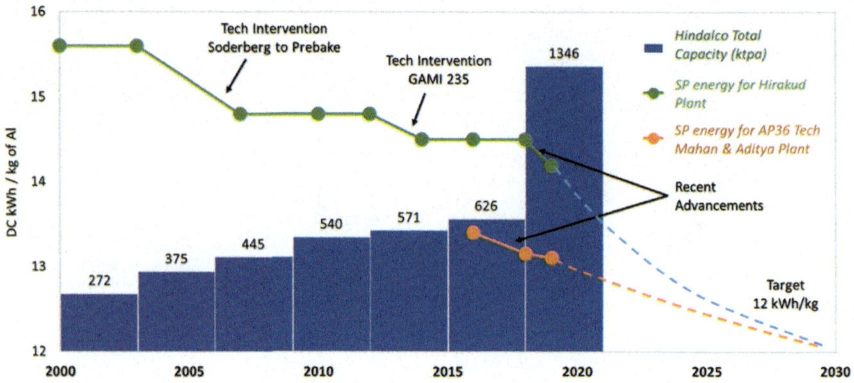

Fig. 6.9 Hindalco smelter production capacity and specific energy

These have helped in achieving some benchmark energy numbers, e.g. 13.6 kWh/kg (DC) in Hirakud 85kA & 235kA pots, 13.1 kWh/kg (DC) in Mahan/Aditya 360 kA pots (Gupta and Basu 2019; Jha et al. 2019). Hindalco has a target of achieving sp. energy of 12 kWh/kg by 2030 in high amperage smelter as shown in Fig. 6.9.

Hindalco and ABSTC have developed '*HiPoT-85kA Technology*' to replace all the old 85kA pots with new design. This technology consists of a new magnetically compensated busbar configuration, new low energy cell lining with novel Cu-insert collector bars along with an improved anode/cathode assembly. It also consists of PLC-based pot control system powered by a new control logic, which has inbuilt advanced analytics algorithms to predict key parameters for operational excellence. After successful replacement of all the pots in 85 kA potlines with the new HiPoT-85kA technology, the energy consumption of potline would achieve 13.6 kWh/kg from the original figure of 14.6 kWh/kg, thereby bringing down the energy level by about 1 kWh/kg. The improved lining has also enhanced the pot life and reduced SPL generation by about 5–6 kg/tonne of Al. These improvements facilitate the sustainability of low amperage smelter along with greater understanding of smelting process and technology (Gupta and Basu 2019; Sahoo et al. 2020; Rajgire et al. 2019).

6.3.4 Downstream Application Developments

Product innovation and responsible business practices are integral to Hindalco's operations. The Company constantly focuses on innovation, excellence and quality in all of its product development efforts. As part of the aluminium industry, Hindalco is among the top five aluminium producers, based on shipments, globally. Hindalco is an integrated producer with a strong base across the value chain. Its diverse downstream offerings, such as extrusions, flat-rolled products, foils, wire rods and billets

Fig. 6.10 Aluminium application development: Freight Trailer and Bulker

find application across various industries, ranging from automobiles, packaging, pharmaceutical, transportation, building and construction to name a few.

Launch of Aluminium Freight Trailer and Bulker

Hindalco had launched India's first complete aluminium freight trailer and bulker in 2019 shown in Fig. 6.10. The trailer is an initiative for India's logistics and freight industry. The trailer is 34-foot long, 50% lighter and weighs 2.5 t less than an equivalent steel trailer. The high strength aluminium alloy ensures that the vehicle is safe, strong, durable, efficient and environment friendly. Each trailer saves over 15,000 L of fuel and emits 25 tonnes less GHGs and helps in achieving BS-VI emission targets. It also has 70% higher scrap value. Aluminium bulker is another such product where steel is replaced by aluminum in the manufacturing process which makes it 1.8 t lighter, thus increasing the fuel efficiency and reducing the GHG emissions (Hindalco Annual Report 2019).

Launch of Aluminium Foil-Laminated Jute Bags at Tirumala Venkateswara Temple, Andhra Pradesh

Tirumala Tirupati Devasthanams (TTD), the trust that manages the temple, has joined hands with Hindalco and the Jute Corporation of India (JCI) to roll out a pilot project on distributing '*prasadam*' in 100% recyclable aluminium foil-laminated jute bags. The jute and aluminium bag for Tirupati *laddoos* is an alternative to plastic bag. This is an Indian innovation that provides the devotees an eco-friendly option. This green initiative also provides employment to thousands of artisans (Hindalco Annual Report 2019).

6.4 Going Global

The vision to be the global best saw Hindalco acquire US-headquartered Novelis Inc. in May 2007, a company four times its size at that time. Novelis added robust value to Hindalco's portfolio—providing access to technology (Novelis being an auto lightweighting pioneer) and ready access to global marquee customers, while de-risking Hindalco's upstream India business. The acquisition resulted in the formation of the world's largest aluminium rolling and recycling enterprise. Today Novelis recycles 74 billion cans a year, enough to circle the globe more than 160 times. The complementary expertise of Hindalco and Novelis is to create and provide a strong platform for sustainable growth and ongoing success (Hindalco Corporate, 2021).

The acquisition of Aleris Corporation through Novelis in April 2020 was another major milestone for Hindalco on its path to global leadership; with a global footprint spanning 49 state-of-the-art manufacturing facilities in North America, South America, Europe and Asia. The Aleris deal, crucially enabled further diversification of Hindalco's metal's downstream portfolio, into other premium market segments, most notably aerospace. It further insulates the Company from global price volatility (Primary aluminium price at London Metal Exchange) and sharpens its focus on the downstream business. Aleris enhanced Hindalco's strategic position in Asia and solidifies its position as a leading global metals player, with a stronger presence across the U.S. and Europe as well (Hindalco Corporate, 2021).

Future Strategy

With focus on transformational, responsible and sustainable change, we have grown from India's largest leading non-ferrous metals company to become a global leader in aluminium business. Over a period, we have transformed our business model by reducing dependency on volatile LME price movement and focusing on a stable portfolio of value-added products. Our purpose of manufacturing products that are Greener, Stronger and Smarter has been at the core of our transformation journey.

6.5 We have set out clear strategic priorities for ourselves as we embark on the next phase of our growth journey:

- Focus on value-added products
- Strong ESG commitment
- Strengthening the balance sheet
- Capital allocation to maximize shareholder returns.

6.5.1 Focus on Value Added Products

Our growth over the medium term will come from our investments in downstream business. This will continue to provide insulation from LME movements as LME price is a pass through in downstream business imparting stability to the overall earning stream. Additionally, it put us closer to the end customers and consumers.

At Novelis, we continue to defend and grow the core business and diversify the portfolio. We will begin first with our China expansion to fully integrate that facility with our downstream auto finishing lines. We will continue to implement our world-class manufacturing, automation, and digital initiatives and advancements in R&D

to unlock further capacity, capture further growth that we see in the marketplace, and to support our sustainability initiatives.

The dominant segments of aluminium downstream, viz. flat-rolled products (FRP) and extrusions are under-represented in India and the per capital aluminium consumption is significantly below global average. Increasing aluminization in industries such as Building & Construction, Transportation and Packaging offers a significant market opportunity to a large and integrated player such as Hindalco.

The refined Copper consumption in India is expected to grow at 7–8% over the next 10 years driven by electrical applications such as wires and cables, motor windings, transformer strips as well as and infrastructure projects such as railways and metro. Today we are the only large, refined Copper manufacturing plant in the country and also the largest manufacturer of Copper rods. We plan to further increase the share of value-added products in our business, which will aid our margins and help us move up the value chain.

We also plan to expand out value-added products portfolio in our specialty alumina business targeting opportunities in newer and high-growth application areas for our products.

6.5.2 Strong Environment Social and Governance (ESG) Commitment

Going forward, Hindalco will continue with the renewed focus on increasing the recycle content. Company has committed to net carbon neutrality, zero waste to landfill, water positivity in all out-mining locations, and no net loss on biodiversity by 2050.

6.6 Concluding Remarks

This paper offers an insight into Hindalco's growth journey over a period of 6 decades. It started with Renukoot in 1958, and consolidated with the acquisitions of INDAL as well as brownfield expansion of Hirakud and greenfield expansions at Utkal, Mahan and Aditya and culminated with the acquisition of Aleris, which has cemented the Company's position as the world's largest flat-rolled products player and recycler of aluminium.

Hindalco's sustainability journey is also highlighted in the paper which gives a brief account of the giant strides which Hindalco has taken in developing a solution for 100% of its bauxite residue generated at its refineries. This development has further added impedance to the implementation of the concept of circular economy. With its DNA embedded in the Company's purpose statement that reads—"We manufacture materials that make the world Greener—Stronger—Smarter", Hindalco has been rated the world's most sustainable aluminium company in the Dow Jones Sustainability Indices (DJSI) 2020. Hindalco got a total score of 75 points as against the industry average of 51.

Acknowledgements The Authors would like to thank the help provided by Charu Gouniyal, Abhishek Rungta, Pramod Koparde, Nageswar Kapuri, BasavrajGangadharmath, Sagar Dhamorikar and Vilas Tathavadkar in preparing this paper. The encouragement and guidance provided by Satish Pai and Bibhu Mishra in writing this article is thankfully acknowledged.

References

ABSTCPL Website, http://www.adityabirlascienceandtechnology.com/. (Online), July (2021)

Gupta, A., Basu, B.: Sustainable primary aluminium production-technology status and future opportunities. Trans. Indian Inst. Met. **72**(8), 2135–2150 (2019)

Hindalco Annual Report. Hindalco Industries Ltd., Mumbai (2019)

Hindalco Corporate Website, http://www.hindalco.com/. (Online), July (2021)

Jha, A., Gupta, A., Tiwari, V., Ghatnatti, S., Pandey, K., Anand, S.: Copper insert collector bar for energy reduction in 360 kA smelter. In: Light Metals (2019)

Pandit, S.S.: Development of Bauxite Residue (BR) as a value added product for cement industry. In: Mineral Processing Technology, Hyderabad (2019)

Pandit, S.S., Jadhav, R.N.: Bauxite Residue utilization–development of method for soda recovery using mechano treatment. In: International Conference on Non Ferrous Minerals and Metals, New Delhi (2017)

Rajgire, S., Jha, A., Gupta, A., Chulliparambil, M., Choudhary, S., Verma, G., Upadhyay, V., Nath, S.: Advancement in control logic of HINDALCO low amperage pots. In: Light Metals (2019)

Sahoo, M., Gupta, A., Badajena, S.K.: Minimizing SPL generation via redesigning pot and life enhancement. In: 38th International ICSOBA 2020, Virtual Conference, 16–18 Nov 2020

Shah, R.P., Gararia, S.N., Singh, R.J.: Alumina refinary brownfield expansion–the HINDALCO experience. In: Light Metals (2004)

Surawar, V.: Hindalco Sustainability Report. Hindalco Industries Ltd., Mumbai (2020)

Tandon, S. Prasad, R.: Twenty years of progress at Hindalco's aluminium smelter. In: Light Metals (2003)

Chapter 7
MECON's Journey of Six Decades—A Saga of Engineering Excellence in Iron and Steel

Kumud Ranjan, Sajal Kumar Bhattacherjee, and U. K. Vishwakarma

Abstract Starting as a small design organization, MECON's six decades of glorious journey has witnessed many benchmarks leading to its emergence as one of the premier Indian consultancy and engineering companies, being a Mini Ratna CPSE under Ministry of Steel, Government of India. Over the years, MECON has taken giant strides forward and grown in stature offering turnkey project execution on single point responsibility. Through business diversification and restructuring into three business verticals, viz., Metals, Energy and Infrastructure, it has transformed itself to the needs of changing economic environment and has become a key player in India's industrial and infrastructure progress.

7.1 Genesis and Profile

Steel Industry was reckoned as an engine of economic growth and prosperity of India by the policy planners in the 1950s, as it plays a vital role for development of industries/manufacturing base, infrastructure, strategic self-reliance and also in creation of employment and revenue to National exchequer due to its strong forward and backward linkages to other sectors of the economy. Accordingly, vast resources were mobilized and allocated to develop integrated capability and self-sufficiency in iron and steel manufacturing, with due consideration for development of backward areas with regional balance and employment generation. Towards this endeavour, today's MECON was born with an objective to assimilate and absorb foreign technology and develop indigenous technology base for the Iron and Steel industry and render consultancy and engineering services for integrated steel plants in India. Since then, MECON has become a key player in India's economic development especially in the areas of mining and metals, energy and infrastructure.

K. Ranjan · S. K. Bhattacherjee · U. K. Vishwakarma (✉)
MECON Limited, Ranchi, Jharkhand 834002, India
e-mail: ukvishwakarma@meconlimited.co.in

R. Divakar et al. (eds.), *Indian Metallurgy*, Indian Institute of Metals Series,
https://doi.org/10.1007/978-981-99-5060-7_7

Fig. 7.1 MECON Limited headquarters at Ranchi, Jharkhand

MECON Limited, the erstwhile Central Engineering and Design Bureau (CEDB) consisting of only thirteen elite engineers then, was born in April 1959, as an independent consultancy outfit within Hindustan Steel Limited (HSL), the first public sector steel organization under the leadership of Late Mr. K. M. George as the Head and Chief Engineer.

MECON renders full range of services required for setting up of Greenfield and Brownfield projects from Concept to Commissioning including Turnkey execution in coke oven, agglomeration plant, mini blast furnace, rolling mill and also R&M/AMR schemes. MECON has collaboration agreements with leading global technology providers for equipment design capability to enhance its cutting-edge technology (Fig. 7.1).

MECON has also developed in-house indigenous design for 1.0 Mtpa Coke Oven Battery and 4250 m^3 Blast Furnace which was launched by the then Hon'ble Minister of Steel on 21st December 2016 at MECON, Ranchi by brand names "ANGARA 7.1" and "LOHA 4250" respectively.

MECON has also strengthened its footprint in the International market by providing World Class Design, Engineering and Consultancy Services for about 130 assignments in different countries.

Besides steel, MECON also caters to the non-ferrous sector including aluminium, copper, zinc, lead, magnesium and precious metals and mining sector including ferrous, non-ferrous, atomic minerals and fossil fuel minerals (coal and lignite). Due to cyclic swing in metals business, MECON diversified into energy and infrastructure sectors with focus on renewable power and green energy and iconic and institutional infrastructure, green buildings, defence infrastructure and aero-space sector. A glance through the decade-wise development of MECON is depicted in Fig. 7.2.

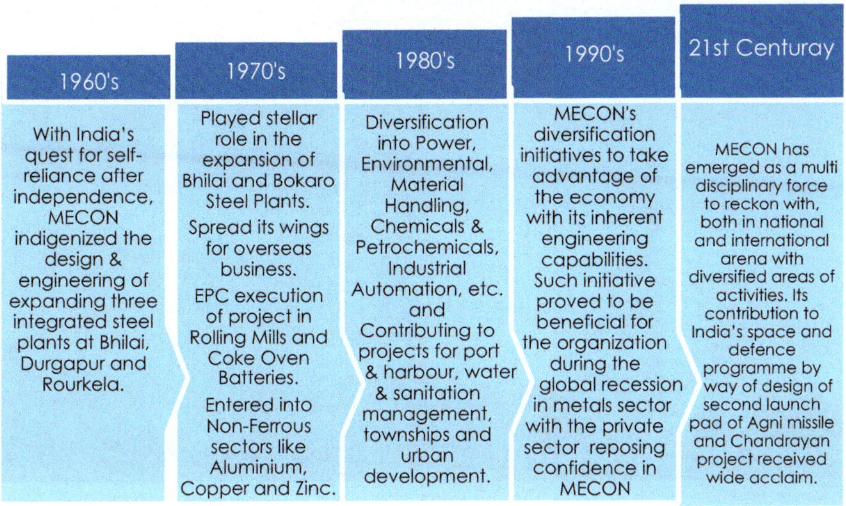

Fig. 7.2 Growth of MECON limited

7.2 Achievements

MECON has executed a host of outstanding projects in all areas of its business operation. Over 3,500+ Consultancy and EPC assignments have been successfully completed. It has three Engineering Offices at Ranchi (HQ), Delhi and Bengaluru and over 100 project/site/inspection offices on pan India basis (Fig. 7.3).

Fig. 7.3 MECON's contribution in complete value chain of steel plants

Fig. 7.4 The second launch pad project executed for ISRO

7.2.1 *Projects in Strategic Sectors*

- Second Launch Pad, ISRO (Fig. 7.4)
- Development of Uranium Mines for UCIL
- Geotechnical Centrifuge, IIT Mumbai
- Project Seabird Phase-IIA Dockyard and Fleet Base Buildings at Naval Base, Karwar
- Solvent Extraction Plant to recover Rare Material (Yellow Cake) for Heavy Water Board at Tuticorin
- Integrated Engine Testing Facility of IPRC—Semi Cryo for ISRO, Mahendragiri, TN
- Naval Aircraft Yard at Goa and Kochi, Ministry of Defence
- Solid Propellant Space Booster Plant (SPROB), Augmentation Facilities for ISRO, Sriharikota
- Payload Fabrication and Testing Facility for ISRO at Bhopal, Ahmedabad
- Environmental Engineering with state of the art environmental lab recognized by MOEF & CC and CPCB (Fig. 7.5)

7.2.2 *Oil and Gas*

MECON has executed several projects for Long Distance Transmission of Oil and Gas through long-distance cross-country pipelines, CNG and CGD Networks (Fig. 7.6) and Storage and Handling of Liquid Petroleum Products in POL Terminals including Loading Gantries both RAIL/Truck.

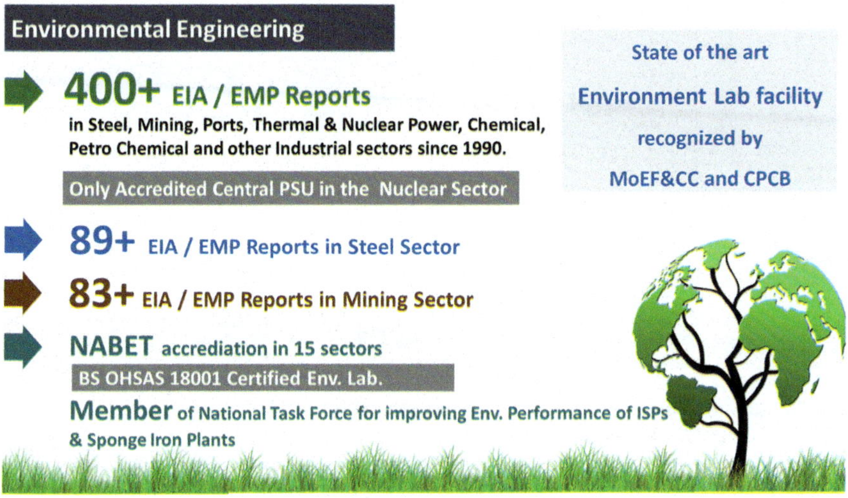

Fig. 7.5 MECON's contributions to environmental engineering

Fig. 7.6 City gas distribution projects

7.2.3 *Power*

MECON has rendered services in Renewable Energy, Thermal and Hydel Power and Transmission and Distribution Projects (Fig. 7.7).

7.2.4 *Infrastructure*

MECON has rendered services for creation of iconic and institutional infrastructure including defence and space (Fig. 7.8).

Fig. 7.7 (top to bottom, left to right) 2 × 500 MW TPP for NTPL, solar power generation for BPSCL and NMDC, FGD projects for TANGEDCO, RLA of HYDEL power for DVC, OHPC, TANGEDCO and APGENCO, DDUGJY and IPDS projects for Govt. of Jharkhand and Odisha

Fig. 7.8 Iconic and institutional infrastructure

7.2.5 Overseas Operations

MECON has expanded its successful business operation to encompass overseas market as well. MECON was actively involved in setting up two integrated Steel Plants and provided consultancy, project management and technical services in Nigeria for Delta Steel Company, Warri and Ajaokuta Steel Company, Ajaokuta. Africa's first Integrated Steel Plant way back in the late 1970s.

7.2.6 Alignment with National Priorities

(See Fig. 7.9).

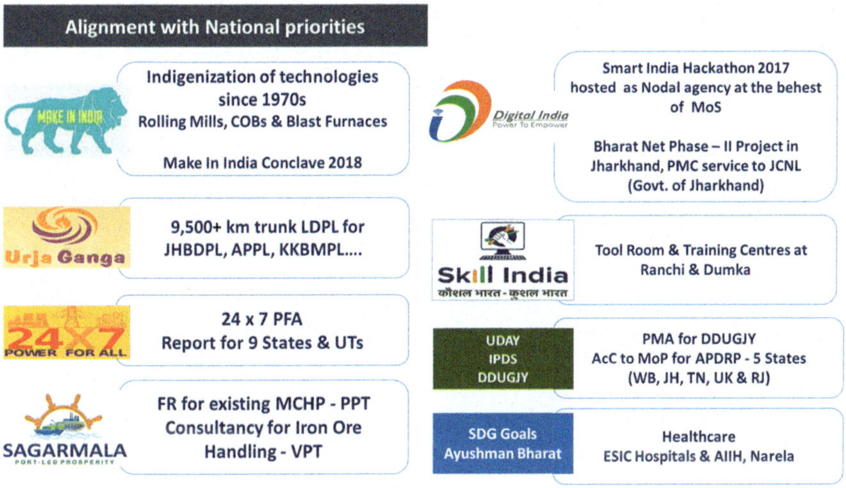

Fig. 7.9 Alignment with national priorities

7.3 Atmanirbhar Bharat

- In-house Technology Development in the areas of Coke Oven Battery, Iron Making, Steel Making and Value Added Steel
- Technology Absorption in Steel Making and Emission Norms
- Indigenization and Creation of Design and Innovation Cell

MECON has been actively involved in giving policy suggestions to Govt. of India for future planning. It has been instrumental particularly in Policy Formulation such as National Steel Policy, Domestically Manufactured Iron and Steel Products (DMI&SP), National Mineral Policy, PLI scheme for specialty steel etc.

7.4 Service Portfolio

(See Fig. 7.10).

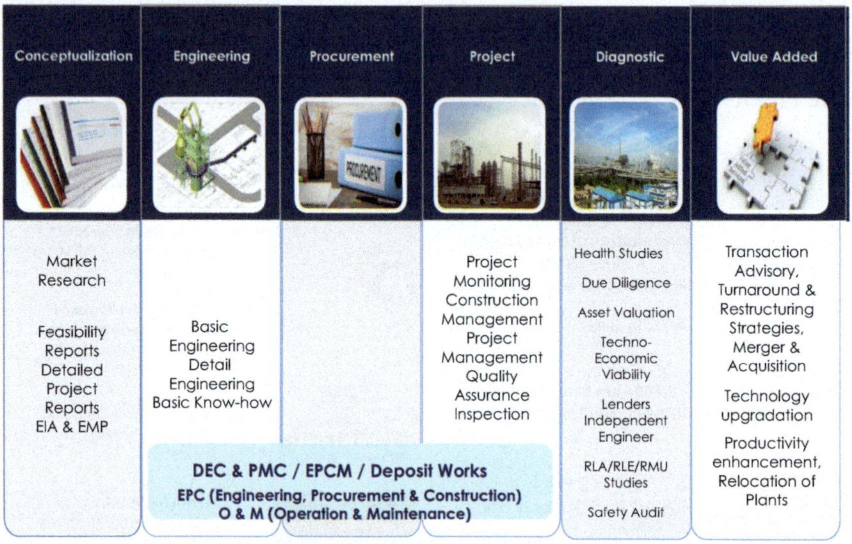

Fig. 7.10 MECON's service portfolio

7.5 Summing Up and Future Vision/Outlook

Over the past six decades since inception, MECON has developed capability to enrich its service portfolio encompassing design, engineering and consultancy, Project Management Consultancy (PMC), project execution on Engineering, Procurement and Construction Management (EPCM)/Deposit Works mode as well as turnkey project execution on Engineering, Procurement and Construction (EPC) mode and to offer services in its three business verticals, viz., Metals (Iron and Steel and non-ferrous metals) and Mining, Energy (Power and Oil and Gas) and Infrastructure. It can also undertake entire project execution based on basic engineering support from the technology suppliers including assistance in capability development to bridge the technology gap. It can also provide complete engineering for balance of plant/overall plant integration based on inputs from technology package suppliers leading to discrete turnkey mode of execution.

MECON has a strong presence in the Iron and Steel sector with basic engineering support from global technology providers especially in the areas of steel making and rolling mill. It has entered into MoU with leading global technology providers to offer end-to-end cost-effective solution for the entire value chain of integrated steel plants with adoption of state-of-the-art technologies and maximum indigenization.

MECON expanded its operation into diversified sectors to de-risk its business from the cyclic swings in the metals sector. In power sector, the Company is capable of rendering engineering and consultancy services in thermal power plants including Flue Gas Desulphurisation (FGD) projects, Remaining Life Assessment (RLA) studies of hydel power plants, other renewable energy including solar and

Government Technology and Development projects, viz., 24 × 7 Power for All, Deen Dayal Upadhyaya Gram Jyoti Yojana (DDUGJY), Integrated Power Development Scheme (IPDS), etc. MECON also has large footprint in the Oil and Gas sector in low to mid-end segments of long-distance cross-country pipelines, CNG and City Gas Distribution (CGD) network. The Company has also rendered services for creation of iconic and institutional infrastructure including green buildings and strategic sectors of aero-space, defence and atomic mineral. MECON also has strong credentials in environmental studies with National Accreditation Board for Education and Training (NABET) accreditation in 15 sectors and the state-of-the-art environmental lab facility which is recognized by Ministry of Environment and Forests (MoEF) and Central Pollution Control Board (CPCB).

MECON plays a vital role in giving inputs to Government in policy formulation especially in the Iron and Steel and Mining sectors. Its business operations are aligned with the National Priorities and Government Flagship Programmes/Policies, viz., National Steel Policy, National Mineral Policy, Make in India, Atmanirbhar Bharat/ Vocal for Local, Urja Ganga, 24 × 7 Power for All, APDRP, IPDS, DDUGJY, National Solar Energy Mission, Digital India, Swachh Bharat, AMRUT, Ayushman Bharat, Skill India, etc.

MECON's future vision is to emerge as an internationally recognized brand in design, engineering, consultancy, project management and Engineering Procurement and Construction (EPC) execution. It has mission to be a global centre of excellence to provide innovative and cost-effective engineering and technological solutions in Metals and Mining as well as diversified sectors (Energy, Infrastructure, Space, Defence, etc.); to leverage its deep domain knowledge in the Metals and Mining sectors to provide solutions from concept to commissioning; to leverage in-house capabilities to provide engineering, technological and project management services to priority sectors of economy; to develop indigenous technological base/promote self-reliance and expand geo-strategic presence and export of services targeting Middle East, North Africa and Asia–Pacific regions. The Company's future outlook also focuses on harnessing/adopting the state-of-the-art newer energy-efficient green technologies suiting to indigenous natural resources with minimum GHG footprint and maximum indigenisation/cost-effectiveness. The company also has vision to nurture O&M businesses.

MECON is committed to promote Atmanirbhar Bharat campaign of Government of India through maximization of indigenous content while adopting state-of-the-art foreign technology in participation with the manufacturing and consuming industries.

Chapter 8
BALCO's Journey: A Story of Growth, Determination, and Transformation

Abhijit Pati

Abstract Bharat Aluminium Company (BALCO) was established as a public sector undertaking in Korba, Chhattisgarh in 1965 and subsequently acquired by Vedanta Group in 2001, going on to emerge as India's leading producer of aluminium. BALCO's growth story has had a multiplier effect on the country's economic progress, with downstream businesses and employment opportunities. BALCO has contributed significantly to empowering the country through its 56 years of developmental journey, and its aluminium production has assisted the country's self-reliance. The article discusses BALCO's progress towards excellence by adopting state-of-the-art technology in its processes.

Keyword BALCO · Aluminium · Power

8.1 Laying the Foundation Stone

Bharat Aluminium Company (BALCO) was established in 1965 as a Public Sector undertaking in Korba, Chhattisgarh. It was conceived by India's first Prime Minister Pandit Jawaharlal Nehru, due to its futuristic valuable contribution towards the development of the nation. Aluminium, the wonder metal, is vital for the nation's growth and progress. Realising the importance of this metal, an aluminium complex with a production capacity of 1 Lakh tonne/annum by the name of Bharat Aluminium Company Limited along with a power generating facility of 270 MW was established.

Later the organisation was acquired by Vedanta Group. This strategic acquisition of BALCO in 2001 by Vedanta, led to a turnaround of an unprecedented scale and BALCO became one of India's privatisation success stories. The Vedanta group has a 51% stake in the company, and the Government of India has a 49% stake. Today, BALCO stands tall as India's iconic producer of a vital raw material for the country—Aluminium.

A. Pati (✉)
Bharat Aluminium Company Limited, Aluminium Sadan, Lodi Road, New Delhi 110003, India
e-mail: abhijit.pati@vedanta.co.in

R. Divakar et al. (eds.), *Indian Metallurgy*, Indian Institute of Metals Series,
https://doi.org/10.1007/978-981-99-5060-7_8

Vedanta Limited is the world's 6th largest diversified natural resources conglomerate with business operations in India, South Africa, Namibia, and Australia. It is a leading producer of Oil and Gas, Zinc, Lead, Silver, Copper, Iron Ore, Steel, Aluminium and Power. Vedanta contributes 1% towards India's GDP, as per IFC.

8.2 The Story of Success

BALCO has grown by leaps and bounds throughout its journey of 56 years since its inception. Producing 20% of the country's Aluminium today, BALCO's initial capacity was 96,000 tonnes/annum (TPA) till 2004. This was followed by 155% volume ramp up to 245,000 TPA in FY 09 and by 30% increment to 325,000 TPA in FY 14 with current capacity of 575,000 TPA.

The heart of the smelter i.e. Potline was started with Soderberg technology with 60 kA current in 1965. In 2002, BALCO took a giant leap in switching over to Hall Heroult process with 100 kA current. At this stage, Chinese non-graphitised anodes were used which yielded a Current Efficiency of 93%. Currently, BALCO is running with fully graphitised anodes with Current Amperage of 340 kA and current efficiency of 95% with a specific energy consumption of 12,862 kWh/MT which is on par with global standards.

The power generation capacity at BALCO has also reached 1740 MW from 270 MW at the time of its establishment. The organisation has also created plenitude of employment opportunities for people from the state and across the nation. BALCO's turnaround has created a 360° multiplier effect for the country. BALCO's growth story has resulted in the economic progress of the regional business community comprising over 5,000 downstream businesses.

BALCO believes that privatisation has the ability to unlock value and realise synergies for businesses in a way that no other means can create. This is proven by many accounts in the past. The magnitude of the turnaround was such that its impact created ripple effects that went beyond the sphere of business to have an effect on the community.

BALCO seamlessly inculcated the values and ethos of Vedanta, with the takeover proving to be a win–win for both. While Vedanta benefited from the growth prospects and commercial extraction of BALCO's aluminium, post-acquisition, BALCO derived tremendous value that infused the once-ailing public giant with a new lease of life. The success on a mammoth scale was possible due to Vedanta's wealth of experience and learnings that it acquired over the years from its endeavours in the metals and energy spaces that transformed businesses and lives the world over.

BALCO, one of India's pioneering Aluminium producers, has always been a front-runner in India's Aluminium production at the global scale. It is also humbling that BALCO has been instrumental in bringing its areas of operations in rural Chhattisgarh, which has impacted the region appreciably and has introduced multidimensional progress to the remote areas in the vicinity by bringing them into the developmental economic mainstream of the country.

8.3 Making India Self-reliant

BALCO's aluminium production assists the country's tread towards self-reliance. BALCO provided Hard Alloys to ISRO between the period of 2006–2014 for defence purposes. The Aluminium sector plays a vital role in nation's economy and is a sector of strategic importance to the country due to its role in energy security, national defence, aerospace, automobile, infrastructure, electricity transmission and distribution, and many other ancillary applications. The metal also has a significant role in Union Government's forward looking and visionary schemes of Make in India, Smart Cities, Power for All, etc.

BALCO, the only aluminium industry in Chhattisgarh, has contributed significantly in empowering the state and the country through its 56 years of developmental journey. In the two decades after disinvestment, BALCO contributed about 10,000 Cr to State Exchequer and provided employment to 15,000 people directly and indirectly. BALCO has created economic opportunities worth thousands of Crores of rupees in the state.

8.4 Stride Towards Excellence

BALCO has always adopted state-of-the-art technology in its processes. Catering to the needs of Defence and Space Research since 1980, the Company has firmly established itself as a front-runner among the producers of Aluminium sheets and plates.

The molten metal (Aluminium) is fabricated into saleable products by adapting different processes. Fabrication Strategic Business Unit (SBU) is broadly divided into three units operating as separate profit centres—Cast House, Foundry, and Sheet Rolling Shop.

Many firsts as an Organisation

- First to produce Alloy Rods for conductors used in power transmission industry
- First to roll material for Aerospace in the Country
- First to produce largest Billet of diameter—800 mm, length—6 MT
- First to set up the widest Hot Rolling Mill

Wire Rods—BALCO is one of the leading wire rod manufacturers in the world. Our mills are fully equipped with in-line degassing and filtration systems to ensure good internal metal quality and cleanliness. Alloy Rods, EC Rods, and Flip Coils all are covered in our portfolio with the very first time production of T4 Rods in October 2021.

Ingots—BALCO produces Primary Aluminium ingots that are re-melted to produce a variety of end products covering the entire spectrum of aluminium applications using the state of the art technology.

Primary Foundry Alloys—BALCO supplies Primary Foundry alloys (PFA) in many industry segments. The casting facility has metal treatment facilities of Degassing, Metal Filtration Unit, and Vertical Chill Casting Technology for AlSi. PFA facility has started from 2018 in Cast House 3 and is currently having a capacity of 90 kTPA. First P1020 T-Ingots were developed in Foundry in April 2020.

8.5 Milestones over Four Decades and Journey or Rolled Products

BALCO is equipped to deliver high quality rolled products in segments with applications in automobiles, insulations, bus bars, power projects, electrical, packaging, etc.

- BALCO is first to set up the widest Hot Rolling Mill in India
- BALCO is first to roll material for Aerospace in the Country
- BALCO specialises in Hot Rolled Products
- BALCO has one of the oldest mills in the world: Technology from Russia, USA, and Italy

Initially, in 1981–82, Hot Rolling Mill * Old cold rolling Mill (NKMZ, Russia) was inducted in BALCO which was followed by induction of Caster and New Cold Rolling Mill (NCRM, Fata Hunter Italy) in 2001. Addressing the market needs preheating furnace was modified in the year 2019. In the year 2020, AA3105 was produced first time through Strip Caster (FATA Hunter Machinery) which was developed by in-house improvements. In the year 2021, manufacturing of circles was done for the first time and 3105 and 19,600 alloy series have also been manufactured through caster route in the same year.

BALCO supports critical emerging sectors like electric vehicles, modern construction and other ancillary segments, as Aluminium is a metal of endless possibilities. BALCO is providing impetus to the vision of 'Make in India', by leveraging the potential of the entire aluminium value chain, from mining to end usage. BALCO continuously strives to adopt advanced and updated technologies and processes to maximise efficiency. On the same lines, several digitalisation initiatives have been integrated in operational areas. BALCO has adopted image and video analytics, contextual analytics, situational awareness, safety, and security risk analysis that is available in real-time with accuracy.

At the same time, BALCO-manufactured value-added aluminium products that found application in several projects of national interest, scientific innovation, and strategic significance like long-range missiles and space exploration equipment.

BALCO is also deploying industry-leading digital technologies, creating super alloys to support the country's emerging applications of aluminium to make aluminium usage comparable to those of the developed countries in the world. In developed countries, aluminium has over 3,000 applications whereas it is limited to only 300 applications in India. Thus, BALCO is dedicated to explore the enormous scope to enhance the aluminium usage in India, as the country's per capita aluminium consumption is a meagre 2.2 kg against the world average of 10 and 25 kg for China with a comparable population. Germany ranks first with per capita aluminium consumption of 42.1 kg followed by Taiwan at 33.3 kg.

India is one of the fastest growing economies in the world and Aluminium is one of the key growth sectors for India which has a huge potential to contribute towards the wealth creation of the nation.

8.6 Towards the Betterment of Society

The robustness that BALCO's financials achieved under Vedanta opened up a bunch of opportunities for the community in the vicinity. Vedanta's priority for a holistic development through the empowerment of its stakeholders triggered an investment of $17.63 million in community initiatives alone. It introduced social welfare programmes and health facilities and created employment opportunities for the BALCO community that comprised of 123 villages.

Presently, Vedanta runs an institute of skills in Korba which benefits over 9,000 students from Chhattisgarh. This is a one-of-its-kind institute of skills that provides technical education to the rural youth to prepare them for a vocation that allows them to become financially self-reliant.

BALCO also runs a Cancer Hospital at Raipur with complete support from the government. The government has always been very proactive and supportive of the industries to sustain their business and grow further. BALCO received complete support from the government for continuous operations and retaining the employment as well.

As is the characteristic of any sustainable venture, the brand has undertaken several development projects under the themes of sustainable livelihood, women empowerment, healthcare, education, afforestation, water conservation, and agriculture programmes benefiting over 1,50,000 people in the region and beyond. The communities are witnessing continuous and commensurate development along with BALCO's growth.

8.7 Environmental, Social, and Governance (ESG) Initiatives

BALCO is credited with planting over 41 Lakh trees in the surrounding regions. To create awareness about the importance of water conservation, the company implemented rainwater harvesting techniques in the plants and in the township. Towards this purpose, an array of modern equipment was installed, which resulted in zero effluent discharge into the water bodies. In bauxite mining areas, there is a special focus to promote the regional flora and fauna.

As BALCO grows in strength, it has been able to explore India's demographic dividend that currently needs to be capitalised for the realisation of our vision of becoming a USD 5 trillion economy. We truly believe that despite the mammoth challenges that industries across the world are faced with, BALCO is on the right path to experience new heights of social, sustainable, and economic growth as we move from one milestone to the next.

BALCO's 1200 and 540 MW Power Plants have undertaken various innovative and improvement projects for energy conservation. In the year 2020–21 the specific water consumption of 1200 MW and 540 MW power plants was 2.1 and 1.95 m^3/MWh respectively. A solar plant of 33 KW has been established to promote renewable energy. The plants have also ensured 100% utilisation of ETP discharge water and almost 8000 Quality Circle and Kaizen projects have been implemented. In its other energy saving endeavours at all the areas of operations, BALCO has also reduced its power consumption in its potline 1 of its smelter as per the global benchmark and has become the best potline in India and Gulf countries in its power consumption and hence reducing its carbon footprint.

BALCO has been the recipient of various awards and accolades for its environmental conservation efforts. BALCO secured the coveted top spot with the highest Energy Saving Certificates, a testimony of its continuous endeavours towards energy conservation for business and environment sustainability. BALCO has won the Golden Peacock Award for Energy Efficiency-2021, 'CII Energy Efficiency Award', and 'Golden Peacock Sustainability Award'. The company has also bagged 'International Green Apple Award' 'CII Energy Excellence Award', 'Sustainable Business of the year Award', 'National Award for Excellence in Energy Management' and 'Energy and Environment Global Environment Award' in recent years.

8.8 Summary

BALCO is sincerely committed towards the goal of maximising its efforts in the direction for the betterment of the society and environment. Through our continuous endeavour, we aim to achieve the larger goal of creating a sustainable future for the generations to come and contribute to the expeditious development of the nation. BALCO has been an early adopter of smart technologies for heightened operational efficiencies, which further bolsters the culture of energy optimisation, safety, and productivity that we have meticulously fostered across the organisation. We are dedicatedly to driving the Environment, Social, and Governance dimensions of sustainability for building a greener tomorrow and, therefore, are actively exploring innovations that can help reduce our carbon footprint as early as possible in the creation process.

Chapter 9
JSW Steel—Journey with Speed and Innovation

Lokendra Raj Singh

Abstract The progressive growth of JSW group in the past three decades has mainly been attributed to the core principles on which the group stands viz. culture and work environment of complete freedom, innovation, empowerment, ethical governance, generous rewards, focus on results and global standards. JSW Steel has been a pioneer in introducing the new technologies in Indian Steel industry, such as Corex technology, setting up of non-recovery vibro-compacting coke ovens, India's largest blast furnace, the largest and widest hot rolling mill, most modern cold rolling mill and continuous annealed tinplate products. This article brings out the journey of JSW Steel and its contributions to the Indian steel industry.

9.1 Introduction

The journey began in 1982, with JSW Group's first steel plant at Vasind, near Mumbai. This project was helmed by Jindal Iron and Steel Company. The next two decades saw several acquisitions and expansion plans being drawn out. This was followed by the merger of Jindal Iron and Steel Company (JISCO) and Jindal Vijayanagar Steel Ltd (JVSL) in 2005. Today, JSW Steel can be found in six locations across India—Vijayanagar in Karnataka, Salem in Tamil Nadu, and Tarapur, Vasind, Kalmeshwar and Dolvi in Maharashtra—with a combined capacity of 18 MTPA. The mother plant, a world class manufacturing facility at Vijayanagar, Karnataka, is the largest single location steel-producing facility in India with a capacity of 12 MTPA. It's India's pride, and that inspires every visitor. With the completion of JSW Dolvi Works' expansion from the existing 5 MTPA to 10 MTPA, it becomes a mammoth 27 MPTA steel conglomerate. At JSW Steel, we are well-placed to address the strong demand growth enabled by the government's thrust on infrastructure, housing and the increasing share of manufacturing in GDP. By 2026, JSW Steel aims to produce 40 MTPA steel annually (Fig. 9.1).

L. R. Singh (✉)
Vijayanagar Works, JSW Steel Ltd., PO Vidyanagar, Dist - Ballari, Karnataka 583275, India
e-mail: lokendraraj.singh@jsw.in

R. Divakar et al. (eds.), *Indian Metallurgy*, Indian Institute of Metals Series,
https://doi.org/10.1007/978-981-99-5060-7_9

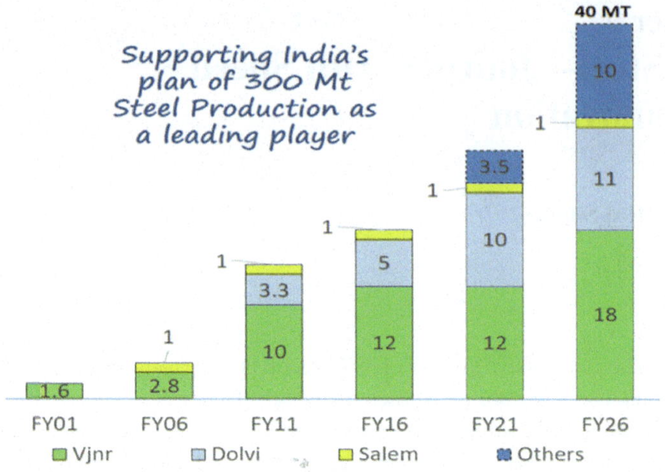

Fig. 9.1 Growth of JSW steel

To stay on the leading edge of technical advancement, JSW has entered into technological collaboration with JFE Steel Corp. Japan, for manufacturing high strength & advanced high strength steel for automobile sector. JSW Steel also entered into a joint venture with Marubeni-Itochu Steel Inc., Tokyo, to set-up state-of-the-art steel processing centers. To strengthen its global network, the Company has acquired a Pipe and Plate making steel mill in USA and JSW Steel Italy, Piombino. By 2030, the company plans to increase its capacity to 10 Million Tons overseas.

JSW Steel is recognized worldwide as a purveyor of value-added steel. JSW offers a wide gamut of steel products that include Hot Rolled, Cold Rolled, Bare & Pre-painted Galvanized & Galvalume, TMT Rebars, Wire Rods and Special Steel. Today, nearly 40% of the company's products are high-value steel, and one-fifth of the products are exported to countries around the world. With a footprint in over 100 countries, JSW Steel has become the largest exporter of coated products in India. The JSW Group also embarked its journey of expanding business landscape from Steel to Energy, Cement, Paints, Infrastructure, Sports and so on. The Company has been at the forefront of state-of-the-art, cutting-edge technology, research and innovation while laying the foundation for long-term growth. Strategic collaborations with global technology leaders to offer high-value special steel products for various applications across construction, infrastructure, automobile, appliances and other sectors.

The quick growth of JSW group within the short span of three decades could not have happened without creating a culture and work environment of complete freedom, innovation, empowerment, ethical governance, generous rewards, results focus and global standards and this is what it has done admirably. Adherence to global environment management standards protects and enriches the ecosystem, while creating more opportunities for those at the bottom of the social pyramid, which is a hallmark of his limitless growth and leadership qualities. JSW Steel is the only Indian Steel company ranked among the top 10 steel producers in the world by World Steel Dynamics for the last 10 consecutive years. Key honors & awards include World Steel Association's Steel Sustainability Champion (2019), Deming Prize for Total Quality Management at Vijayanagar (2018) and Prime Ministers Trophy-2012 for the best steel plant among others. With the cutting-edge technology and constant innovation, it has emerged as a steel plant with one of the most efficient conversion costs globally.

9.2 Introducing New Technologies and Unique Initiatives

JSW Steel has been a pioneer in introducing the new technology in Indian Steel industry, viz., bringing Corex technology (an alternative ironmaking process) first time in India in 1998, setting up of non-recovery vibro-compacting coke ovens, India's largest blast furnace (4019 m^3 volume) in 2009, the largest and widest hot rolling mill in 2012, and of late the most modern cold rolling mill (CRM-II) in 2013, which is India's answer to high grade automotive Steels. JSW Steel is the first Indian steel company to introduce continuous annealed tinplate products in the country which can help reduce carbon footprint as India moves towards reducing plastic consumption by eliminating single-use plastics over the next few years. These are the few bold initiatives of JSW in the recent past.

- Adopted Corex technology First time in the country—1999
- Largest blast furnace in the country—2009
- Largest and widest hot rolling mill—2011
- Largest State of the art cold rolling mill—2014
- Country's largest beneficiation plant (20 MTPA)—2011
- World's largest Corex gas based DRI Plant (1.2 MTPA)—2014
- India's First Pipe Conveyor for iron ore transportation—2019

On research and development front, JSW Steel has set-up a world class state-of-the-art steel research facility at Vijayanagar Works, Ballari (Karnataka) and has sub-centers at Dolvi (Maharashtra) and Salem (Tamil Nadu). In order to scale up the research outcome to a commercial scale, pilot plants for processing BHQ iron ores, coal briquetting, slag to sand making, steam aging of steel slag, plastic injection in EAF and coke ovens etc. have been commissioned and stabilized.

At JSW Steel, adopting innovative technologies has helped in cutting down the operational cost, mitigating the environmental concerns and conservation of natural resources. JSW is India's only steel plant with zero effluent discharge. JSW Steel has set-up the Asia's largest iron ore beneficiation plant of 20 MTPA capacity to upgrade the low to medium grade iron ores for producing agglomerates to feed the iron making units. For optimal utilization & conservation of natural resources as well as 100% recycling of solid waste generated in steel plants, a number of technology

Fig. 9.2 Unique initiatives at JSW steel

have been adopted, i.e. micropelletization technology, mill scale briquetting Technology, DRI fines briquetting, Iron recovery from slime, dust and sludge, usage of slag as alternative construction materials and in agriculture etc. JSW Steel has an integrated strategy towards efficient waste management that ensures optimal utilization of resources and minimal disposal of solid waste. Some first of its kind initiatives are shown in Fig. 9.2.

9.3 Human Resources and Corporate Social Responsibility

Trained and experienced human resource is the key to efficiently run the plant and achieve the operational excellence. A rigorous induction training of 3½ months to the youngster and continuous learning opportunities are provided to every workforce to develop skills and functional expertise. In every area of manufacturing, learning centers are set-up and classroom as well as job based training are regularly provided. Leadership development is derived as one of the key process which ensures the development of employee as potential leader.

The CSR activities through JSW Foundation mount a number of initiatives. The company through its CSR activities facilitates access to water and sanitation to nearby villages/talukas. It also promotes skill development through vocational training and education. JSW partners with government and various civic organizations for providing outreach to the community. JSW Steel, through JSW Foundation, is committed to contribute 2% of its net profit towards community development activities. The fund is allocated to empower women, impart primary education, mitigate infant mortality and nurture maternal health. The Company also aims to safeguard water, using renewable energy and promoting hygienic sanitation.

9.4 Summary

JSW Steel firmly believes that continued innovative research with an emphasis on Nature and Society is the driving force for long-term growth, competitiveness and sustainability of the steel industry. The company reiterates its commitment to align all the initiatives to achieve the overall business objectives and to put India higher on the map of Global Steel Production.

Part II
Research and Development Organisation

Chapter 10
Seven Decades of Metallurgical Excellence—CSIR-NML

I. Chattoraj and Mita Tarafder

Abstract The National Metallurgical Laboratory was established in the year preceding independence, for research and development in all areas of minerals, metals and materials. It has played a significant role in the industrial revolution of India starting from 1950, especially in the areas of mineral processing, iron and steel making, ferroalloys and extraction of non-ferrous metals. It has pioneered the production of ferroalloys, sponge iron and magnesium in the country. Its recent achievements include several developments to improve the quality of steel produced by secondary melters, coal beneficiation, strategic metals such as tungsten, sodium, gadolinium and lithium, and effluent management, several developments for metal artisans, and various skill development initiatives.

10.1 Introduction

CSIR-National Metallurgical Laboratory (CSIR-NML) is a premier Indian research organisation dedicated to various facets of Minerals, Metals and Materials including scientific research, technology development, industrial services and human resource development. Since its inception, CSIR-NML has diversified its research areas, which now encompass extractive metallurgy, alloy development, mathematical and physical modelling of metallurgical processes, minerals research, advanced materials and materials tailoring, integrity evaluation of critical industrial components, surface engineering and cleaner and sustainable metals production.

The foundation stone for National Metallurgical Laboratory (NML) was laid by Hon'ble Sri C. Rajagopalachari on 21st November 1946. It was formally inaugurated and dedicated to the nation on 26th November 1950 by Pandit Jawaharlal Nehru "in a spirit of hope and in a spirit of faith in the future". The laboratory was an element of Sri Shanti Swaroop Bhatnagar's vision of providing India with a network of research institutions for taking the country ahead in science and technology. CSIR-NML played a significant role in the industrial revolution of India starting from 1950,

I. Chattoraj (✉) · M. Tarafder
CSIR-National Metallurgical Laboratory, Jamshedpur, India
e-mail: ichatt@nmlindia.org

R. Divakar et al. (eds.), *Indian Metallurgy*, Indian Institute of Metals Series,
https://doi.org/10.1007/978-981-99-5060-7_10

especially in the areas of mineral processing, iron and steel making, ferroalloys and extraction of non-ferrous metals. Asia's largest creep testing facility was also set up at CSIR-NML in the early 1970s and, even today, it ranks as the second largest creep testing lab in Asia. A historical account of past achievements (1950–2010) of CSIR-NML is preserved in the Diamond Jubilee commemorative volume '*la vintage metallurgie: 60 years of marriage of science to industry*'.[1]

The stated vision of the laboratory is "to become a global leader and an internationally benchmarked laboratory in mineral and metallurgical research and development; to become a self-sustained technology centre in minerals, metals and materials."

10.2 Legacy Achievements

NML's history is filled with groundbreaking innovations for the country, especially in the early days after independence. Some pioneering and impact making developments of the yesteryears are presented below.

10.2.1 Ferroalloys Production (Circa 1960–65)

Development of ferroalloys was essential for the growth of iron and steel industry in India. NML had initiated work on producing low carbon and carbon free ferrochrome through the aluminothermic route to meet the urgent requirements of ordnance establishments. Special emphasis was placed on ferroalloys in the third Five Year Plan. Ferroalloys development commenced in full earnest after the commissioning of a 500 kVA submerged arc furnace (SAF), the first of its kind in India. While the furnace provided an ideal test facility for a large number of ferroalloy producers in the country, the aluminothermic production technology was exploited by dozens of small and medium entrepreneurs across the country. The submerged arc furnace also served as a national facility for the smelting of ores in general.

Since the early sixties, NML's contributions to the development of the ferroalloy industry were phenomenal. Calcium-Silicide was produced for the first time in India (1975–76) at NML in a 500 kVA submerged arc furnace. Substantial amount of this alloy, which conformed to ISI Standard 1387/1967, was sent for industrial use to Hindustan Machine Tools (HMT, Bangalore) and Heavy Engineering Corporation (HEC, Ranchi). Industrial scale trials were conducted at Visveswaraiah Iron and Steel Works to produce vanadium rich slag in a 1500 kVA submerged arc furnace. The slag is a source for the production of ferrovanadium. NML's relationship with Ferroalloys Corporation (FACOR), which started in 1962, grew stronger. Based on the extensive pilot scale trials done at NML, FACOR established a 12 MVA furnace for high carbon ferrochrome and silico-chrome production and an 8 MVA slag furnace for

[1] http://eprints.nmlindia.org/4360/.

low carbon ferrochrome in 1967. Besides these successful pilot scale investigations, FACOR sponsored many other investigations in the seventies, including macro- and micro-structure of low carbon ferrochrome samples, utilisation of ferromanganese slag, sintering of manganese ores, beneficiation of low grade chrome ore samples, analysis of nitrogen content of low carbon ferroalloys, and use of ferrochrome slag as hardener for sodium silicate bonded sand in foundries. Interestingly, FACOR utilised the magnesium metal flakes generated as waste during the production of magnesium crowns at the Magnesium Pilot Plant of NML for producing inoculants.

10.2.2 Pioneering Sponge Iron Production in India (Circa 1970–75)

Sponge iron making with the use of coal and other low grade fuels was vital for developing the steel economy of the country to replace scrap in electrical steel making. In tune with the research priorities of the Planning Commission and the Ministry of Steel, a rotary kiln facility (200–300 kg/h) was created for larger scale trials for the production of sponge iron. The then Union Minister of Steel, Shri Charanjit Singh Yadav lauded the efforts in this direction: "National Metallurgical Laboratory has shown exemplary courage in taking up the challenging initiative for promoting the direct reduction techniques in the production of sponge iron. This can result in establishing a large number of mini steel plants in the country which will serve the steel requirements of the areas nearby utilising wherever possible, the deposits of iron that are available locally. These mini steel plants, using sponge iron technique, will be able to provide 20–25% of the steel requirements in the years to come" The iron ores from Industrial Development Corporation of Odisha Limited (IDCOL), Bhilai Steel Plant (BSP), State Industrial Investment Corporation of Maharashtra (SIICOM), Andhra Pradesh Industrial Development Corporation (APIDC) and National Mineral Development Corporation (NMDC) were tested at pilot scale in the rotary hearth furnace for the production of sponge iron. Under a UNIDO sponsored study, reduction tests were carried out on Aswan (Egypt) ore (NML's 1st international sponsored research project). Nepal Bureau of Mines sponsored yet another study to test their raw materials. India's first commercial sponge iron plant with a capacity of 25,000–30,000 tonnes per year based on the technology of solid reduction process developed by NML was commissioned (1975–76) at the plant site of Andhra Cement Company at Vijayawada and inaugurated by the Minister of Steel and Mines. Dr. Ing. H. P. Mishra, former Chairman, Industrial Promotion & Investment Corporation of Odisha (IPICOL), Bhubaneswar acknowledges (2010) NML's contributions in this area: "From 1970 to 1974, I took the help of NML for our sponge iron and Ferro-vanadium project. This work helped us to comprehend the Orissa sponge Iron & Tata Sponge Iron projects".

10.2.3 Pioneering Magnesium Production in India and Related Developments (Circa 1970–1987)

The most significant development in the early seventies was the commissioning and commencement of successful trial runs in the Magnesium Plant. The first production of sponge magnesium in the laboratory happened in the late 1960s. The magnesium team led by Mr. M. P. Menon produced the first magnesium ingot in the country. The magnesium pilot plant was commissioned in February 1972 and it was dedicated to the country by Sri Mohan Kumaramangalam, Union Minister for Steel and Mines, in November 1972. High purity magnesium (99.9%) was successfully produced in tonnage quantities and the sponge magnesium was subsequently melted and alloyed at a pilot scale. The magnesium metal produced during the development period was supplied to a large number of users, such as Ordnance Factories, Hindustan Aeronautics Limited, Government Mint, Bharat Aluminium, Heavy Engineering Corporation, Vikram Sarabhai Space Centre, Defence Research Laboratories and Nuclear Fuel Complex. These organisations confirmed that the magnesium supplied was of purity good enough to replace the imported magnesium used by them. By 1975, the know-how was ready for transfer to a commercial producer. However, the government of that time was very insistent that the technology should be transferred only to the public sector and not to any private organisation. Despite the several proactive measures of NML, it took twelve years for the transfer of technology because of the bureaucratic hold-ups and finally the plant was sold to Southern Magnesium in August–September 1987, who became the only magnesium producer in the country for more than a decade. The magnesium technology development did have its positive spin-offs. While the extraction of magnesium was being focussed on, in parallel, the Laboratory developed a strong programme on magnesium based cast and wrought alloys in collaboration with Bhabha Atomic Research Centre (BARC) and the Indian Space Research Organisation (ISRO). Technology for magnesium inoculants to produce nodular cast iron was developed and successfully transferred to half a dozen companies.

10.2.4 Establishment of the Largest Creep Testing Laboratory (Circa 1970)

The establishment of Asia's largest creep laboratory (at that time) for uninterrupted long-term testing at elevated temperatures was a major milestone in the history of NML. The engineering support for its entire installation and commissioning was carried out by the engineering division at NML. The Creep facility at National Metallurgical Laboratory was a cynosure for all the visitors. Since its inception, the creep testing facility has served hundreds of industrial users from all major sectors, atomic energy, oil and petrochemicals, steel, power, fertilisers, and many others. By 1980, the activities in the areas of creep, remaining life assessment and failure analysis had reached substantial maturity. Development of creep resistant steels and long-term creep tests for BHEL, continued. The materials qualification activity for the Fast Breeder Reactors of Reactor Research Centre, now the Indira Gandhi Centre for Atomic Research (Department of Atomic Energy) was also at its peak. On an average, 20–30 failure analysis investigations, several of these as consultancy assignments, were pursued every year for the power plants, petrochemical industries, cement industry, paper mills, Indian Railways, state electricity boards and so on.

10.3 Recent Scientific and Technological Achievements

CSIR-NML continues to play a vital role in the quest of the country towards scientific and technological leadership and in providing scientific solutions to the industries in the areas of minerals, metals and materials. CSIR-NML is also carrying out major activities for creating awareness among the common masses on issues relating to health, environment, rural technology and sustainable development. The Laboratory has kept pace with the changing research scenarios and needs of the country. In the last few years, greater emphasis was given to industry sponsored research and on alignment with government programmes, namely, Make in India, Innovate in India, Strategic sector needs, Swachch Bharat, Societal and Skill India, etc. The activities of the Laboratory touch upon several major sectors relevant to the growth of India, including iron and steel, power and energy, oil and gas, automotive, railways, strategic, societal, and others. The following is a brief sectorial/topical account of the achievements and efforts of the laboratory.

10.3.1 Steel

In India, nearly thirty per cent of steel is produced by secondary steel producers using induction furnace route. High phosphorous content makes the steel substandard and unsuitable for structural applications. CSIR-NML in association with All India Induction Furnace Association (AIIFA) and Ministry of Steel has developed a flux to reduce the phosphorus level from 0.07–0.09% to BIS prescribed limits of <0.05%. The flux which is tested under industrial conditions has the potential to benefit a large number of secondary steel producers. Efforts are on to fine tune the flux chemistry and its usage with high lining life. Research pursuits of CSIR-NML are also aligned with the material needs of the various sectors. A wear resistant steel was developed by the Laboratory to address the silt erosion problem during hydel-power generation in the Himalayan region. An abrasion resistant steel was also developed for grinding media and was successfully utilised by the client.

The present thrusts are on the development of advanced automotive steels including steels for e-vehicles. Quench and partitioned steels, very high strength steels and multi-phase steels are being researched in collaboration with different steel majors. A substantial thrust is also on the reuse of iron and steel plant wastes including steel slags, slimes, fines, mill scales, etc., by conversion to value added products.

10.3.2 Coal

Significant efforts are directed on coal and the activities range from prospecting using 'coal core analysis' to beneficiation for ash removal and obtaining value added products from fly ash. One of the first efforts in the country on dry beneficiation of coal was successfully carried out by CSIR-NML. CSIR-NML column flotation technology, which was earlier exploited in commercial operations for beach sand minerals, limestone and barite, is now successfully tested for coal, on site. CSIR-NML is also involved in an effort to convert non-coking coal to coking coal.

10.3.3 Strategic Metals

Specific and sustained focus of the Laboratory is on the extraction and processing of strategic metals, from primaries as well as secondaries. For most of the strategic metals, the primary resources are either absent or insignificant in our country. The thrust of research at CSIR-NML has been on extraction of these metals from secondaries and wastes, along with some efforts towards exploitation of low grade indigenous resources.

To address the immediate needs of the country, especially defence needs, NML has developed and commercialised a technology for the recovery of tungsten from a large variety of tungsten-bearing scraps. The developed technology is superior in terms of tungsten recovery (>90%), co-recovery of associated metals (e.g. cobalt, nickel), process economics and environmental considerations. Indigenous technologies for sodium and gadolinium, essential for atomic energy programme were successfully developed. Sodium is produced using electrolysis cells from 50 to 500 kVA capacity, and further scale up of the process by industry is in progress. Similarly, high purity gadolinium was produced by fused salt electrolysis.

Lithium is another metal with huge future demand and India does not have much primary resources. CSIR-NML is involved in the exploitation of primary resources and is spearheading lithium extraction from secondary resources (end of life batteries) with significant success. Similarly, in the matter of Rare Earth elements, substantial efforts have been made to exploit secondaries, including a variety of urban ores like discarded computers, cell phones, magnets, compact fluorescent lamps, etc. Additionally, strategic metals valorisation from diverse industrial wastes, like spent catalysts, are being carried out. A dedicated urban ore recycling centre has been created and is in operation. Several technology transfers in valorisation of urban wastes, especially with regard to the extraction of lithium, cobalt, nickel and rare earth elements, have happened over the last five years.

10.3.4 Industrial Wastes Valorisations and Effluents Treatment

Solid wastes and effluents are of major concern in metallurgical industries from the point of view of resource conservation and environmental considerations. Large quantities of solid wastes (e.g. slimes and fines during iron ore mining, slags, mill scales) and effluents (e.g. pickle liquor) are generated during iron and steel making operations. NML has developed a number of processing options and value added products to address the problems. Some of these have been tested in real plant conditions and include: DRI from slimes/mill scale using tunnel kiln; mill scale briquettes as coolant in LD converter steel making; pig iron from the smelting of self-reducing briquettes (slime and Jhama coal) in a Low shaft furnace; and, magnetite for heavy media separation using high purity haematite produced during pickling operation as raw material. Strategies for the recycling and reuse of LD slag have been developed using smelting in a 175 kVA furnace and the processes are available for further probing on industrial or semi-industrial scale. In yet another significant development, technologies have been developed and tested for the production of a wide spectrum of value added iron oxide pigments. In a NitiAyog inspired programme, CSIR-NML along with three industries and two research organizatios, is spearheading the holistic utilization of red mud, an aluminium industry waste.

10.3.5 Devices and Equipment Development

In recent times, the Laboratory has paid increasing attention to the design and development of equipments which are tailor-made for specific metallurgical research and industrial applications. The objective is also to minimise dependence on imported equipment, which are often priced exorbitantly. Under one such initiative, an annealing simulator was developed which offers exciting possibilities to simulate batch and continuous annealing processes commonly practised in steel industry (e.g. IF steel with desirable dew point setting, ultrafast cooling for dual phase or complex phase steel annealing cycles, etc.). Similarly, several devices were developed for on and off site NDT of industrial components and materials. One of these, MagStar, a portable magnetic sensing device for Non-destructive Evaluation of Steel Structures/ Components measures Magnetic Hysteresis Loop (MHL) and Magnetic Barkhausen Emissions (MBE). MagSys, a Giant Magneto Impedance (GMI) based magnetic sensing device is also ready for commercialisation as a structural health measuring tool. The device is capable of detecting very low magnetic fields and magnetic phases.

An NDT device for detecting defects in the wire during fabrication has been developed and installed at the sponsor's site. In collaboration with our sister laboratory, CSIR-CGCRI, various applications of FBG sensors including temperature profiling of continuous casting moulds, have been perfected. The use of in-situ FBG sensors, for breakout detection in billet casters, was possibly tried out successfully for the first time anywhere.

10.3.6 Functional Materials

Several new alloys and materials have been developed based on the specific needs of our clients and collaborators. Several biomimetic products useful for bio-applications have been commercialised. A collagen-graphene composite was developed for supercapacitive applications for a Sri Lanka based multinational. Several advanced coatings have been developed at the behest of aerospace multinationals with green protocols and safe and sustainable materials in mind. Amorphous alloys have been developed and used in sensors and other applications. Present focus is on 2D materials and MXenes, and magnetic shape memory alloys. Various ceramic and metallic bio-implant materials have been developed and commercialized.

10.3.7 Structural Health Monitoring

The laboratory has been a nodal organisation in structural health monitoring (SHM). The excellent state of the art facilities in Materials characterisation and evaluation, capabilities for NDT evaluations, coupled with the availability of nationally and internationally renowned experts in these areas, have enabled CSIR-NML to be a leader in this country in SHM of critical industrial components and related efforts in remaining life assessments, fitness for service evaluations, and failure investigations. Over the last thirty years, more than three hundred projects and programmes have been carried out in this area, with a very varied clientele from Oil & Gas, Chemicals, Power, Aerospace, Automotive, and other sectors. The Indian Air Force as well as the Indian Navy has benefited from NML provided services and recommendations in this area.

10.4 Working for the Society

10.4.1 Assisting Artisans

Brass-ware artisans across the country are in a crisis because of stiff competition from China. The home based traditional pit furnaces used by the artisans are fuel inefficient and polluting. As part of a National Innovation Council initiative, NML has developed a cost effective, fuel efficient and eco-friendly coke based brass melting furnace for the metal artisans utilising waste heat recuperating system and arresting suspended particulate matter (SPM) inside the pit. A few such furnaces were installed in Moradabad, Uttar Pradesh and in Balasore, Odisha. Simultaneous efforts are on to enhance awareness and training of the artisans. CSIR-NML has also developed an efficient, low cost anti-tarnishing lacquer to prevent tarnishing and to maintain the metallic cluster of brass handicrafts. The developed lacquer is technically and cost-wise superior to the lacquers available in the market. The technology has been transferred to a private entrepreneur for its commercial production.

NML has been entrusted with the responsibility to generate detailed project reports for multiple silver artisan clusters in West Bengal, which have been completed successfully.

10.4.2 CSIR Integrated Skill Initiative

In 2016, CSIR launched an Integrated skill training initiative to equip young minds with the necessary technological skills through exposure to research laboratories at national facilities. Under CSIR integrated skill initiative, CSIR-NML offers skill training ranging from basic training for rural youths, summer/winter internship for undergraduate students, M.Tech./Ph.D. dissertation guidance for postgraduate students, apprentice training for ITI/Diploma/Graduate engineers and specialised training for professionals from the industries.

10.4.3 Interactions with Students and Schools

CSIR-NML has had a running programme to enhance the scientific temper of school students. NML has, over the last ten years, allowed student visits on a weekly basis along with interactions with NML scientists during such visits. Additionally, a special programme for Central Schools, initiated by CSIR, the Jigyasa programme, has been made operational. These programmes have benefited nearly 18,000 students of more than 200 schools and have been widely appreciated.

10.5 Epilogue

Since its inception, CSIR-NML has been striving to achieve the sustainable development goals (SDG) through technology development for the utilisation of natural resources and by translating them to the industries. The vision set forth during the first decade of our independence, to make India "AtmaNirbhar" remains relevant today. The performance of this laboratory as gauged through patents filed, technologies transferred, revenues generated, research publications, global visibility and other metrics has trended up over the last decade. The current trends and future directions of research and development at NML are in concert with the national aspirations, and with sustainability, inclusion and equity.

Chapter 11
Molten Salt Electrowinning of Metals and Materials: Opportunities and Challenges

M. Jayakumar, Naveen Chandrasekaran, R. S. Prasannakumar, James Ebenezer, N. Mohanapriya, C. Andrew, B. Subramanian, and L. John Berchmans

Abstract Conventional metallurgical processes to produce metals involve carbothermal reduction of their compounds and suffer from massive carbon emissions. Electrowinning using aqueous electrolytes is a demonstrated approach to obtain many metals. However, aqueous electrolytes are not suitable to produce several classes of metals such as alkali, alkaline earth, refractory, rare earths, and actinides due to their high negative reduction potentials. Molten salt electrolysis is a commercially viable and feasible route for the production of these metals. The present chapter details on the molten salt electrowinning of alkali, alkaline earth, rare earths and refractory metals with a special focus on the efforts at CSIR CECRI in this niche area of science and technology.

Keywords Molten salt · Electrowinning · Pyrometallurgy · Light metals · Rare earths · High temperature

11.1 Introduction

The overall development of a country is directly related to the flow of materials, utilization of commodities, imports and exports. It is propelled by the population growth and rapid industrialization which further steers innovative technologies for the production of advanced materials. Mobilization of rural population to urban areas leads to higher utilization of lifestyle driven consumer products.

M. Jayakumar (✉) · N. Chandrasekaran (✉) · R. S. Prasannakumar · J. Ebenezer · N. Mohanapriya · C. Andrew · B. Subramanian · L. John Berchmans
Electroplating and Metal Finishing Division, CSIR Central Electrochemical Research Institute, Karaikudi, Tamil Nadu 630 003, India
e-mail: jayam@cecri.res.in

N. Chandrasekaran
e-mail: naveen@cecri.res.in

R. Divakar et al. (eds.), *Indian Metallurgy*, Indian Institute of Metals Series,
https://doi.org/10.1007/978-981-99-5060-7_11

In recent years, the consumption of minerals, metals and fossil fuels has steadily been increasing in our country which steers the development of improved ferrous and non-ferrous materials for sustainable growth in aviation, automotive, space, construction and heavy engineering sectors. As critical players, metallurgical industries have a compelling demand to maximize the production of metals and materials through improved process technologies. Traditional metallurgical processes involve carbothermal reduction of metal compounds to their respective metals which suffers from colossal carbon footprint (Springer and Hasanbeigi 2016). Electrowinning is a demonstrated approach to obtain metals by the passage of electrical charge into a molten solution containing the metal ion. Several classes of metals such as alkali, alkaline earth, refractory, rare earths and actinides cannot be produced in metallic form by electrolysis of their aqueous solutions due to the limited electrochemical stability of water (~1.23 V). These metals are characterized by their high negative reduction potentials that make their electrowinning into their metallic forms extremely difficult since their respective compounds are very stable. Molten salt electrolysis is the commercially viable and feasible route for the production of these metals. The present chapter dwells on the molten salt electrowinning of alkali, alkaline earth, rare earths and refractory metals with a special focus on the efforts at CSIR-CECRI in this niche area of science and technology.

11.2 Principles of Metal Extraction Processes and Electrometallurgy

The extraction of metals depends on the difference in chemical potentials which makes the separation of metal values from the compounds difficult. Metallurgical processes are classified into pyrometallurgical and hydrometallurgical processes. Pyrometallurgical processes generally involve beneficiation of minerals followed by the reduction of purified minerals such as metal oxides, sulphides and chlorides at high temperatures by using a reductant like carbon (C), hydrogen (H), carbon monoxide (CO) to produce the respective metals. The Ellingham diagram shows the relative thermodynamic stabilities of the various oxides as a function of temperature, thus enabling an understanding of the thermodynamics of the reduction process. In hydrometallurgical processes, the minerals are subjected to various ore processing steps like crushing, grinding, sieving, magnetic separation and calcining followed by dissolution in a suitable aqueous liquid from which the metals can be extracted.

In electrometallugical processes, the purified ore is leached with a suitable solvent and subsequently decomposed by applying an electrical potential, for example, in electrowinning of Cu, Ni, Zn, Cr etc. Mostly acids, alkalis and organic solvents are used to trap the metal values from the mineral particles. The process of leaching enables the dissolution of metal values in to the leached medium. The leached solution is then electrolyzed by applying a current to deposit metal at the cathode. Noble metals like copper, silver, gold etc. which exhibit reduction potentials more positive than hydrogen in the electrochemical series could be conveniently electrodeposited from aqueous solution (Vol'skii and Sergievskaia 1971). Even though certain metals such as manganese (Mn), zinc (Zn), chromium (Cr), iron (Fe), cadmium (Cd), cobalt (Co), nickel (Ni), tin (Sn) and lead (Pb) have negative reduction potentials, they could be electrowon from aqueous solutions taking the advantage of hydrogen over-voltage under appropriate conditions of electrolysis (Ray 1985).

In an electrolytic process, electrical current is passed through an ionic media resulting in non-spontaneous chemical reactions at the electrode surfaces. This concept is widely applied for the deposition and purification of a variety of metals from their primary and secondary resources.

The essential components required to carry out the electrodeposition process are,

1. An electrolyte either in liquid or solid form having mobile ions.
2. A power source, which provides energy through a voltage and current for the decomposition of soluble metal compound.
3. Electrodes are made of electronically conducting materials and current distribution system with control devices.
4. An electrolytic cell called the electrolyzer comprising of anodes and cathodes which are dipped into the electrolyte to complete the electrical circuit connected to a DC power source. The electrical energy supplied results in the discharge of the ions at the electrodes which are immersed in the electrolyte.

During electrolysis, the positively charged ions (cations) move towards the negative electrode called as cathode thereby depositing the respective metal, whereas the negatively charged ions, anions migrate towards the positively charged anode. At the electrode surface, the electrons are absorbed or released by the atoms and ions. Those atoms that gain or lose electrons to become charged ions pass into the electrolyte whereas those ions that gain or lose electrons become uncharged atoms and usually separate from the electrolyte. Other components include electrolytic vessel to hold the reactants and products, electrical circuitry and control devices of the electrochemical reactor (Yan and Fray 2008).

Electrolytic processes may be subdivided into Electrowinning, Electrorefining, Electroplating and Electroforming.

1. In electrowinning process, the metal is deposited in a pure or partially impure form from a compound of the metal dissolved in the electrolyte.
2. In electrorefining, the pure metal is transferred from the impure anode to the cathode by the application of a suitable voltage. The impurities may reside at the anode itself or go in to the liquid electrolyte.
3. In electroplating, a thin uniform coating of the metal is formed on the surface of an inexpensive substrate to protect the substrate from corrosion or oxidation.
4. Electroforming is performed to produce an article on a disposable substrate through an electrolytic process (Mamantov and Marassi 2012).

During electrowinning of electropositive metals from aqueous solutions, electrolyte decomposition precedes metal reduction and the cathodic product is invariably H_2 and results in the formation of respective hydroxides or oxides of the metal.

Therefore, metals with more negative reduction potentials can be electrowon only from non-aqueous electrolytes. Among them, molten salt electrolysis is a feasible and scalable strategy for the electrowinning of alkali, alkaline earth, refractory, rare earths and actinide metals.

11.3 Molten Salt Electrolysis

Molten salt electrolysis involves passage of electric current into molten metal halides which act as the solvent and the electrolyte (hence termed as fused salt electrolysis) to enable efficient migration of the ions to the respective counter electrodes (Thonstad 1992). Majority of the molten salt systems comprise of alkali and alkaline earth halides such as chlorides and fluorides. They possess high negative Gibbs free energies of formation and therefore, they possess very high decomposition potentials at their operational temperature (Mamantov and Marassi 2012; Gale et al. 1983). The lowering of the decomposition voltage may be possible, when the activity of the deposited solvent metal is lowered either through interaction with the solvent cations or through interactions with the electrode surfaces or both. The energy necessary in terms of voltage and current may be small to deposit any metal from its compound and directly relates to the efficiency of the process. The molten alkali and alkaline earth salts and their mixtures have good ionic conductivities thereby the energy consumption can be reduced by lowering the temperature of the operation. Molten salt electrolysis reduces the number of preliminary steps and can be performed in compact electrolytic cells as compared to aqueous electrolysis. A very high current density can be applied in fused salt electrolysis for the deposition of various metals from their compounds (Thonstad 1992). Electrowinning of metals by fused salt electrolysis has been attempted to metals that cannot be directly electrodeposited from aqueous baths. Among them, the most important metals such as lithium (Li), sodium (Na), potassium (K), rubidium (Rb), caesium (Cs), beryllium (Be), magnesium (Mg), calcium (Ca), strontium (Sr), barium (Ba), radium (Ra), aluminium (Al), tantalum (Ta), thorium (Th), tellurium (Tu), titanium (Ti), vanadium (V), zirconium (Zr) and many rare earth metals and alloys, are produced by fused salt electrolysis (Viswanathan 1985; Lovering 1982; Mouron 1997; Osaka et al. 1997; Sunshot 2016). In particular, aluminium and magnesium are the important metals produced in tonnages by fused salt electrolysis (Kannan et al. 1986).

11.3.1 Important Parameters in Designing Efficient Molten Salt Systems

11.3.1.1 Chemistry of Molten Salt Baths

An ideal molten salt bath should have low vapour pressure and melting point, high decomposition potential and ionic conductivity, non-corrosive, and constitutes environmental friendly and low-cost materials. Even though a wide range of molten salt mixtures are available, only some of the salts possess the above mentioned properties. Thus the possible range of molten salts is wide ranging from the low melting structurally complex and predominantly covalent system such as $KAlCl_4$ to the high melting structurally simple and ionic systems such as NaCl (Viswanathan 1985).

The fused baths can be classified as,

i. Chloride melts
ii. Fluoride or Fluoride-Chloride melts
iii. Oxides dissolved in Fluoride-Chloride melts

Chloride baths are generally chosen for alkali and alkaline earth metals, because their respective chloride salts can be easily purified from their raw materials. Chlorides are preferred electrolyte melts for the purpose of increasing the conductivity and lowering the melting point of the bath. The chloride mixtures could be chosen in such a way that they should have a melting point above that of the metal to be electrodeposited. Hence, the electrodeposited metal could be obtained in the molten form. Chloride salts are inexpensive, less corrosive and have lower melting points than fluoride salts. However most of the chloride salts are hygroscopic which requires dehydration and handling in inert atmosphere (Gale et al. 1983; Mishra and Olson 2005).

In case of refractory metals such as Zr, Nb, Mo Hf, Ta and W with high melting points and reactivity, the electrolysis is performed at temperatures below the melting point of the metal and the metal is obtained in the form of crystals. The refractory metal fluorides are stable compounds and are comparatively non-hygroscopic. Hence, the electrolyte baths chosen for the recovery of refractory metals are a mixture of double fluorides and alkali metal fluorides (Osarinmwian 2019). Refractory metal chlorides have high volatility and in some cases, they possess poor conductivity such as titanium chloride which becomes disadvantageous for easy electrolysis. This problem can be resolved either by complexing their chlorides with appropriate solvents such as BeCl–NaCl wherein the melting point of the chloride salts is lowered resulting in less volatility.

Whenever oxides are employed, they are generally dissolved in fluoride melts. Hall-Herault process for the electrowinning of aluminium is an outstanding example in this category (Choate and Green 2000). The application of similar procedures in other cases has considerable difficulties because of the poor solubility of oxides in fluoride or fluoride-chloride melts. Electrowinning of tantalum from its oxide dissolved in $KCl–KF–K_2TaF_6$ is an example, which is industrially practised. Some of the important factors affecting the deposition process are solutes, volatility of the feed, formation of cluster compounds, formation of insoluble oxides and oxy-halides and stable oxy-cations.

(a) **Solutes**

The solutes should have physical properties such as common anion which facilitate the dissolution and high degrees of miscibility of the mineral in the molten melt. For example, the appreciable solubility of alumina in cryolite melt is a critical factor for the successful electrowinning of aluminium (Choate and Green 2000). The strength of the interactions between the cations of the solute and the anions of the solvent and between the anions of the solute and the cations of the solvent are considered for the better solubility of the solute in the melt.

(b) **Fusibility**

Fusibility of mixtures of salts employed in molten salt electrolysis varies widely depending on the components of each system. Eutectic mixtures whose melting points are very low compared to individual components of the salts are used in molten salt electrolysis. For example, eutectic mixtures can have a melting point as low as 425 °C for the salt mixtures of NaCl–KCl–$MgCl_2$ whose individual melting points are 804, 790 and 712 °C, respectively. Practical molten salt baths necessarily do not employ a universal eutectic composition, as efficient electrolysis requires only a stable and clear molten melt and it may vary depending on the metal to be deposited (Vol'skii and Sergievskaia 1971).

(c) **Volatility of the Feed**

Fluoride solutes have less volatility in fluoride melts than in chloride melts. In some cases, a mixture of chlorides and fluorides is used in the electrowinning of refractory metals. In such conditions, the fluoride ions will complex with high oxidation state refractory metal ions more strongly than the chloride ions and are stabilized in the melt. For example, in the case of electrowinning of aluminium using aluminium chloride, its volatility can be reduced by the addition of fluoride in the melt which forms complexes such as AlF^{4-} and AlF^{6-} (Choate and Green 2000). In certain cases, this effect can be complicated by disproportionation reactions as observed during the addition of fluoride ions to stabilize Ta(IV) ions in LiCl–KCl eutectic melt; $5Ta^{4+} \rightarrow 4Ta^{5+} + Ta$ (Thonstad 1992). This reaction is possible due to higher valency of tantalum resulting in the formation of a stable complex. Further, the refractory metal ions can exist in different oxidation states; for example, tantalum can exist in oxidation states +5 to -1. As a result, tantalum chloride can undergo disproportionation reactions in the pure state at relatively low temperature and both homogeneous and heterogeneous disproportionation equilibrium are possible in chloride melts (Thonstad 1992).

$$2Ta^{4+} \rightarrow Ta^{5+} + Ta^{3+}$$

$$3Ta_6Cl_{18}^{4+} + 16Cl^- \rightarrow 4Ta + 14TaCl_5^{2-}$$

Similar reactions have been reported during titanium metal deposition in molten electrolytes in the presence of fluoride ions in the melt (Chamelot et al. 2000). While the disproportionation nature of the higher oxidation state of refractory compounds and their importance in electrolysis has been studied, refractory metal ions in lower oxidation states can occur as intermediates of the overall cathodic reduction process. The formation of such insoluble and soluble compounds with such diverse chemistries and their consequent disproportionation may or may not inhibit the overall reduction process.

(d) **Formation of Cluster Compounds**

Cluster compound formation is a particular feature of the lower oxidation states of the heavier refractory metals. While the cluster formation is thermodynamically favoured, it may be kinetically hindered. Tantalum can be produced from $TaCl_5$ at positive electrode potentials with reasonably high current efficiency due to the anodic formation of clusters in the molten melts; $20MCl + 16Ta + 14TaCl_5 \rightarrow 5M_4Ta_6Cl_{18}$ (Thonstad 1992; Chamelot et al. 2000; Sehra and Vijay 1998) where $M = Na^+$, K^+, Li^+. On the other hand, niobium metal is produced with a low current efficiency because of the formation of clusters in the NaCl melt (Sehra and Vijay 1998).

(e) **Formation of Insoluble Oxides and Oxy-Halides and Stable Oxy Cations**

While the presence of alkali and alkaline earth metal oxides as impurities in the product is inevitable, they can have deleterious effect on the product formation. For example, oxypolynuclear anions or insoluble oxides may be formed as insoluble films on the electrode surfaces by the reactions between oxide ions and metal ions in a lower oxidation state (Jeong et al. 2008). On the other hand, the deposition of metal oxides from stable oxy-cations can be an inherent feature of molten salt electrolysis. For example, when UO_2Cl_2 is dissolved in halide melt, the UO_2^{2+} cation is formed and the cathodic reduction produces UO_2 (Dubrovskiy et al. 2018). On the other hand, there is an evidence for the formation of stable oxy anions during the aluminium electrowinning bath ($AlOF_5Cl^{4-}$, $AlOF_3^{2-}$) which are discharged at the anode (Sehra and Vijay 1998).

11.3.1.2 Electrochemical Aspects of Molten Salt Systems

(a) **Conductivity**

Molten salts are good conductors of electricity. Their conductivities are considerably greater than the conductivities of the same salts in their aqueous solution. For example, molten KCl at 800 °C has a specific conductance (K) of 2.24 mho/

cm, whereas for an aqueous solution of KCl at 20 °C, K is 0.1 mho/cm. Increase in temperature decreases the viscosity, and increases the conductivity in most of molten salt systems (Redkin et al. 2012). While an increase in temperature is expected to increase the conductivity linearly, fused salt mixtures do not usually follow such a trend but deviate from the law of mixtures.

(b) **Electrode and Decomposition potentials**

In the case of aqueous electrolysis of pure salt solutions, the potential (E_{cell}) of the cell depends upon the activity of the conducting ion, and the question of concentration or activity does not generally arise. Since, all the reactants are in pure state, they have unit activities and hence the electrode potentials are standard potentials. However, in the case of mixtures of salts, the deposition potential will depend on the concentration of the salt corresponding to the metal used as the electrodes (Castrillejo et al. 2002).

$$E_{deposition} = E_0 + RT/nF + \log A$$

wherein A is the activity of the conducting ions

(c) **Kinetic factors**

Kinetic limitations in the processes are minimal at high temperatures. Three main factors are considered with regard to the formation of metal deposition from soluble solution species.

- Solute mass transport
- Electron transfer
- Coupled chemical reactions and nucleation and growth

Electron transfer reaction of metal ions leading to metallization is generally rapid at interfaces in molten salt systems due to high exchange current densities. However, this may not be the case for many refractory metals. For example, the electron transfer reaction in the reduction of Mo(III) to Mo(0) in NaCl–KCl melt is characteristic of a slow charge transfer process as the overall rate is controlled by the rate of the preceding chemical reaction leading to the formation of Mo polynuclear complex ions and not by the electron transfer step (Andrew et al. 2020). Nucleation overpotential is another critical factor controlling the metal deposition process. In many cases, the deposit of metal is in dendritic form and is occluded with solidified melt which is subsequently removed by a leaching step. Powdery deposits can arise due to secondary processes like disproportionation reactions near the surface or the chemical reactions between the deposited alkali metal and metal ions as the alkali metals diffuse away from the electrodes (Inman and Lovering 1983; Redkin et al. 2012; Castrillejo et al. 2002; Gleb Mamanto 1987; Senderoff 1966; Inman and White 1978). Coherent metal deposition from fluoride melts may be partly attributed to the interfacial effects .

(d) **Current Efficiencies and Critical Current Densities**

The productivity of an electrolytic cell is expressed by current efficiency in terms of Faraday's laws of electrolysis. It is defined as the ratio of the number of equivalents of metal produced to the number of moles of electric charge passed into the cell from external power source. Current densities of about 1 A/cm^2 are common in molten salt electrolysis and the cathodic current efficiencies remain in the range 60–80% (Choate and Green 2000). In fused salt electrolysis, the deposition of the metal at the cathode is realized only at current densities above a critical value. When the current density is raised subsequently the rate of the cathode metal production is greater. The current efficiency may be zero at low current densities if the losses are greater or equal to the rate of the formation of electrode products. The maximum limit of the current density is set by heating effects, increase in IR drop, anode effects and metal formation. The factors which lead to lower current efficiencies during electrolysis are;

a. The mechanical losses occurring during the product formation are carried away in the slimes or entrapped in the cell lining.
b. Distillation, volatilisation or sublimation of the products may be controlled by adjusting the temperature of electrolysis.
c. Mixing up of the electrode products, such as the recombination of anodic chlorine and cathodic product may be minimized by employing a suitable diaphragm between anode and cathode.
d. Undesired reactions between the product of electrolysis and the materials of construction, or the reaction with gases at the vicinity of the electrodes.
e. The current efficiency may be reduced due to the formation of lower compounds at the cathode (e.g. $CaCl_2$ instead of Ca) or higher compounds at anodes (e.g. $SnCl_4$ instead of Cl_2).
f. The current efficiency may also be reduced by simultaneous deposition of several ions at the electrode surface, for example in the case of Mg extraction, the lower current efficiency may be due to the presence of moisture in the cell feed which leads to the decomposition of H_2 and O_2.
g. Dissolution of deposited metal in fused salt melts:

   There is a loss in current efficiency by the dissolution of the deposited metal in the fused salt itself. For example, in the case of Na metal extraction using fused sodium hydroxide, the metal may dissolve completely in the molten electrolyte if the temperature exceeds 350 °C. In some cases, the formation of sub-salts may lead to the dissolution of metal in the fused salts. The dissolved metal is totally dispersed in the bath which causes colour change in the melt and is termed as "metal fog". The solubility of metals is more in pure salt melts than in a mixture of salts, hence mixed salts are employed in many of the electrowinning processes.
h. Anode effect:

   The anode effect often occurs during the electrolysis of fused baths and is frequently seen in the production of aluminium by Hall-Heroult process (Haupin and Frank 1981), titanium and tantalum metals. The effect will abruptly increase the cell voltage wherein the cell current reduces to almost zero. Such a situation arises due to the formation of a gas film surrounding the anode surface offering high resistance during electrolysis. The anode effect will disappear when the surface of the anode is cooled or stirred. In addition, the current density can be minimized thereby the gas film collapses and the evolution of the gas reappears.

11.4 Extraction of Alkali and Alkaline Earth Metals Through Molten Salt Electrolysis

In general, the conventional metallurgical processes such as carbothermic reduction of metal salts to their respective metals are considered energy intensive and emit large amounts of greenhouse gases as by-product (Springer and Hasanbeigi 2016; WS 2008). Availability of electricity from renewable energy sources can change this scenario. Molten salt electrolysis is considered in the large scale production of alkali and alkaline earth metals such as lithium, sodium, magnesium and calcium. These metal salts exhibit extremely negative Gibbs free energies of formation, enabling high electrochemical decomposition potential to produce alkali and alkaline earth metals (WS 2008; Sohn and Geskin 1990). In molten salt electrolysis, steel and graphite are used as cathodes and anodes, respectively. The primary cathodic and anodic reactions are the reduction of metal and chloride evolution, respectively (Mishra and Olson 2005; WS 2008; Fray n.d.; Minh 1985). The annual production of lithium is 25,000 tonnes per year, while sodium accounts for 100,000 tonnes per annum. On the other hand, 20,000 tonnes of calcium is produced annually and owning to their structural applications in diverse technologies, magnesium accounts for the highest production of about 766,000 tonnes per year. China is one among the major producers of these reactive metals. This section will summarize the recent advancements in the chemistry and engineering aspects for the production of lithium, sodium, magnesium and calcium through molten salt electrolysis.

11.4.1 Electrowinning of Sodium Metal Through Molten Salt Electrolysis

Sodium (Na) is a highly reactive metal next to lithium and potassium and is well known that sodium metal reacts vigorously with water and tarnishes when exposed to the environment. Na is used as a reducing agent and as a coolant in heat exchangers in nuclear industries (De NoraPlacido et al.; Thayer 2008).

In 1888, Hamilton Castner introduced the production of sodium through Castner's cell which operates at a temperature of 330 °C to produce Na metal from sodium hydroxide. The Castner cell primarily consisted of anode and cathode either made of iron or nickel. Na metal is formed at the cathode and oxygen evolution occurs at the anode. The chemical reaction is given below,

$$4Na^{+} + 4e^{-} \rightleftharpoons 4Na$$

$$4OH^{-} \rightleftharpoons O_2 + 2H_2O + 4e^{-}$$

The only side reaction at the cathode is hydrogen evolution due to the diffusion of water into the electrolyte resulting in the reduction of the cathodic efficiency of the process (De Nora et al.; Hine 1985; Gilbert 1952). In 1922, the Castner cell was replaced by Down's cell for the production of Na metal. Down's cell primarily consisted of four anodes and cathodes each with a steel or iron gauze diaphragm preventing the recombination of the products (Cl_2 gas and the metal). Unlike, Castner's process, the anodic reaction is Cl_2 evolution, since sodium chloride was used as the raw material. Graphite rods were used as anodes and steel plates were used as cathodes. The old-style Down's cell was operated at 50 KA in a molten melt of $CaCl_2$ and NaCl at a temperature of 600 °C. Modified Down's cell employed a molten salt electrolyte consisting of NaCl (28 wt%), $CaCl_2$ (25 wt%) and $BaCl_2$ (47 wt%) at a temperature of 600 °C. The cell was operated with an optimized current density of 1 A/cm^2 and a cell voltage of 7 V. The specific energy consumption was about 10 kWh/Kg of Na produced. The current efficiency of the process was as high as 90% (Ghareh Bagh et al. 2013; Kepler et al. 2001; Stewart and Michael 1961). The electrolyzer comprises of a metal collection chamber to collect the floating metal. The cell consists of electrolyte, electrodes and an absorption system, made of three packed columns containing NaOH solution, to collect the liberated chlorine gas. The main challenge for successful production of Na metal relies on preventing the recombination of Na metal and Cl_2 gas. Calcium chloride was used as the supporting electrolyte due to an increase in the conductivity of the electrolyte (Hine 1985; Thompson et al. 2002). In addition to the above, Ca can reduce the Na from NaCl electrolyte during electrolysis.

11.4.2 Electrowinning of Lithium Metal Through Molten Salt Electrolysis

Being one of the most anodic (or least noble) metals in the electromotive force series, lithium metal is the primary candidate as electrode material for the fabrication of new generation energy applications. Moreover, Li is also utilized in the production of organolithium compounds and also as an alloying addition to Al and Mg. Unlike Mg, Li is not a structural material but is widely employed as a chemical reagent in various chemical reactions. Li metal had many historical applications in the production of lithium greases in World War II and in the fabrication of nuclear weapons. In the modern days, the demand for Li metal is steadily increasing due to its applications in the energy sector. It is expected that Li metal will capture the market share similar to Mg metal widely employed in structural applications. Li is used to reduce the melting temperature of glass and aluminium oxide in the electrolytic production of aluminium. Australia, Chile and China are the major producers and exporters of lithium metal (USGS 2021). Due to its high reactivity, the metal does not exist in elemental form. It is disseminated in hard rocks and as brine in sea water. It is estimated that 230 billion tonnes of lithium content is available in sea water. Lithium aluminium silicate (Spodumene) is an example of a hard rock source of lithium (Dessemond et al. 2019). The silicate based material is transformed to carbonate form by alkali roasting and treating with acid, followed by calcination and carbonation (Margarido et al. 2014). The potassium lithium sulphate salt obtained from the brine of sea water is converted to carbonate after removal of K^+ ions. Finally, the obtained lithium carbonate is converted to chloride salt by treating with chlorinating agents such as HCl or Cl_2 gas (Yan et al. 2012). Similar to Na metal production, Lithium metal is extracted from the purified LiCl using the modified Down's cell. Here, the electrolyte employed comprises of LiCl and KCl eutectic melt with an operating temperature between 400 and 460 °C. Anhydrous LiCl is employed as the cell feed with high density graphite as the anode and mild steel as cathode. Chlorine (Cl_2) gas evolves at the anode and Li metal floats at the top of the melt. Due to low density of Li, the metal is collected in the bell shaped reaction chamber. The current density and energy consumption for ideal production of Li have been optimized to be 2 A/cm^2 and 35 KWh/Kg respectively. In order to reduce the melting point of LiCl (606 °C), KCl is employed as a supporting electrolyte. Li is generally stored in liquid petroleum, due to its recombination with chlorine. The current efficiency is generally between 50 and 60%. The other production processes employed, reduction of lithium oxide and hydroxide using carbon or iron carbides as the reductant has also been reported (Kipouros and Sadoway 1998). Metallothermic reduction of oxides and hydroxides under vacuum using Mg, Al, Ca and Si can be employed to obtain Li metal (Gourishankar and Karell 1999).

11.4.3 Calcium Metal Production Through Molten Salt Electrolysis

Calcium (Ca) metal is the fifth most abundant metal constituting 3.64% of the earth's crust. The most important source of the metal and its compounds are calcium carbonate, limestone, marble, calcite and aragonite. Like Mg, Ca is an effective reducing agent and its widely used for the reduction of other metals like Ti, V, Cr, Zr, Ta, W, Th and U. At elevated temperatures, it reacts with metallic oxides or halides to form its corresponding metal. It also combines with many metals forming a wide range of alloys and intermetallic compounds (Ca–Mg, Ca–Pb, Ca–Cu and Ca–Ni alloys). Calcium and its alloys find wide applications in a variety of processes including ferrous and non-ferrous metallurgy (Heilig 1994; Panigrahi et al. 2013; Ono et al. 2009). Demand for Ca is rising steadily, China is believed to be the world's largest producer and most important exporter. World production of Ca is estimated to be about 2500 tonnes per annum. The metallurgical applications account for about

90% of total Ca consumption. The steel industry uses Ca as a ladle addition to assist desulphurization and deoxidizer to produce killed steels. Total consumption in the steel industry is estimated to be about 650 tonnes per annum. The growing market of Ca metal accounts for 400 tonnes per annum.

Two major methods for the production of calcium metal are,

1. Thermal reduction of lime with Al under vacuum and
2. Electrolysis of fused calcium chloride.

Since nineteenth century, strenuous efforts have been made to produce calcium through molten salt electrolysis. The solubility of calcium in the melt seriously hampered the efficiency of the process. USA and Germany successfully conducted experiments to produce calcium ingots. The lower energy efficiency and impure product hindered the commercialization of the process. Zaikov et al. demonstrated an electrolytic process with the operational parameters to produce calcium with high current efficiency. The same group developed an improved version of the electrolyzer (0.5–3.0 A) to produce high purity calcium (Zaikov et al. 2014a, b).

An improved molten salt electrolytic process for the production of calcium metal was developed by CSIR-CECRI, using calcium chloride as the molten electrolyte. High purity Ca metal was produced in a single stage reactor with new cell design using commercially available raw materials. Anhydrous $CaCl_2$ with the addition of alkaline and alkaline earth salts was used for the efficient production of Ca metal. The electrolyte comprises of 85% $CaCl_2$, 15% NaCl, KCl and CaF_2. Mild steel cylindrical cell with fixed circular anode was used for the process. Cylindrical graphite with suitable arrangements was indented for the electrolysis (Lukasko and Murphy 1990). During electrolysis, calcium metal was deposited at the cathode and it was periodically collected and kept in high density diesel oil. The electrolyte level and the concentration were maintained with regular additions of $CaCl_2$. The metal obtained was purified by vacuum melting techniques depending on its application. An optimized electrolyte process was developed at CSIR-CECRI with different concentrations of electrolyte, current densities and cell voltage. The electrolytic cell was operated at 1000–1500 A capacity producing 10–12 kg/day (Berchmans 2003).

11.4.4 Production of Magnesium Metal Through Molten Salt Electrolysis

Magnesium metal because of its light weight, superior strength and its tendency to form alloys with many metals, finds several applications in defence, aerospace and nuclear sectors. Even though the resources for magnesium metal are widely distributed in nature, the cost of the metal is high due to high energy involved in its extraction. The world's production of magnesium has been growing steadily because of its demand both in domestic and industrial sectors. Over 50% of magnesium is consumed for alloying with aluminium to increase the strength and other properties for structural applications. Die cast magnesium alloys find applications in automotive and computer industries. Global magnesium metal market is expected to grow significantly owing to increasing demand in automotive sector and for the fabrication of Al–Mg alloys. The demand for Mg metal has steadily increased because of its applications including new batteries, light weight metal alloys, and wind turbines, capturing of CO_2, production of iron and steel, Ti and Zr metals. Mg scrap, wastes from metallurgical and agro-chemical industries are the sources of raw materials for the production of Mg metal.

Commercially, electrowinning of Mg metal is carried out by two processes namely, I.G. Farben and Dow's process. The primary difference in the processes lies in the Mg feed. I.G. Farben process employs anhydrous $MgCl_2$, whereas, Dow's process works with hydrated salts of $MgCl_2$ ($MgCl_2{\cdot}1.5–1.7H_2O$). The precursor for Mg is from brine which approximately comprises of 0.13 wt% of Mg. In general, Mg^{2+} ions are precipitated by the addition of calcium oxide (CaO). The precipitation of insoluble $Mg(OH)_2$ is acidified to yield $MgCl_2$. Dow's process employs 25%$MgCl_2$, 15%$CaCl_2$ and 60%NaCl, with the cell operating temperature of 700 °C. In Dow's process, the graphite anode is consumed by the reaction of graphitic carbon with water to form H_2 and CO_2. Electrowinning of Mg through I.G. Farben and Dow's process consumes ~10.5–18 Kwh/Kg of Mg with a purity of 99.9%. Technological improvements has been attempted to reduce the cell voltage, and in the development of alternate inert anodes as the replacement for conventional reactive carbon based anodes. The molten salt electrowinning experiments were performed at temperatures between 680 and 750 °C, using electrolyte melt comprising of $MgCl_2$ (20.5 wt%), NaCl (39 wt%), $CaCl_2$ (39 wt%) and KCl (1.5 wt%). During electrolysis, H_2 gas was purged into the anode compartment. The overall cell reaction of the modified Mg electrowinning cell is given below,

$$2Cl^- + H_2(g) \rightleftharpoons 2HCl + 2e^-$$

$$Mg^{2+} + 2e^- \rightleftharpoons Mg$$

$$MgCl_2 + H_2 \rightleftharpoons Mg + 2HCl(g)$$

Yen et al. developed inert niobium doped TiO_2 electrodes as a replacement for conventional reactive graphite anodes. The fabricated electrodes with 3 mol% of Nb_2O_3 sintered at 1300 °C were found to be more efficient for Mg electrolysis (Yan and Fray 2010). Sharma et al. investigated an electrowinning process, where MgO was dissolved in neodymium chloride to form $NdOCl_3$ and $MgCl_2$. The overall reaction of the process is the formation of metallic Mg and CO_2 (Sharma 1996). Lu et al. reported a process of dissolving MgO in $LaCl_3$–$MgCl_2$ electrolyte using steel and carbon as cathode and anode respectively (Lu et al. 2016). The main products were metallic Mg and CO_2 respectively with current efficiency of >85% and energy consumption of 12 kWh/kg of Mg. Cost of the rare earth salts is the primary and crucial disadvantage of the process. Pal et al. developed a synthetic route for the preparation of Mg and other metals using solid oxide membrane (Pal et al. 2001). In this process, solid-oxygen-ion-conducting stabilized ZrO_2 electrolyte separates the inert cathode and anode. The main products of this process are $H_2O_{(g)}$ and metallic Mg. Krishnan et al. reported the production of Mg using solid oxide membrane, from MgO dissolved in fluoride salts at 1300 °C. The cell was operated at a current density of 1 A/cm^2, with a cell voltage of 4.5 V and power consumption of 9–12 kWh/kg of Mg (Krishnan et al. 2005).

CSIR—Central Electrochemical Research Institute, Karaikudi (CECRI) has carried out significant work for more than two decades on the production of magnesium (Mg) metal using various raw materials like magnesite, sea bittern and anhydrous magnesium chloride.

1. Mg metal production from Low grade Magnesite ore as the primary source.
2. Sea bittern is a waste liquor produced from salt panes.
3. By-product Magnesium chloride ($MgCl_2$) from Titanium (Ti) and Zirconium (Zr) manufacturing plants.
4. Magnesium hydroxide $(MgOH)_2$ from chemical industries.

Pilot plant investigations were undertaken in all aspects of Mg metal production by molten salt electrolytic process.

1. **Production of Mg Metal from Low Grade Magnesite Ore**

CSIR-CECRI has tried for the production of magnesium metal from low grade Magnesite ore, which is extensively available in our country. In this process, finely ground magnesite ore was treated with HCl, resulting in the production of slurry and a clear solution of $MgCl_2$. This filtered solution was spray dried to produce fine powder of anhydrous magnesium chloride. The $MgCl_2$ powder was used as the cell feed for the electrolyzer. The molten electrolyte consisted of $MgCl_2$, NaCl and KCl salts. Electrolysis was performed by using mild steel cathode and graphite anode. Eutectic salt mixture was periodically fed into the electrolyzer and the molten magnesium metal was produced and casted in steel moulds.

2. **Mg Metal Production Using Sea Bittern**

Sea bittern a by-product from salt industries is considered a waste and disposed off into the sea. This solution was used as raw material for the production of Mg metal. Bittern solution consists of a fairly good concentration of magnesium chloride, obtained after the crystallization of common salt. The vast establishment of a number of salt farms around the coastal area is one of the major sources for the production of Mg metal. The solar heat available in plenty is used to concentrate the sea bittern.

CSIR-CECRI undertook the task of beneficiation of sea bittern to make it suitable for the production of magnesium metal. The concentration of the bitterns was identified as 36°Be. CSIR-CECRI has evolved a process for the removal of sulphate and boron which are considered as harmful impurities present in the concentrated brine. The process involved the separation of sulphates as calcium sulphate ($CaSO_4$) by treating with calcium chloride ($CaCl_2$) solution. The filtered solution has a concentration of $MgCl_2$ around 380 g/l, which is sufficient for the production of anhydrous magnesium chloride by spray drying process. Spray dryer was operated to produce anhydrous $MgCl_2$, having 85–90% $MgCl_2$ with 4–5% MgO and the rest being H_2O. The spray dried $MgCl_2$ was used as the raw material for the electrolytic cell. CSIR-CECRI also undertook the task of upgrading the electrolysers and developed the modular cells which brought down the energy consumption up to 16 kWh/kg.

Based on the CSIR-CECRI process, Tamil Nadu Magnesium and Marine Chemicals Ltd (TMML), Govt. of TamilNadu has put up a 600 TPA production plant at Valinokkam. The commercial plant started its operation in 1989. However, TMML could not be in operation after an year due to high humid and high wind velocities at their site. Due to high moisture content in the feed, large quantity of sludge was formed which lead to lower efficiency in metal production. These problems lead to passivation of cathodes which resulted in poor coalescence of the metal. Tiny magnesium metal globules recombined with anodic chlorine gas thereby the current efficiency was lowered. All these factors made the operation of plant difficult and due to these issues, TMML stopped its operations in 1993.

3. **By-Product Magnesium Chloride ($MgCl_2$) from Titanium (Ti) and Zirconium (Zr) Manufacturing Plants**

Magnesium is used as a reducing agent in Kroll's process for the manufacture of zirconium, titanium metals etc. $MgCl_2$ is produced as a by-product after the reduction of $TiCl_4$ and $ZrCl_4$. CSIR-CECRI and Nuclear Fuel complex (NFC), Hyderabad had developed a process for the production of Mg metal by using anhydrous magnesium chloride, which is a byproduct generated from the zirconium plant. Magnesium metal was produced with modular type, bipolar and multi-polar cells with active support from DMRL and NFC. DMRL-Hyderabad had set up a large scale Ti-sponge production unit for which, a captive Mg plant was jointly established by CSIR-CECRI and DMRL consisting of two 30 and 42 KA electrolytic cells at their site. Ti-sponge plant and the Mg plant had been operated for more than six years. The DMRL team had tried with different raw materials for the production of magnesium metal.

4. **Magnesium Hydroxide $(MgOH)_2$ as Raw Material from Chemical Industries**

Magnesium metal production was also tried using Magnesium hydroxide $(MgOH)_2$ as the raw material supplied from Birla Periclase. Concentrated $MgCl_2$ solution was prepared by dissolving magnesium hydroxide $(MgOH)_2$ in concentrated HCl solution. From this solution, anhydrous $MgCl_2$ was prepared by spray drying process. Then, it was used as the raw material for producing magnesium metal through molten salt electrolytic process.

Bipolar Technology Developed at CSIR-CECRI Karaikudi

There has been a tremendous interest in bipolar electrode configuration for the production of strategic metals like Aluminium (Al) and Magnesium (Mg). The advantages of the cell are compactness, high energy efficiency and flexibility in cell voltage and current. This bipolar cell allows with smaller bus bar without sacrificing the voltage. The production of Mg metal comprises of facing anode and cathode surfaces defining an inter-electrode space, through which the electrolyte is caused to flow. Chlorine is evolved at the anode surface molten metal is generated at the cathode surface and it is collected in a metal collection zone. If the metal and chlorine come into contact, they recombine and reduce the current efficiency. Recombination reaction can be reduced or prevented by separating the facing anode and cathode surfaces by introducing diaphragms. But the introduction of diaphragms increases the internal resistance of the cell.

A proto-type ALCAN tapered anode cell of the capacity of 80 KA (ALCAN Company) with small anode cathode distance could be used to advantage for the reduction of specific energy consumption. But the basic problem was the excessive heat generated at the electrodes. The most direct way to obviate the heat generated when operating at low specific energy consumption is by adopting a multi-polar cell design. Bipolar electrodes are valuable to increase the effective cathode area, where the metal formation can take place without either increasing the size of the cell or increasing the heat and power loss. Because of these bipolar electrodes, the productivity of the cell is nearly double. But in practise, a portion of the current bypasses the bipolar electrode and flows around it. The loss becomes one of the inherent factors, and can be minimized by inserting additional bipolar electrodes between anode and cathode to increase the cell productivity.

Dow had explored cell designs of this type in 1940's after that, the first ALCAN multi-polar cell of capacity 40 KA was installed in 1980 at Ocska Titanium company. Several configurations were tested before operating the cell at 80–140 KA capacity by the same company in 1982. The production of Magnesium 1000 MT/cell/year with a specific energy consumption of 9.5–10 KWh/Kg was tried. Comparative assessment of the tapered anode and multi-polar cell revealed that the later showed both capital and operating costs lowered by about 1/3.

The Ishizuka Research Institute Japan has developed a bipolar cell in 1983 and operated by Showa Titanium, Toyama. Five bipolar electrodes (Steel-Graphite) were used in the cell. The cell was operated at 670 °C and carried out 50 KA corresponding to 300 KA in a mono-polar cell. Current density was 5600 A/m^2 and the inter-polar gap of 4 cm. The power consumption was 11 KWh/Kg of Mg using molten $MgCl_2$ from Kroll's Titanium production. Ishizuka preferred electrolyte composition of 50% NaCl, 30% KCl and 20% $MgCl_2$. A European patent filed by Ishizuka Hiroshi described a method for electrolytically obtaining Mg metal. The electrolytic bath is composed of $MgCl_2$ and additional ingredients, with a bath density of 0.02–0.10 g/cm^3. The electrical conductivity of the bath was 2.4 Ω^{-1} cm^{-1}, holding magnesium metal at the surface of the electrolyte. Six intermediate bipolar electrodes were used. Each electrode consisted of an iron plate joined to a graphite slab with several bolts of iron implanted at one end in the graphite and welded to the iron at the other end.

In 1989, Hiroshi Matsunami et al. of Toho Titanium company filed a patent where, the electrolytic cell comprised of an anode, cathode and plurality of bipolar electrodes. A first partition wall is provided with partition openings situated beneath the level of the electrolyte. A second partition wall is comprised of an intermediate chamber between the first and the second partition walls for the removal of chlorine gas. A metal collection chamber was formed between the second partition wall and the front wall of the electrolytic cell and a control plate for preventing the deposition of sludge.

CSIR Central Electrochemical Research Institute (CECRI) Karaikudi, has undertaken the task of upgradation of the production of Mg metal by utilizing $MgCl_2$ (obtained from Zirconium and Titanium production) through adopting bipolar and multi-polar cell technology. Research activities had been focused on the development of stable heterogeneous bipolar electrodes, design and construction of different configurations of cells and the role of current density and energy efficiencies, etc. In CSIR-CECRI, bipolar cells of different capacities ranging from 0.5 to 6 KA were operated. These included graphite plates coated with electrodeposited or plasma sprayed iron on one side and snugfit type graphite-mild steel plates on the other side. Graphite-mild steel plates were connected with mild steel spacers in between the electrode configuration. Inter-polar gaps ranging from 1.5 to 3.5 cm and current densities ranging from 0.8 to 1.25 Acm^{-2} were tried. Electrolytes containing ternary $NaCl–KCl–MgCl_2$ and quaternary $NaCl–KCl–BaCl_2–MgCl_2$ systems were studied in bipolar electrolytic cells. Multi-polar cells with two and four modules were also designed and operated with varying capacities.

11.5 Molten Salt Electrowinning of Refractory Metals

The nine transition elements of Groups IVA, VA and VIA of the periodic table, comprising of Ti, Zr, Hf, V, Nb, Ta, Cr, Mo and W, are known as the refractory metals because of their high melting points and the refractory nature of their compounds (Inman and White 1978; Konings et al. 2014). Refractory metals exhibit attractive mechanical, physical, corrosion resistance and nuclear properties. There has been a renewed interest in these metals in recent years (Tyrer and Gibbon 2012). Specifically, emerging applications in nuclear and aerospace industries under extreme conditions have increased the demand for titanium, niobium and zirconium. Many of these applications demand metal of very high quality whereas the extraction of refractory metals remains difficult due to their complex chemistries and ability to exist in multiple oxidation states (Industrial and Processes 2021; Sehra and Suri 1993).

Refractory compounds exhibit a fascinating array of properties (Liddell and Sadoway 1990). For example, the borides and nitrides are excellent electronic conductors; the electrical conductivities of TiB_2 and TiN exceed that of titanium metal itself (Liddell and Sadoway 1990). Due to their heat and erosion resistance, the carbides find usage in cutting tools. The silicides are used as heating elements in electrical resistance furnaces for service at elevated temperatures under oxidizing conditions. The compounds comprising the borides, carbides, nitrides and silicides of the refractory metals are also termed as ‘refractory hard metals’. Due to their unique properties suitable for specific critical applications, many of the refractory metals are considered strategically important.

The primary extraction of refractory metals can be accomplished in two ways; thermo-chemical and electrochemical methods (Inman and White 1978; Al-Jothery et al. 2020). The thermo-chemical extraction, practised especially with metals of Groups IV and V, involves metallothermic reduction of a refractory metal compound. The refractory compounds used are oxides, fluorides or chlorides; the reductant is primarily chosen from among aluminium, magnesium, sodium and calcium, although other reactive metals have also been used. For example, titanium metal is produced by the Kroll process which is based upon the reaction of titanium tetrachloride with magnesium.

Electrochemical extraction involves the electrolytic reduction of a refractory metal compound dissolved in an ionic medium. Aqueous solutions are unsuitable as electrolytes for the electro-reduction of the refractory metals, with the exception of chromium, because of the lower stability of water. Therefore, there is a need to use non-aqueous electrolytes, viz., organic solvents and molten salts. Due to their poor ionicity, low electrical conductivity, low flash points and violent reactivity with water, organic electrolytes are unsuitable for electrowinning of refractory metals (Huang et al. 2016). Therefore, molten salt eutectics are the only class of electrolytes suitable for the electrochemical extraction of refractory metals.

Because of high melting points and high densities of some of the refractory metals, preparing them in the liquid state involves high temperature operation above 1923 K and tapping at the bottom of the electrolytic cell, unlike light metals which are generally tapped at the top of the cell (Chenyao et al. n.d.). Therefore, the production of these metals in liquid state is not advisable; instead, the metal deposition is performed in solid state. Oxides, halides and double fluorides have been suggested as functional electrolytes for the electrodeposition of refractory metals (Chamelot et al. 2000). Oxides, except for MoO_3, have very high melting temperatures and cannot be electrolyzed directly. Suitable solvents, called carrier electrolytes, are required to dissolve them. The selection of a carrier electrolyte depends on the choice of the functional electrolyte. For example, if the oxides are functional electrolytes, the carrier electrolytes must possess high solubilities (Jeong et al. 2008).

The fused salt electrolytic process is a promising route for the preparation of refractory metals and offers the possibility of metals being produced continuously at a much lower cost than the metallothermic processes. During the electrowinning of the refractory metals, two kinds of reactions occur in the melts; homogeneous reactions of the type: $M^{(n+z)+} + \text{ne}^- \rightarrow M^{z+}$ where $M^{(n+z)+}$ and M^{z+} are two oxidation states of the same metal ion; heterogeneous reactions leading to the metal deposition: $M^{n+} + \text{ne}^- \rightarrow M^0$ (Dubrovskiy et al. 2018). The potentials corresponding to each of these reactions are linked to the Gibbs free energy of formation of salt solutions (Lantelme and Groult 2010). Table 11.1 compiles representative literature on molten salt eutectics and electrowinning parameters for obtaining high purity refractory metals (Sehra and Vijay 1998). In the electrolytic production of molybdenum (Mo) and tungsten (W), oxyionic baths are preferred than the chloride and fluoride based baths due to the volatile nature and they do not have the tendency to form stable complexes with alkali or alkaline earth metal chlorides (Lantelme and Groult 2010).

Table 11.1 Molten salt electrowinning of refractory metals: representative bath compositions and electrowinning parameters

Metal	Type of electrolyte	Operating temperature (K)	Molten eutectic system	Operating voltage (V)	Cathode current density (A/ cm^2)
Ti	Chloride	1073	$NaCl–KCl–TiCl_4$	9.5–9.8	3
	Fluoride	1123–1223	$NaCl–K_2TiF_6$	5.7	0.95
Zr	Chloride	1073	$NaCl–NaF–ZrCl_4$	5.3	0.323
	Fluoride	1073–1123	$NaCl–KCl–K_2ZrF_6$	2.0–2.5	5
	Oxide	1273	$NaF–ZrF_4–ZrO_2$	2–3.8	2
Hf	Chloride	838	$NaCl–KCl–HfCl_4$	2.4–7	0.19–1.15
	Fluoride	1123	$NaCl–K_2HfF_6$		
Nb	Fluoride	1173	$KF–K_2NbF_7$		
	Oxide	1025	$KCl–KF–K_2NbF_7–Nb_2O_5$	2.5	
Ta	Fluoride	1023	$NaCl–KCl–K_2TaF_7$	2.7–3.4	0.75
	Oxide		$K_2TaF_7–Ta_2O_5$		0.17
Mo	Chloride	1323	$CaCl_2–CaMoO_4$		0.5–5
	Oxide	1273	$Na_2B_4O_7–NaCl–NaF–MoO_3$	8–12	0.38
W	Fluoride	1073	$NaCl–NaF–KAlF_4–WO_3$	5.0–7.5	3.5
Cr	Oxide	1173–1193	$NaCl–Cr_2O_3–C$	3.5–4.5	1.6–5.5

Recently, one of the authors (M.J.) reported electrodeposition of tantalum based superalloys using deep eutectic solvents (Andrew et al. 2020).

11.6 Rare Earth Metals

Rare Earth Elements (REEs) or Rare Earth Metals are the collection of 17 elements comprising of lanthanides (atomic numbers 57–71) along with scandium and yttrium (Lucas et al. 2015; Zhu 2014; Gupta and Krishnamurthy 1992; Jha et al. 2016; Haque et al. 2014). REEs are silver, silvery-white, or grey metals, characterized by high density, high melting point, high electrical and thermal conductivities, and readily oxidize in air. Due to their similar chemical properties, they are found together in minerals and it is difficult to separate or even distinguish one from another. REEs are classified into lighter REEs from atomic numbers 57 through 63 (La, Ce, Pr, Nd, Pm, Sm and Eu) and heavier REEs with atomic numbers from 64 through 71 (Gd, Tb, Dy, Ho, Er, Tm, Yb and Lu) (Ganguli and Cook 2018). Generally, lighter REEs are more common and more easily extracted than heavier REEs (Coey 2020).

The lanthanide ions are the first row of the f-block in the periodic table, often called the inner transition elements, due to their placement between the s- and the d-block. Owing to an increase in relativistic and non-relativistic nuclear attraction, atoms with heavy nuclei, starting with the third row of the periodic table, experience contraction of the s and p orbitals, while the d and f orbitals, already diffuse and less penetrating, further expand owing to increased shielding from the nucleus. This leads to an electronic configuration in which all 15 lanthanide elements share the xenon core, [Xe], corresponding to $1s^2 2s^2 2p^6 3s^2 3p^6 3d^{10} 4s^2 4p^6 4d^{10} 5s^2 5p^6$. The 4f orbitals, while part of the valence configuration, are shielded from the coordination environment by the filled 5s and 5p orbitals, and that is the case for lanthanides in their atomic state as well as higher oxidation states and this phenomenon is called as 'lanthanide contraction' (Lucas et al. 2015). The remaining electrons, the valence electrons, are then placed in the 6s orbital and, with the exception of La, Ce, Gd, and Lu, in the 4f orbital, since the energy difference between the 4f and 5d orbitals is large. Rare earths are generally found with non-metals, usually in the 3+ oxidation state and there is little tendency to vary the valence, such as Eu^{2+} and Ce^{4+}.

REEs are considered key drivers of today's technologies. Their applications are spread across several areas such as catalysis in petroleum refining, automobile catalytic convertors (La, Ce), magnets (NdFeB, SmCo), polishing (CeO_2), batteries—NiMH (La), metallurgical additives and alloy preparation (Ce, La, Nd), lamp phosphors (Eu, T, Y), capacitors and ceramics (Bharadwaj et al. 2017). Emerging technologies include anodes of solid state fuel cells (Sc, Y), high temperature superconductors and lasers (Y) among others (Behrsing et al. 2014).

In India, monazite is the principal source of rare earths and is a prescribed substance as per the notification under the Atomic Energy Act, 1962. It occurs in association with other heavy minerals, such as ilmenite, rutile, zircon, etc. in concentrations ranging from 0.4 to 4.3% in beach sand and inland placer deposits of the country (Bharadwaj et al. 2017). It is estimated that 10.21 million tonnes of monazite are available in the beach and inland placer deposits (Vogel 2015) and India has almost 35% of the world's total beach sand mineral deposits rich in LREE (Krishnamoorthy and Guptha 2008). Generally, the processing of REE ores leads to the formation of pure REE or mixture of REEs, primarily in their oxide forms.

11.6.1 Electrochemistry of Rare Earths

Rare Earths are generally characterized by their high negative reduction potentials ($E^0(RE^{3+}/RE^0) = -1.99$ to -2.60 V vs. SHE) (Lide 2006) and are among the metals considered to be most difficult to produce in pure form. Therefore, electrowinning through molten salt electrolysis is the only viable route to obtain rare earths in metallic form. Electrochemistry and electrodeposition of REEs in chloride and fluoride molten salt eutectics using inert Ta, Mo and W electrodes have been studied by several research groups (Table 11.2, 82–93). Majority of the studies reported detailed reaction mechanism, nucleation phenomena, thermodynamic stability and transport properties. Most of the lanthanides undergo reduction reactions through a single-step, but sluggish three-electron charge-transfer, except, Nd(III), Sm(III),

Table 11.2 Electro-reduction of REEs in molten salt electrolytes

Element	Electrode	Reduction steps	Reference
La, Ce, Er	Mo and W	$RE^{3+} + 3e^{-} \rightarrow RE^{0}$	Vandarkuzhali et al. 2012; Kondo et al. 2012)
Pr, Gd, Tb, Dy, Ho, Sc	W		Castrillejo et al. 2005; Caravaca et al. 2007; Rayaprolu and Chidambaram 2014; Saïla et al. 2010; Han et al. 2017; Castrillejo et al. 2012)
Lu, Y	Mo		Bermejo et al. 2008; Li et al. 2011)
Nd, Eu	Mo and W	$RE^{3+} + e^{-} \rightarrow RE^{2+}$	Abbasalizadeh et al. 2015; Kuznetsov and Gaune-escard 2011)
Sm, Tm, Yb	W	$RE^{2+} + 2e^{-} \rightarrow RE^{0}$	Gao et al. 2016; Castrillejo et al. 2009; Castrillejo et al. 2011)

Eu(III), Tm(III), Yb(III) which exhibit a two-step reduction (Table 11.2). It was widely reported that the reduction and reoxidation of RE^{3+}/RE^{2+} redox couple is a reversible diffusion-controlled reaction.

11.6.2 Electrowinning of Rare Earths

Several processes have been developed for the production of rare earth metals of different purities. These include: (i) reduction of anhydrous chlorides or fluorides, (ii) reduction of rare earth oxides, and (iii) fused salt electrolysis of rare earth chlorides or oxide–fluoride mixtures. RE metals are highly reactive, particularly at the high temperatures required for their preparation.

The advantage in employing fused salt electrolysis is the avoidance of a number of preliminary steps normally involved in aqueous chloride electrolysis and in having very compact electrolytic cells as compared to aqueous electrolysis. REE extraction through molten salt electrolysis is strongly governed by key electrowinning parameters such as electrolysis temperature, concentration of the species, decomposition potential, current density and period of deposition (Naboychenko et al. n.d.). Fused-salt electrowinning of rare earths can be operated continuously, except for "tapping". The first recorded attempt at the preparation of a RE metal is credited to C. G. Mosander (1827).

Lanthanum, Cerium, Praseodymium, and Neodymium have melting points that permit their recovery in the liquid state by electrolysis of relatively inexpensive chlorides at temperatures less than 1100 °C (Mishra and Olson 2005). Winning of rare earth elements in the liquid state facilitates slag (electrolyte) metal separation, minimizes contamination of the reduced metal, and enables continuous operation and high volume levels of production. The technical feasibility of preparing limited quantities of high melting rare earth metals such as Gadolinium, Dysprosium, and Yttrium in consolidated form has also been demonstrated by using fluoride electrolytes in place of chlorides (Han et al. 2016). The simplest type of rare earth metal electrowinning is through chloride electrowinning. The metal is denser than the electrolyte. It sinks to the bottom of the furnace where it is periodically syphoned out of the furnace via a vacuum pipe or drained out of the furnace through a submerged closable tap hole (Zhu 2014).

$$RE^{3+} + 3e^{-}\underrightarrow{\Delta(1050^{\circ C})}RE^{\circ}(\text{liquid})$$

Industrial molten salt electrowinning is always designed to produce molten metal. A molten product is easily tapped, poured, or syphoned from the electrowinning furnace. It allows almost continuous electrowinning. Electrowon RE product is as pure as its rare earth chloride feed. Industrial chloride electrowinning started in the beginning of the twentieth century, mostly in China and Japan. Large (up to 50,000 amperes) electrolytic cells were used, producing several tonnes of metal or alloy per day (Gupta and Krishnamurthy 1992).

Molybdenum was recognized as a suitable cathode material after Guertler (1923) reported that molybdenum is insoluble in the REs at high temperatures (Inman and White 1978; Cao et al. 2011). Molybdenum was used as the cathode in the experiments for the preparation of high purity REs. A carbon crucible was used as the anode, and the cathode in the form of a spinning rod was employed for the electrolysis. However, the chloride process was subsequently replaced with oxide-fluoride electrolyte by many industries because of ease and safe handling. Rare earth metal preparation by electrolysis of a fluoride–oxide bath was first reported in 1907 (Gupta and Krishnamurthy 1992; Wang et al. 2013; Lee et al. 2015).

Electrolytic processes are inherently simpler and cheaper compared to complex metallothermic reduction procedures. The technical feasibility of preparing limited quantities of high melting rare earth metals such as gadolinium, dysprosium, and yttrium in addition to neodymium, praseodymium, and didymium (Nd + Pr) in a consolidated form has also been demonstrated by using the fluorides as electrolytes in place of chlorides (Yasuda et al. 2014). For cerium and lanthanum, cells have been designed and operated at U.S. Bureau of Mines for continuous production of these metals by oxide–fluoride electrolysis (Wang et al. 2013; Dysinger and Murphy 1994). Instead of pure metals, if the REs are recovered as binary alloys having melting points in the range of 600–1000 °C, the electrolytic process has been found applicable to all the rare earth metals with the added advantages of liquid metal recovery and low temperature operation (Takeda et al. 2014). By this process, all the REs, including those with stable divalency, could be recovered as alloys. If a relatively more volatile metal such as cadmium, zinc, or magnesium is used as the alloying component with a relatively less volatile high melting rare earth, e.g. neodymium, gadolinium, and yttrium, the alloy can be vacuum distilled to yield pure RE metal.

11.6.3 Electrowinning of Rare Earths at CSIR-CECRI

Molten Salt Metallurgy is a niche area of CSIR-CECRI not only among national laboratories, but among international research community as well, since its inception in 1953. Particularly, CSIR-CECRI has a seamless legacy of electrowinning of rare earths in metallic and alloy forms, and as compounds (Fig. 11.1). Molten salt electrowinning of lanthanum, cerium and neodymium has been successfully carried out. CECRI carried out a collaborative project on the electrowinning of cerium metal using ceria as raw material. Using chloride and fluoride baths, attempts were made to understand the electrochemical behavior of Ce(IV) using cyclic voltammetry. The team also developed various baths and optimized electrowinning parameters to obtain metallic cerium. Electrowinning of neodymium metal is currently being carried out under DAE-IREL sponsored project at CECRI.

Molten salt electrowinning of rare earth alloys such as $NdFe_x$ and $LaNi_y$ has been demonstrated at a scale of 100 g/batch. Continuous electrowinning of lanthanum-nickel alloy was carried out at 75 A capacity using chloride melt and soluble nickel cathode under CSIR sponsored project. Neodymium iron alloy was prepared from

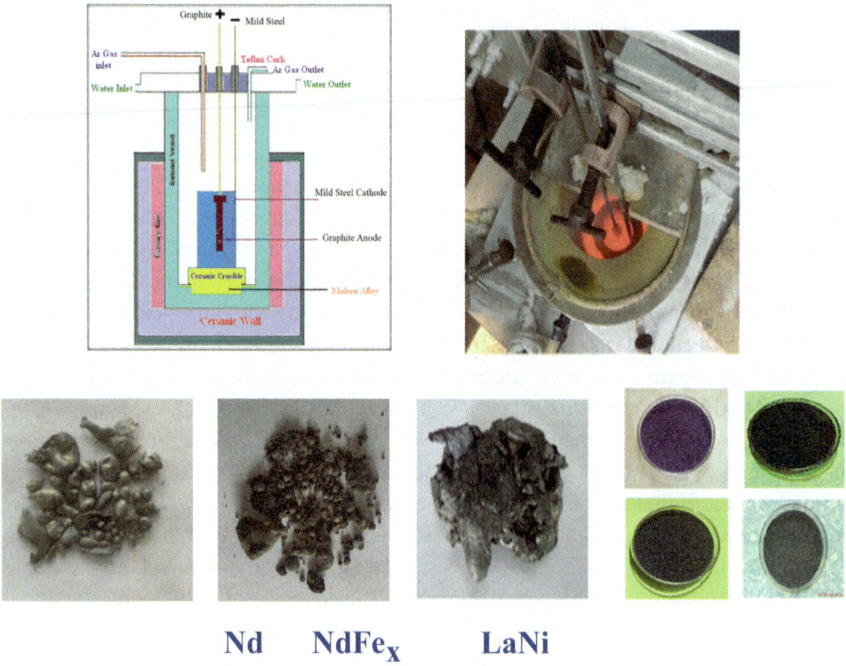

Fig. 11.1 **a** Schematic diagram of molten salt setup for electrowinning of rare earths **b** electrowinning cell with electrode assembly under operation (top open only for photograph) **c** Samples of neodymium, neodymium-iron, lanthanum-nickel and rare earth borides electrowon through molten salt electrolysis at CSIR-CECRI

chloride melts at various current capacities, 50–100 A. Among compounds, rare earth borides of the type, REB_6 (La, Ce, Nd, Sm, Eu) have been prepared at a scale of 250 g/batch. Further, CECRI has proven capability and strong interaction with strategic agencies such as DAE, BARC, IREL and IGCAR in producing strategic metals, particularly, rare earth metals, alloys and compounds. Recently, novel electrolyte systems based on Ionic Liquids and Deep Eutectic Solvents are also being developed at CECRI for studying the electrochemistry and electrodeposition of rare earths at near ambient temperatures (Andrew et al. 2021).

11.7 Challenges in Molten Salt Electrolysis

In addition to their merits, molten salt electrolytes have certain challenges as well. First, the melts have to be maintained in the liquid form at high temperatures throughout and the energy required is an added burden. The materials of construction should withstand the high temperature and have negligible or nil corrosion at the operational conditions. Fire, explosion, pyrophobicity, radioactivity (for nuclear materials) and toxicity are potential hazards when working with molten salts (Janz 1988). As most of the alkali and alkaline earth salts are hygroscopic, they are difficult to handle during the operations. Some of the salts are volatile at high temperatures. Prior to feeding into the electrolyzer, the salts must be purified for efficient melting and to form a uniform eutectic melt. Whenever there exists chance of the recombination of the cathode product with the anode gas, it is necessary to separate the anode and cathode by using high temperature ceramic diaphragms. The deposit at the cathode in the dendritic form is generally governed by mass controlled electrochemical process and such powdery deposits are difficult to separate from the frozen melt post-electrolysis (Fray 1988).

11.8 Scope for Molten Salt Metallurgy

Molten salt technology is a unique and potential area for the production of strategically important light metals, refractory and rare earth metals, and some of the transition metals like Al from their mineral and secondary resources. Even though significant amount of work has been carried out worldwide, many processes have not been commercialized to produce tonnage of metals to meet the requirements for various applications due to certain unresolved issues which mandate scalable and commercially viable solutions. In our country, only a few research groups are working in this important area of metallurgy. To meet the growing demand for speciality metals and alloys, indigenous technological approaches should be focused particularly in the production of special alloys in both ferrous and non-ferrous materials. While pilot scale production of metals such as lithium, magnesium and calcium metals have been established by CSIR Institutes like CSIR-CECRI, CSIR-NML and Defence

Laboratories like DMRL and DAE-BARC, due to various reasons their operations are not continuous to meet our requirements. Impetus research scope is there and it should be focused to evolve innovative technologies with automation and continuous production of these metals, alloys and materials. While India is blessed with abundant resources of rare earth minerals, bulk production of rare earth metals like La, Ce, Nd, Gd, etc. and their alloys using indigenous resources using molten salt technologies have not been taken up on a bigger scale.

There are several key areas of research that Indian research community can focus on in the field of molten salt technology. First, environment friendly technologies which can minimize the emission of toxic gases like CO_2 and NO_X from the metal production industries remain unsolved. Second, the development of conducting inert anodes to replace conventional carbon based materials, particularly, for the bulk molten salt production of aluminium metal is another key area. Some of the complex conducting oxides are to be developed and retrofit as inert anodes in the metal production technologies. Third is the development of bulk production technologies for refractory metals like Ti, Ta, Nb, Zr and V from their oxide raw materials by direct electrolysis. Fourth, molten salt electrodeposition of platinum group metals like Ir, Re, Rh and their alloys for the fabrication of complicated engineering devices like thrust chambers and exhaust lines for rocket nozzles is needed for strategic sectors. Considering the great demand for renewable energy sources, production of solar grade materials such as 6N Si (99.9999%), InAs and GaAs is another key area. Another important area of research is the development of strategies to recover critical metals such as lithium, cobalt, neodymium and samarium from end-of-life products such as permanent magnets, lithium ion batteries through molten salt processes. While this area has gained widespread interest from government and MSMEs in the country, the last-leg of the processing, i.e. getting the metallic products with high purity remains a coveted domain controlled by major foreign players. To thrust research in molten salt technologies, young researchers must be encouraged to work in this niche area along with domain experts and Centres of Excellence focused on R&D in molten salt technology need to be established under the sponsorship of government agencies and private partners involved in strategic metals production. To conclude, indigenous, commercially viable and sustainable processes using molten salt technologies are key cornerstones in India's march towards a responsible global power.

11.9 Author Contributions

M.J. and L.J.B. conceived the chapter and designed the plan. L.J.B. and M.J. contributed to Sect. 1 and 7. L.J.B. contributed to Sect. 2 (with R.S.P.) and Sect. 3. C.N.K., J.E. & N.M. contributed to Sect. 4. M.J. and R.S.P. contributed to Sect. 5 and Sect. 6 (with C.A.). M.J. and L.J.B. reviewed and supervised the overall manuscript.

Acknowledgements The financial support by IRELTDC via GAP 40/19 (M.J., R.S.P.) SERB-ECR via Project No. ECR/2018/002075 (M.J.). CSIR-TWAS Postgraduate Fellowship via Award No. 22/FF/CSIR-TWAS/2017 (C.A.) is gratefully acknowledged.

References

Abbasalizadeh, A., Teng, L., Sridhar, S., Seetharaman, S.: Trans. Inst. Miner. Metall. Sect. C Miner. Process. Extr. Metall. **124**, 191–198 (2015)

Al-Jothery, H.K.M., Albarody, T.M.B., Yusoff, P.S.M., Abdullah, M.A., Hussein, A.R.: IOP Conf. Ser. Mater. Sci. Eng. **863** (2020)

Andrew, C., Jayakumar, M., Chithambara Thanu, V.R.: Surf. Interfaces **19**, 100539 (2020)

Andrew, C., Dhivya, M., Jayakumar, M.: J. Electroanal. Chem. 115398 (2021)

Behrsing, T., Deacon, G.B., Junk, P.C.: The Chemistry of Rare Earth Metals, Compounds, and Corrosion Inhibitors. Woodhead Publishing Limited (2014)

Berchmans, L.J.: Improved Process with New Cell Design for the Production of Calcium Metal, 225/NF/2003 (2003)

Bermejo, M.R., Barrado, E., Martínez, A.M., Castrillejo, Y.: J. Electroanal. Chem. **617**, 85–100 (2008)

Bharadwaj, M.D., Kumar, P.: Natl. Conf. "Critical Strateg. Mater. Adv. Technol. (CSMAT-2017)" (2017)

Cao, P., Zhang, M., Han, W., Yan, Y., Wei, S., Zheng, T.: J. Rare Earths **29**, 763–767 (2011)

Caravaca, C., de Córdoba, G., Tomás, M.J., Rosado, M.: J. Nucl. Mater. **360**, 25–31 (2007)

Castrillejo, Y., Bermejo, M.R., Pardo, R., Martínez, A.M.: J. Electroanal. Chem. **522**, 124–140 (2002)

Castrillejo, Y., Fernández, P., Bermejo, M.R., Barrado, E., Martínez, A.M.: Electrochim. Acta **54**, 6212–6222 (2009)

Castrillejo, Y., Fernández, P., Medina, J., Vega, M., Barrado, E.: Electroanalysis **23**, 222–236 (2011)

Castrillejo, Y., Hernández, P., Rodriguez, J.A., Vega, M., Barrado, E.: Electrochim. Acta **71**, 166–172 (2012)

Castrillejo, Y., Bermejo, M.R., Díaz Arocas, P., Martínez, A.M., Barrado, E.: J. Electroanal. Chem. **575**, 61–74 (2005)

Chamelot, P., Taxil, P., Oquab, D., Serp, J., Lafage, B.: J. Electrochem. Soc. **147**, 4131–4137 (2000)

Chenyao, S., Che, Y., Li, J., Shu, Y., He, J., Song, J.: Integr. Med. Res. **9**, 9341–9347 (n.d.)

Choate, W.T., Green, J.A.S.: US, Department Energy Effic. Renew. Energy 12–24 (2000)

Coey, J.M.D.: Engineering **6**(2), 119–131 (2020)

De Nora, V., Spaziante, P.M., Nidola, A.: Molten salt electrolysis. US4187155A

Dessemond, C., Lajoie-Leroux, F., Soucy, G., Laroche, N., Magnan, J.F.: Minerals **9** (2019)

Dubrovskiy, A., Makarova, O., Kuznetsov, S.: Coatings **8**(12), 442 (2018)

Dysinger, D.K., Murphy, J.E.: 1–8 (1994)

Fray, D.: J. Miner. Metals Mater. **40**, 420 (1988)

D.J. Fray, October 60 (n.d.)

Gale, R.J., White, S.H., Lovering, D.G. (eds.): Molten Salt Technique. Plenum Press, New York (1983)

Ganguli, R., Cook, D.: Rare earths: a review of the landscape. MRS Energy Sustain. **5**, E9 (2018)

Gao, Y., Shi, Y., Liu, X., Huang, C., Li, B.: Electrochim. Acta **190**, 208–214 (2016)

Ghareh Bagh, F.S., Mjalli, F.S., Hashim, M.A., Hadj-Kali, M.K.O., Alnashef, I.M.: Ind. Eng. Chem. Res. **52**, 11488–11493 (2013)

Gilbert, H.N.: J. Electrochem. Soc. **99**, 305C (1952)

Gleb Mamanto, R.M.: Molten Salt Chemistry, An Introduction and Selected Applications. O. Reidal Publishing Company (1987)

Gourishankar, K.V., Karell, E.: Report in TMS Annual meeting, San Diego, California (1999)
Guertler, W.: Molybdenum as a constituent in alloys, Z. Metallkd. *15*, 151–154 (1923)
Gupta, G.K., Krishnamurthy, N.: Extr. Metall. Rare Earths (1992)
Han, W., Li, M., Zhang, M.L., De Yan, Y.: Rare Met. **35**, 811–825 (2016)
Han, W., Li, Z., Li, M., Hu, X., Yang, X., Zhang, M., Sun, Y.: J. Electrochem. Soc. **164**, D934–D943 (2017)
Haque, N., Hughes, A., Lim, S., Vernon, C.: Resources **3**, 614–635 (2014)
Haupin, W.E., Frank, W.B.: Compr. Treatise Electrochem. 301–325 (1981)
Heilig, M.L.: Personal communication 131–134 (1994)
Hine, F.: Fused salt electrolysis and electrothermics. In: Electrode Processes and Electrochemical Engineering. Plenum Press, New York (1985)
Huang, Z., Liu, J., Deng, X., Zhang, H., Lu, L., Hou, Z., Zhang, S.: Int. J. Refract. Metals Hard Mater. **54**, 315–321 (2016)
Petrovic, S.: Industrial electrochemical processes. In: Electrochemistry Crash Course for Engineers. Springer, Cham (2021)
Inman, D., Lovering, D.G.: Compr. Treatise Electrochem. 593–640 (1983)
Inman, D., White, S.H.: J. Appl. Electrochem. **8**, 375–390 (1978)
Janz, G.J.: Thermodynamic and Transport Properties for Molten Salts Correlation Equations for Critically Evaluated Density. American Chemical Society and the American Institute of Physics for the National Bureau of Standards, New York (1988)
Jeong, S.M., Yoo, H.Y., Hur, J., Seo, C.: J. Alloys Compd. **452**, 27–31 (2008)
Jha, M.K., Kumari, A., Panda, R., Rajesh Kumar, J., Yoo, K., Lee, J.Y.: Hydrometallurgy **165**, 2–26 (2016)
Kannan, G.N., Dandapani, K.S., Srinivasan, K.S., Srinivasan, L.K., Subramanian, P., Augustin, C.O., Selvin Devasakayam, T., Rajagopalan, N.: Report in Proc. Electrometallurgy Confs. India (1986)
Keppler, S., Messing, T., Proulx, K., Jain, D.: Molten salt electrolysis of alkali metals. US20010045365A1 (2001)
Kipouros, G.J., Sadoway, D.R.: Jom **50**, 24–33 (1998)
Kondo, M., Tanaka, T., Muroga, T., Tsujimura, H., Ito, Y.: Plasma Fusion Res. **7**, 7–9 (2012)
Konings, R.J.M., Beneš, O., Kovács, A., Manara, D., Sedmidubskỳ, D., Gorokhov, L., Iorish, V.S., Yungman, V., Shenyavskaya, E., Osina, E.: J. Phys. Chem. Ref. Data **43** (2014)
Krishnamoorthy, N., Guptha, C.K.: Miner. Process. Extr. Metall. Rev. **22**, 477–507 (2008)
Krishnan, A., Lu, X.G., Pal, U.B.: Metall. Mater. Trans. B Process Metall. Mater. Process. Sci. **36**, 463–473 (2005)
Kuznetsov, S.A., Gaune-escard, M.: J. Nucl. Mater. **414**, 126–131 (2011)
Lantelme, F., Groult, H.: Trans. Inst. Miner. Metall. Sect. C Miner. Process. Extr. Metall. **119**, 82–87 (2010)
Lee, G.G., Jo, S.K., Lee, C.K., Ryu, H.Y., Lee, J.H.: TMS Annu. Meet., pp. 249–252. 15–19-March (2015)
Li, Y., Wang, F., Zhang, M., Han, W., Tian, Y.: J. Rare Earths **29**, 378–382 (2011)
Liddell, R.G.B., Sadoway, R.: Refractory metals extraction, processing and applications. Int. Nucl. Inf. Syst. (1990)
Lide, D.: Handbook of Chemistry and Physics, 87th edn. CRC Press, Boca Raton, Fl (2006)
Lovering, D.G.: Molten Salt Technology. Springer, United States (1982)
Lu, H., Jia, W., Liao, C., Yuan, W.: Pilot experiments of magnesia direct electrolysis in a 5KA magnesium reduction cell. In: Essential Readings in Magnesium Technology, pp. 101–105. Springer, Cham (2016)
Lucas, J., Lucas, P., Rochelle, L., Davenport, W.: Rare Earths Rare Earths Production and Use. Elsevier Ltd., (2015)
Lukasko, J.J., Murphy, J.E.: Personal communication (1990)
Mamantov, G.E., Marassi, R.: Molten Salt Chemistry: An Introduction and Selected Applications. Springer Science & Business Media (2012)

Margarido, F., Vieceli, N., Durão, F., Guimarães, C., Nogueira, C.A.: Comun. Geológicas **101**, 795–798 (2014)
Minh, N.Q.: JOM **37**, 28–33 (1985)
Mishra, B., Olson, D.L.: J. Phys. Chem. Solids **66**, 396–401 (2005)
Mosander, C.G.: The preparation of cerium by reduction of CeCl with potassium, Pogg. Ann. **11**, 406–416 (1827)
Mouron, L.: Comput. Chem. **21**, 431–435 (1997)
Naboychenko, S.S., Murashova, I.B., State, U.: Production of Rare Metal Powders. Elsevier Ltd. (n.d.)
Ono, Y., Yamaguchi, M., Application, F., Data, P.: **1** (2009)
Osaka, T., Ito, Y., Aladjov, K., Stoyanova, A., Sefman, J.R., Lin, Y.P., Masuko, N. (eds.): Electrochemical Technology: Innovation and New Technologies. CRC Press (1997)
Osarinmwian, C.: ChemRxiv, **48** (2019)
Pal, U.B., Woolley, D.E., Kenney, G.B.: Jom **53**, 32–35 (2001)
Panigrahi, M., Iizuka, A., Shibata, E., Nakamura, T.: J. Alloys Compd. **550**, 545–552 (2013)
Ray, H.S., Sridhar, R.: Extraction of Nonferrous Metals. East-West Press Pvt Ltd (1985)
Rayaprolu, S., Chidambaram, D.: ECS Trans. **58**, 51–66 (2014)
Redkin, A.A., Nikolaeva, E.V., Dedyukhin, A.E., Zaikov, Y.P.: Ionics (kiel) **18**, 255–265 (2012)
Saïla, A., Gibilaro, M., Massot, L., Chamelot, P., Taxil, P., Affoune, A.M.: J. Electroanal. Chem. **642**, 150–156 (2010)
Sehra, J.C., Suri, A.K.: High Temp. Mater. Process. **11**, 255–292 (1993)
Sehra, J.C., Vijay, P.L.: Miner. Process. Extr. Metall. Rev. **19**, 523–556 (1998)
Senderoff, S.: Metall. Rev. **11**, 97–112 (1966)
Sharma, R.A.: Jom **48**, 39–43 (1996)
Sohn, H.Y., Geskin, E.S.: Jom **42**, 32–34 (1990)
Springer, C., Hasanbeigi, A.: Emerging Energy Efficiency and Carbon Dioxide Emissions-Reduction Technologies for Industrial Production of Aluminum, pp. 1–35 (2016)
Stewart, P.D., Michael, C.: Production of sodium. US3257297A (1961)
Sunshot: Molten Salt—Concept Definition & Capital Cost Estimate, Proj. No. 042839. (2016)
Takeda, O., Nakano, K., Sato, Y.: Mater. Trans. **55**, 334–341 (2014)
Thayer, A.M.: Having the mettle for sodium markets. Chem. Eng. News **86**(43), 20–21 (2008)
Thompson, J.S., Blank, H.M., Simmons, W.J., Bergmann, O.R.: Low temperature alkali metal electrolysis. US20020088719 (2002)
Thonstad, J.: Miner. Process. Extr. Metall. Rev. **10**(1), 41–55 (1992)
Tyrer, M., Gibbon, A.: Strat. Important Metals **69**, 12–13 (2012)
USGS: J. Chem. Inf. Model. 2020–2021 (2021)
Vandarkuzhali, S., Gogoi, N., Ghosh, S., Prabhakara Reddy, B., Nagarajan, K.: Electrochim. Acta **59**, 245–255 (2012)
Viswanathan, S.: Bull. Electrochem. **1**, 561–567 (1985)
Vogel, H.: Proceeding EMC, pp. 1–13 (2015)
Vol'skii, A.N., Sergievskaia, E.M.: Theory of Metallurgical Processes: Pyrometallurgical Processes, 2nd edn. MIR Publishers, Moscow (1971)
Wang, S.D., Li, Q., Ye, X.S., Sun, Q.G., Wu, Z.J.: Trans. Nonferrous Metals Soc. China (English Ed.) **23**, 3104–3111 (2013)
WS: Materials Performance Indices (2008)
Yan, X.Y., Fray, D.J.: Molten salt electrolysis for sustainable metals extraction and materials processing—a review. In Electrolysis: Theory, Types Appl., pp. 46–49. Nova Science Publishers (2008)
Yan, X.Y., Fray, D.J.: Molten salt electrolysis for sustainable metals extraction and materials processing—a review. In: Electrolysis: Theory, Types and Applications, pp. 255–301. Nova Science Publishers, Inc. (2010)
Yan, Q.X., Li, X.H., Wang, Z.X., Wang, J.X., Guo, H.J., Hu, Q.Y., Peng, W.J., Wu, X.F.: Trans. Nonferrous Metals Soc. China (English Ed.) **22**, 1753–1759 (2012)

Yasuda, K., Kondo, K., Nohira, T., Hagiwara, R.: J. Electrochem. Soc. **161**, D3097–D3104 (2014)
Zaikov, Y.P., Shurov, N.I., Batukhtin, V.P., Molostov, O.G.: Metall. Mater. Trans. B Process Metall. Mater. Process. Sci. **45**, 968–974 (2014a)
Zaikov, Y.P., Batukhtin, V.P., Shurov, N.I., Ivanovskii, L.E., Suzdaltsev, A.V.: Metall. Mater. Trans. B Process Metall. Mater. Process. Sci. **45**, 961–967 (2014b)
Zhu, H.: Rare earth metal production by molten salt electrolysis. In: Encyclopedia of Applied Electrochemistry. Springer, New York (2014)

Chapter 12
Contribution of Bhabha Atomic Research Centre to the "Metals Journey of India"

Sanjib Majumdar, Raghvendra Tewari, Vivekanand Kain, and Ajit Kumar Mohanty

Abstract Bhabha Atomic Research Centre (BARC) was established by Dr. Homi Jehangir Bhabha in January 1954 as Atomic Energy Establishment, Trombay (AEET) for pursuing multidisciplinary research activities to support the nuclear power programme of India. Since its inception, BARC is deeply involved in the adoption, ingestion and development of new technologies in various fields—be it electronics, computations, nuclear, radiation, chemical, biological/health/agriculture and materials. Among these fields, developments of materials have taken a centre stage right from the inception of BARC. Materials and processes required for every stage, from ore processing to extraction, alloy making, fabrication, support during operation, degradation modes to spent fuel reprocessing and waste management have been indigenously developed and successfully implemented with complete understanding. BARC is providing materials related support not only for Indian nuclear power reactors but also for heavy water plants, Indian Rare Earths Limited (IREL), Uranium Corporation of India Ltd (UCIL) and various other plants of Department of Atomic Energy (DAE).

12.1 Introduction

Genesis of the materials development programme in BARC got initiated by a request from Dr. Homi Bhabha to Prof. Brahm Prakash, then a faculty at IISc Bangalore, to take up the challenge of establishing a facility for fabrication of fuel assemblies at the BARC campus. Prof. Brahm Prakash and his team at BARC met the challenge of fabrication of fuel assemblies in a short time. One fundamental aspect in meeting this challenge pertained to beta quenching of the uranium rod. Uranium rods on rolling acquire texture but for better irradiation response, it is necessary to have a random texture, which is obtained by quenching the rolled rod from the high temperature beta phase to room temperature. For this purpose, a unique in-house induction heating

S. Majumdar · R. Tewari · V. Kain (✉) · A. K. Mohanty
Bhabha Atomic Research Centre, Trombay, Mumbai, India
e-mail: vivkain@barc.gov.in

R. Divakar et al. (eds.), *Indian Metallurgy*, Indian Institute of Metals Series,
https://doi.org/10.1007/978-981-99-5060-7_12

facility was developed. Two fuel elements that were sent to Canada for qualification when subjected to irradiation at NRX reactor withstood radiation levels of 2000 MWD/t as compared to the normal level of about 1000 MWD/t. This surprised everyone who was probably not expecting a better performance from the Indian fuel elements. Project Faggots now known as Atomic Fuels Division (AFD) delivered the initial half charge as well as all the subsequent periodic replacement fuel elements for CIRUS research reactor until its recent decommissioning. This experience reinforced the drive in Prof. Brahm Prakash to set up facilities at BARC to study and understand the science required to develop components for plant applications. BARC follows this legacy and works to fully understand the science as well as develop a fundamental understanding and uses it to develop materials and processes to support the robust nuclear programme of the country.

BARC, right from its inception, build the concept of 'Atmanirbhar Bharat' in its DNA. Although Dr. Homi Bhabha knew at that time that Indian uranium ores are too lean to give a commercially viable product, yet, with the idea of India to become a sovereign nuclear state, he went ahead with the exploration and extraction of uranium from Indian mines with U_3O_8 content as low as 0.04%. The same concept was applied in the case of extraction of zirconium from zircon sand. These decisions paid off in later years when sanctions were imposed on India, but all the Indian PHWRs continued functioning with fuel made in the country. Since then the spectrum of materials activities in BARC have widened their scope and several entities flourished (e.g. Uranium Corporation of India Ltd. (UCIL), Indian Rare Earths Ltd. (IREL), Nuclear Fuel Complex (NFC) and Heavy Water Board (HWB)) using BARC developed technologies for targeted production. From the lightest metallic element beryllium (Be) to the heaviest naturally found uranium (U), nearly all the metallic elements of interest to Department of Atomic Energy (DAE) were successfully extracted and processes were developed for utilisation in BARC.

As the primary resources/ores of most of the elements present in India are very lean, the challenges to selectively take out the metal values associated with the unwanted gangue led to the development of unique extraction processes for each metal, involving various steps, like innovative physical ore beneficiation, selective chemical ore-breakdown techniques, ion-exchange, solvent extraction, halogenation, metallothermic reduction, hydrogen reduction, reduction-diffusion, molten salt electrolysis etc. Expertise was developed on related technologies for melting-consolidation and vacuum refining using arc and electron beam melting technologies, ultra-purification using thermal decomposition, iodide refining and electron beam zone refining etc. Along with these, the fabrication technologies e.g. powder metallurgy, rolling, extrusion, pilgering, welding, coating, and ceramic metallurgy were pursued vigorously and strong groups were developed that contributed significantly. The development of fundamental understanding of materials, microstructural and mechanical properties characterisation at different length scales, structure–property correlations and degradation mechanisms specifically corrosion has been a strong point of materials activities in BARC. The fundamental understanding thus developed helped to address materials related issues in plants and gave impetus to develop newer applications with a strong basis.

Described below are some salient achievements in the journey of materials activities in BARC.

12.2 Technologies Developed at BARC

The technologies developed for important materials at BARC which were later transferred to produce at industrial scale are described below:

12.2.1 Zirconium, Hafnium and Titanium, and Their Alloys and Applications

The principal source of zirconium is zircon (zirconium silicate ($ZrSiO_4$)) which is found in the coastal beach sand in India. This sand also contains many other valuable minerals because of which a special physical beneficiation technology for the separation of the individual minerals was first established at BARC, Trombay at a pilot plant scale and subsequently, a plant was established in Tamil Nadu by Indian Rare Earths Limited (IREL) for the processing of Manavalakurichi beach sand. The zircon contains about 2.5% Hf, an element chemically similar to Zr and it was successfully separated at BARC (Prakash and Sundaram 1955) using vapour phase dechlorination technique. A process for separation of Zr and Hf was developed by chlorination of Zircon, purification to obtain $ZrCl_4$ and $HfCl_4$ mixed salt, selective dechlorination of $ZrCl_4$ from the vapour phase. This led to the formation of pure ZrO_2 in the residue by adjusting the ratio of the flowing gas mixture composed of chlorine and oxygen at high temperatures leaving the volatiles enriched with $HfCl_4$. Solvent extraction process was developed later for a more effective and larger scale separation of the two elements. Chlorination of ZrO_2 was further optimised using static-bed reactor technology. Magnesio-thermic reduction of $ZrCl_4$ producing pure zirconium (Kroll process), was used for making zirconium sponge at BARC, Trombay and a schematic of the initially designed assembly having different heating zones used for carrying out Kroll process is shown in Fig. 12.1a (Subramanyam and Sundaram 1966). Iodide refining technology (Van Arkel-De Boer Process) was developed and demonstrated at BARC for making ultra pure zirconium from impure zirconium and zirconium alloy scraps. Figure 12.1b shows the outlook of the zirconium crystal grown by iodide refining process, and the inset of the figure represents the overall view of the crystal bar zirconium formed on U-shaped zirconium filament. The zirconium produced by Kroll process was subsequently vacuum arc melted to produce various zirconium alloys, such as Zircaloy 2, Zircaloy 4, Zr-2.5Nb, Zr-2.5Nb-0.5Cu, for structural components in nuclear reactors. A dedicated plant was further established in 1971 for the production of Zr and its alloys, and for the fabrication of different reactor components at NFC, Hyderabad.

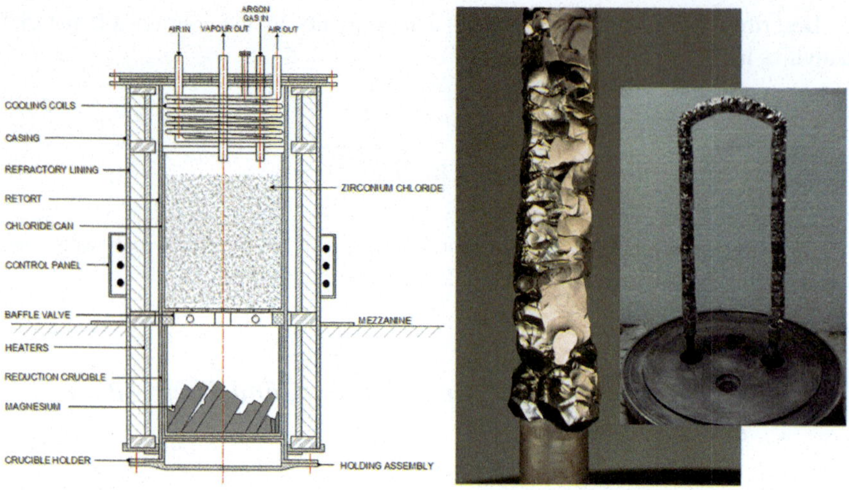

Fig. 12.1 **a** Schematic of the Kroll reduction assembly for zirconium sponge production designed at Bhabha Atomic Research Centre, Trombay, **b** Zirconium crystal bar grown by iodide refining technique at BARC campus and the inset shows the crystal bar zirconium formed on U-shaped zirconium filament

India adopted a different route involving extrusion and pilgering based forming processes for Zr alloys to produce various tubular products required for various structural applications in Indian Pressurised Heavy Water Reactors (IPHWR). For this purpose, detailed deformation processing maps were developed. Detailed studies on the evolution of phases, texture, microstructure, irradiation effects and corrosion behaviour were carried out. Various processing routes were tried and eventually, an optimised route giving products with uniform and desirable properties has been adopted. The basic research carried out in BARC solved many production issues faced by NFC when the expected strength of zircaloy tubes was not achieved initially. Subsequently, a major role in establishing the flow sheet for the production of Zr–Nb tubes was successfully executed with the help from BARC material scientists.

The major drawbacks of Zr based alloys are nodular corrosion and hydride induced embrittlement which have been studied under various conditions. Reactor conditions were simulated and the basics of nodular corrosion and hydride embrittlement have been studied. The oxidation behaviour, corrosion resistance and irradiation response of Zr alloys under various conditions have been studied. BARC provided support in monitoring the healthiness of the materials in service. From hot conditioning during the start-up of a reactor to establishing root cause analysis for components that showed degradation during its service are a few of the services rendered by BARC engineers and scientists.

Hafnium metal was mainly prepared by molten salt electrolysis at 800–950 °C using NaCl-KCl-$HfCl_4$ salt. Figure 12.2 shows the as-deposited hafnium flakes on molybdenum cathode plate. Iodide refining process was used for making crystal bar hafnium. The technology for the production of titanium metal was developed

Fig. 12.2 Hafnium metal produced by molten salt electrolysis

and demonstrated (Sridhar Rao et al. 1969) using Kroll reduction and pyro-vacuum distillation processes as used for Zr. $TiCl_4$ being liquid is fed from a separate chamber, and the $MgCl_2$ formed after magnesio-thermic reduction is progressively removed from the reduction crucible. This technology was demonstrated in pilot plant scale at NFC, Hyderabad, and subsequently, the technical know-how was given to Defence Metallurgical Research Laboratory (DMRL), Hyderabad.

12.2.2 Materials for Fuel: Uranium, Plutonium and Thorium

The detailed flow sheet for the preparation of metallic uranium starting from the lean grade ores was developed at BARC, Trombay. Depending upon the chemistry, the ore was leached with acid or alkali followed by ion-exchange and solvent extraction to separate impurities and to produce uranyl nitrate solution. For the uranium ore at Thummapalle, BARC developed an alkali process and it has been used for the first time in the country. The ammonium diuranate precipitated out from the uranyl nitrate solution was calcined to produce UO_3. UO_3 was further converted to UO_2 by hydrogen reduction followed by hydrofluorination to form UF_4. Metallothermic reduction of UF_4 using calcium or magnesium has been successfully demonstrated to prepare uranium metal. Figure 12.3 shows the reactor used for metallothermic reduction of uranium metal and an example of the uranium ingot produced at BARC.

Plutonium is a man-made element produced during nuclear reaction between neutron and uranium. The detailed flow sheet was developed for reprocessing of

Fig. 12.3 Reactor used for metallothermic reduction of UF4 for making uranium metal (inset: U metal ingot)

nuclear waste for separation and extraction of plutonium metal. The solvent extraction process was developed to separate plutonium from other radio-nuclides in the form of pure plutonium nitrate which was converted to plutonium oxide or peroxide. Halogenation studies for preparing plutonium fluoride and plutonium trichloride were conducted. Metallic plutonium was produced by calico-thermic reduction of plutonium halides and oxide and subsequently by vacuum induction melt consolidation (Roy and Mahajan 1975). Normally, the metal is produced in small quantities (known as 'micro-metallurgy' for microgram level production) to avoid the uncontrolled nuclear chain reaction inside the plutonium matrix.

Thorium metal extraction technology was developed starting from the Monazite ore processed through several stages of solvent extraction, precipitation and calcination to pure thorium oxide. The process of metallothermic reduction of thorium oxide by calcium producing thorium metal powder was standardised at a pilot plant scale.

Development of metallic or ceramic based fuels for nuclear reactors was a challenge as strict restrictions and imposed embargoes did not allow sharing of any information. Without any support, entire laboratories, materials and technologies were developed from scratch at BARC. Being radioactive materials, the fuels for nuclear reactors needed additional precautions and remote handling technologies. In spite of such restrictions, optimising sintering parameters for the (U, Pu)-oxide fuels, development of (U, Pu)-carbide fuels and other fuels were successfully carried out. In this connection, development of carbide fuel needs special mention as the composition of carbon in the materials has to be very strictly monitored. Excess carbon would carburize the wrapping structural material, whereas hypo-stoichiometric carbon leads to the formation of low melting eutectic of U-Fe-Pu. Such a tight control over carbon chemistry has been mastered at BARC and is not an easy task to achieve in commercially produced products. The fuel and clad developed with BARC technologies are produced today to provide the required fuel assemblies for all the Indian nuclear power and research reactors.

12.2.3 Refractory Metals: Niobium, Tantalum, Vanadium, Molybdenum and Tungsten

The technology for aluminothermic reduction of the respective oxides (Nb_2O_5, Ta_2O_5) to produce massive forms of niobium and tantalum metals was developed at BARC. The metal thus produced by the thermit reaction is further refined and consolidated using electron beam melting in which purification is done by vacuum degassing, carbon de-oxidation and sacrificial de-oxidation mechanisms (Gupta and Jena 1968). Technology for the production of capacitor grade tantalum powder by reducing K_2TaF_7 with sodium was established (Jain et al. 1969). All these BARC developed technologies were transferred to NFC, Hyderabad for plant scale production. Aluminothermic reduction reactor was specially designed for making thermit vanadium with low amount of nitrogen, and subsequent electron beam melt refining. These metals were also produced by molten salt electro-extraction using metal carbide as anode feed. The process flow sheet has also been established on laboratory scale at BARC to recover molybdenum from low grade indigenous sources and various secondary sources. The molten salt electro-metallurgical processes for producing pure molybdenum from its carbides and sulphides are also reported earlier (Suri et al. 1974; Mukherjee and Gupta 1974). The technology for the production of molybdenum metal powder by hydrogen reduction of molybdenum trioxide has also been demonstrated (Majumdar et al. 2008a). The flow sheet for low grade wolframite concentrate was developed with the objective of recovery of tungsten and other valuable associates. Process know-how was established for producing tungsten metal powder by hydrogen reduction of WO_3 (Majumdar et al. 2016).

12.2.4 Cobalt

There are no primary resources of cobalt in India. The secondary resources were explored to recover cobalt. BARC has established a flow sheet for producing cobalt oxalate intermediate starting from spent ammonia cracker catalyst (SACC) by hydrometallurgical route followed by thermal decomposition of cobalt oxalate to produce cobalt metal powder (Majumdar et al. 2008b) in kg scale batches. The technology for recovering cobalt from the alloy scraps containing cobalt has been successfully developed and demonstrated recently at BARC. Further, the conversion of cobalt powder into required shapes of cobalt slugs and cobalt pellets, used for cancer therapy

Fig. 12.4 Neodymium metal ingot produced by calcio-thermic reduction of its fluoride (NdF_3)

and other applications, was successfully demonstrated through powder metallurgical route by optimising various process parameters.

12.2.5 Rare Earth Metals

Technologies for the separation and purification of rare earth (RE) oxides (Sm, Ce, La, Pr, Nd etc.) from monazite were developed at BARC and transferred to IREL which is currently producing and supplying various rare earth oxides and compounds. Presently, the technologies are being developed for the separation of rare earths from xenotime, from newly discovered ore deposits and also from various secondary resources. The technologies for the preparation of some of the rare earth metals such as Ce, La, Nd (Fig. 12.4) using molten salt electrolysis and metallothermic reduction of RE-fluoride or oxide and their alloys e.g. $SmCo_5$ using reduction-diffusion technique have been developed. In order to make the country self-reliant in the production of high purity heavy rare earth oxides, rare earth metals, alloys and phosphors, an LoU has been signed between BARC and IREL to develop and transfer the technologies for demonstration at pilot plant scale at the upcoming IREL Theme Park, Bhopal. A plant to produce $SmCo_5$ and Sm_2Co_{17} permanent magnets is under construction at Vizag campus of BARC.

12.2.6 Alkali Metals: Lithium, Magnesium and Calcium

Process flow sheets were developed for the preparation of various lithium metal compounds such as carbonate, hydroxide and chloride, from the lithium bearing ore Spodumene, Lepidolite etc. Technology for the preparation of lithium metal from lithium chloride using fused salt electrolysis process was established. The same technique was also employed to produce magnesium from $MgCl_2$, a by-product produced in the Kroll process. The know-how for magnesium production was provided to DMRL. The Central Electrochemical Research Institute, Karaikudi, developed electrolytic processes for the extraction of crude magnesium and crude calcium. The pyro-vacuum distillation of electrolytic crude calcium was done at BARC to produce pure calcium metal.

12.2.7 Light Elements: Beryllium and Boron

Technologies to extract beryllium from the Indian Beryl ore to produce the intermediates such as oxide and fluoride were developed. Magnesio-thermic reduction technology was established for producing beryllium metal from its fluoride, and also for making Cu-Be alloys. A plant was set up at Navi Mumbai for the production and fabrication of Cu-Be alloy, beryllium metal and BeO for meeting the requirements of space and nuclear industries.

The technology for the preparation of elemental boron from potassium fluoborate (KBF_4) by molten salt electrolysis using KCl-KF-KBF_4 electrolyte was developed and transferred to Indira Gandhi Centre for Atomic Research. The fused salt electrowinning technology is currently being used at Heavy Water Plant, Manuguru for the production of elemental enriched boron. Molten salt electro-extraction technology (Majumdar 2020) using boron carbide as soluble anode was also established for making elemental boron and is shown in Fig. 12.5a, b presenting the particle size and morphology of the electro-extracted boron produced from B4C scraps. This technology was transferred to Ordinance Factory at Khamaria. Figure 12.6 shows a representative image of the molten salt electro-extraction cell designed and fabricated at BARC for making molybdenum, tantalum, vanadium and boron using their carbides as anode feed.

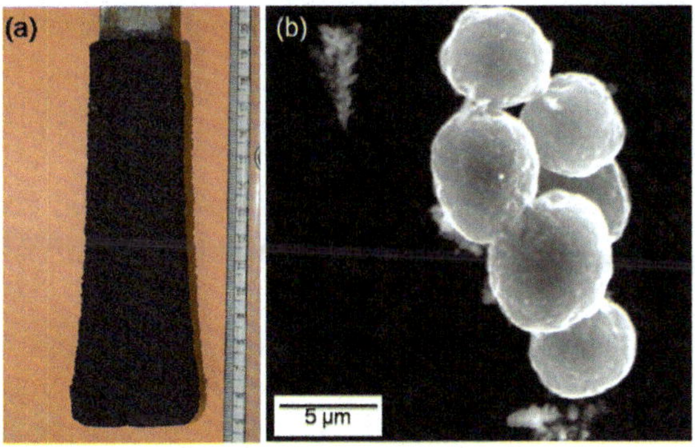

Fig. 12.5 **a** Outlook of the electro-extracted boron deposited on cathode plate and **b** Scanning electron microscopic image showing the morphology of individual boron particles

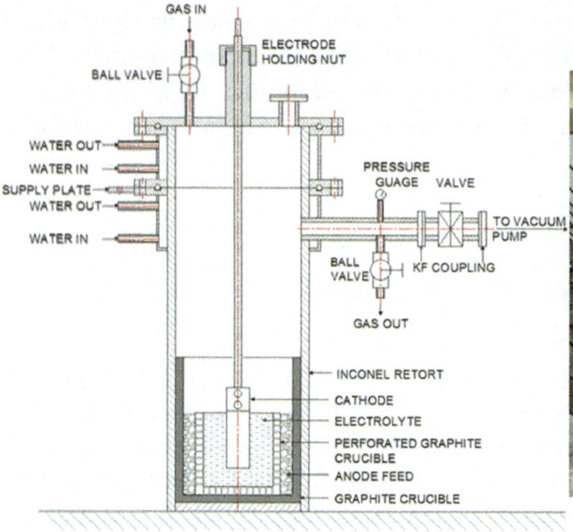

Fig. 12.6 Schematic of the molten salt electro-extraction cell and the set up developed at BARC

Technologies for the synthesis of borides of various reactive and rare metals such as ZrB_2, TiB_2, EuB_6, NdB_6 etc. using carbo-thermic and boro-carbo-thermic reduction at high temperature and high vacuum were developed at BARC. These borides are produced in larger quantities and used for departmental (DAE) applications.

12.2.8 Ferroalloys

Aluminothermic co-reduction technology for the preparation of low carbon ferroalloys such as ferrochromium (Majumdar et al. 2020), ferroniobium, ferromolybdenum, ferrovanadium, ferroboron was developed and transferred to private industries.

12.3 Alloy Development Programme: Evaluation of Physical, Mechanical and Corrosion Properties

In addition to various technological developments, BARC has paid special attention to fundamental scientific research work on structure-processing route-property correlations. Considerable amount of work has been done in understanding phase transformations occurring in various alloy systems (Dey et al. 2004; Tewari et al. 2008, 1999), deciphering the underlying mechanisms of degradation of materials and thermodynamic studies of phase and phase formation reactions. These studies paved the way for better understanding of the performance of materials under various conditions. Detailed diffusion studies on various systems not only generated enormous data on diffusion of species which helped in correcting few of the phase diagrams, but also helped in evolving a new diffusion bonding technique for joining dissimilar materials. Detailed crystallographic work led to the understanding of texture evolution in U, Zr and their alloys, understanding the martensitic phase transformations in Zr (Banerjee and Krishnan 1971), Ni and Fe based alloys which successfully led to tailor the microstructures for specific applications. Among various alloy development work, development of shape memory Ni–Ti alloys (Madangopal et al. 1994) needs special mention, as the coupling produced using this alloy became an integral part of 'Light combat aircraft'.

The physical metallurgy of superalloys and evolution of various phases in this alloy system solved many issues related to ammonia cracker tubes in the production of heavy water. The study of formation of precipitates in various generations of superalloys addressed very long term issues in these alloys. The detailed characterisation studies of various alloys with respect to phase transformation and mechanical properties (Chakravartty et al. 1995) including tensile, fracture, fatigue, creep, embrittlement effects due to hydrogen pick up (Singh et al. 2004), and material degradation behaviour in different corrosive and oxidative environments paved the way for the most optimum use of various alloys in operating plants. Understanding the structure-processing path property including corrosion behaviour remained always an integral part to develop fabrication process flow sheets for many alloys. The materials science activities were extended to understand the physical, mechanical and corrosion properties of nickel based alloys (Inconel 690, Alloy 718, SuperNi etc.), D9 and RPV steels, refractory metal based alloys (Nb-1Zr-0.1C, Mo-0.5Ti-0.1Zr-0.02C, Mo-Ti-Si etc. (Majumdar et al. 2021; Majumdar et al. 2018)) etc. Corrosion behaviour of stainless steels, Zr based, and Ni based alloys has been established at BARC and used to guide the successful deployment of these alloys to ensure safety and plant availability in nuclear power and research plants as well as other plants of DAE. The key strength of BARC is in the fields of stress corrosion cracking (Kain et al. 2002), flow accelerated corrosion (Kain et al. 2008) and various localised forms of corrosion and the developed knowledge and expertise is used to provide metallurgical and corrosion related support to operating plants. Environmental barrier coatings were developed by treating the metal or alloy surfaces using different chemical/physical vapour deposition processes. The powder metallurgical processes are developed to synthesise and consolidate the ceramic materials for instance boron carbide, oxides, carbides and borides of reactive and rare metals.

12.4 Future Perspectives

BARC is currently extending its research and development activities on developing various processes for the preparation and fabrication of metals and alloys. The Centre is extending its support for plant scale production of different rare earths and rare metals technologies for being taken up by public sector companies and private entrepreneurs beyond the family of Atomic Energy. New alloys and coatings are being developed for high temperature and advanced technological applications. Needs for materials in new fields including composites, meta-materials, bio-materials and electronic materials are being addressed for sensor, functional and structural applications. BARC is poised to take up research on newer projects of DAE e.g. molten salt reactors, pressurised water reactors and also on advanced fabrication processes e.g. 3-D printing.

Acknowledgements The dedicated efforts of Dr. Brahm Prakash, Shri H. C. Katiyar, Dr. Kondala Rao, Shri T. K. S. Murthy, Dr. C. V. Sundaram, Dr. P. R. Roy, Shri K. Balaramamoorthy, Dr. C. K. Gupta, Dr. S. Banerjee, Dr. A. K. Suri, Dr. J. K. Chakraborty, Dr. G. K. Dey, Dr. Madangopal Krishnan, and all the past and present members of metallurgical/materials fraternity at BARC helped in the growth of nuclear science technology enriching Indian Metallurgy.

References

Banerjee, S., Krishnan, R.: Martensitic transformation in zirconium-niobium alloys. Acta Metall. **19**(12), 1317–1326 (1971)

Chakravartty, J.K., Dey, G.K., Banerjee, S., Prasad, Y.: Characterization of hot deformation behaviour of Zr-2.5Nb-0.5Cu using processing maps. J. Nucl. Mater. **218**(2):247–255 (1995)

Dey, G.K., Tewari, R., Banerjee, S., Jyoti, G., Gupta, S.C., Joshi, K.D., Sikka, S.K.: Formation of a shock deformation induced ω phase in Zr-20 Nb alloy. Acta Mater. **52**(18), 5243–5254 (2004)

Gupta, C.K., Jena, P.K.: Production of tantalum metal by alumino-thermic reduction of its pentoxide. J. Metals 25–28 (1968)

Jain, S.C., Dose, D.K., Gupta, C.K.: Pilot plant production of capacitor grade tantalum powder. In: Paper Presented at the 23rd Annual Technical Meeting of the Indian Institute of Metals, Bombay, Nov. 14–17 (1969)

Kain, V., Prasad, R.C., De, P.K.: Testing sensitization and predicting susceptibility to intergranular corrosion and intergranular stress corrosion cracking in austenitic stainless steels. Corrosion **58**(1), 15–37 (2002)

Kain, V., Roychowdhury, S., Mathew, T., Bhandakkar, A.: Flow accelerated corrosion and its control measures for the secondary circuit pipelines in Indian nuclear power plants. J. Nucl. Mater. **383**(1–2), 86–91 (2008)

Madangopal, K., Banerjee, S., Lele, S.: Thermal arrest memory effect. Acta Metallurgica et Materialia **42**(6):1875–1885 (1994)

Majumdar, S.: Recovery of elemental boron from boron carbide using KCl-KF-KBF4 Electrolyte in fused salt electroextraction process. J. Electrochem. Soc. **167**(6), 062507 (2020)

Majumdar, S., Sharma, I.G., Samajdar, I., Bhargava, P.: Kinetic studies on hydrogen reduction of MoO3 and morphological analysis of reduced mo powder. Metall. Mater. Trans. b. **39**, 431–438 (2008a)

Majumdar, S., Sharma, I.G., Bidaye, A.C., Suri, A.K.: A study on isothermal kinetics of thermal decomposition of cobalt oxalate to cobalt. Thermochim. Acta **473**, 45–49 (2008b)

Majumdar, S., Kishor, J., Paul, B., Kain, V., Dey, G.K.: Demonstration of production of tungsten metal powder and its consolidation into shapes. BARC Newsletter 30–33 (2016)

Majumdar, S., Paul, B., Singh, P.K., Kishor, J., Kain, V.: Effect of Si content on microstructure, mechanical and oxidation properties of hot pressed Mo-Ti-Si alloys. Intermetallics 100:126–135 (2018)

Majumdar, S., Paul, B., Kishor, J., Kain, V.: Technology for preparation of low-carbon ferrochromium from Indian chrome ore by aluminothermy process. BARC Newsletter 1–5 (2020)

Majumdar, S., Kishor, J., Behera, A.N., Vishwanadh, B., Borgohain, A., Kapoor, R., Tewari, R., Kain, V., Krishnan, M., Dey, G.K., Banerjee, S.: Preparation and characterization of Nb-1Zr-0.1C alloy suitable for liquid metal coolant channels of high temperature reactors. ASME J. Nucl. Eng. Radiat. Sci. 7:011603-1–011603-14 (2021)

Mukherjee, T.K., Gupta, C.K.: Molybdenum extraction from molysulfide. Metall. Trans. **5**, 707–713 (1974)

Prakash, B., Sundaram, C. V.: Separation of hafnium from zirconium by vapor phase dechlorination. In: Proceedings of the International Conference on peaceful uses of Atomic Energy, Geneva United Nations, New York, pp. 8768, 554–558 (1955)

Roy, P.R., Mahajan, V.K.: Preparation of plutonium metal and its alloys. In: Proceedings of INSA-BARC Winter School of Chemistry and Metallurgy of Rare Metal Extraction, BARC, Trombay, pp. 889–920 (1975)

Singh, R.N., Kishore, R., Singh, S.S., Sinha, T.K., Kashyap, B.P.: Stress-reorientation of hydrides and hydride embrittlement of Zr–2.5 wt% Nb pressure tube alloy. J. Nucl. Mater. **325**(1):26–33 (2004)

Sridhar Rao, Ch., Kulkarni, A.P., Sundaram, C.V.: Studies on the production of titanium sponge. Trans. Ind. Ins. Metals 22:9–14 (1969)

Subramanyam, R. B., Sundaram, C.V.: Pilot plant production of nuclear grade zirconium spongefrom Indian zircon. Trans. Ind. Ins. Metals 9–20 (1966)

Suri, A.K., Bose, D.K., Gupta, C.K.: Electro-extraction of molybdenum from Mo2C-type carbide. Metall. Trans. **5**, 451–455 (1974)

Tewari, R., Mukhopadhyay, P., Banerjee, S., Bendersky, L.A.: Evolution of ordered ω phases in (Zr3Al)–Nb alloys. Acta Mater. **47**(4), 1307–1323 (1999)

Tewari, R., Srivastava, D., Dey, G.K., Chakravarty, J.K., Banerjee, S.: Microstructural evolution in zirconium based alloys. J. Nucl. Mater. **383**(1–2), 153–171 (2008)

Chapter 13
Galvanizing Industry in India—Past, Present & Future

L. Pugazhenthy

Abstract Materials degradation due to corrosion is a global issue that disrupts infrastructure, impacting the overall economy. India loses as much as 3-4% of its GDP every year on the account of corrosion. Maintenance is expensive, creates additional logistical problems and disrupts regular activities. India has a long coastal line and the long-term answer to this far-reaching problem is the use of zinc protected steel during the structure's initial construction stages. In 1836, Sorel in France received a patent for the process of coating a cleaned steel by dipping it in molten zinc, which he names as 'galvanizing'. Galvanized steel plays a vital role in our everyday lives and is widely used in construction, transport, agriculture and power transmission where excellent corrosion protection and long life are essential. Although India has become world's second-largest crude steel producer, it still lags behind major developed economies in the uptake of galvanized steel. India's per capita demand for zinc-coated steel is 7 kg, which is far behind the world average of 22 kg. In order to increase the domestic steel production, the government has come out with National Steel Policy 2017, to facilitate faster growth and development of steel industry. In order to achieve this target, it is essential that we protect the steel structures by using galvanized steel, which will ensure a long life, safety, and security of the public. This article presents the details of the galvanizing industry in India, how it has developed over the last several decades and the future outlook.

Keywords Sacrificial · Barrier · Maintenance free

L. Pugazhenthy (✉)
Executive Director, India Lead Zinc Development Association, New Delhi, India
e-mail: ilzda.info@gmail.com

Past President, The Indian Institute of Metals (2008-09), New Delhi, India

R. Divakar et al. (eds.), *Indian Metallurgy*, Indian Institute of Metals Series,
https://doi.org/10.1007/978-981-99-5060-7_13

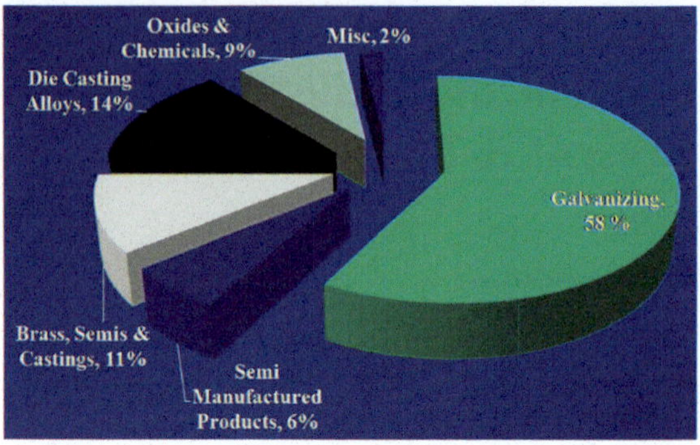

Fig. 13.1 End uses of Zinc—World

13.1 Introduction

Galvanizing industry is an important downstream activity in our economy, playing a significant role in preventing corrosion of steel products and structures. The galvanizing industry had a humble beginning in post-independent India, expanded in the nineties, has come a long way and is poised for a greater role in the coming years in meeting India's ever-growing demand, besides catering to the export markets.

13.2 Global and Indian Scenarios

Of all the coated materials in the world, Zinc-coated steel products would easily be the largest share; zinc is applied through batch hot dip galvanizing, continuous galvanizing, electro galvanizing, thermal spraying, electroplating, zinc anodes, zinc-containing paints as well as powders. Even among zinc-coated steel sheets, there are galvanized, galvalume, galvannealed, galfan, color-coated sheets, etc. Out of the 13.2 MT of Zinc produced globally, 58% goes for galvanizing (Fig. 13.1); in India, out of 0.821 Mt of zinc produced during 2022–23, 72% has gone to the galvanizing sector (Fig. 13.2).

13.3 Why Galvanize?

Steel is the most widely used engineering and construction material. But unfortunately, it corrodes gradually, the corrosion losses in India amount to 3 to 4% of our GDP every year. Hence there is an imperative need for protecting exposed steel structures; public infrastructure, created with huge outlays, are permanent national assets and hence they need to be protected for a long service life. Zinc, well known

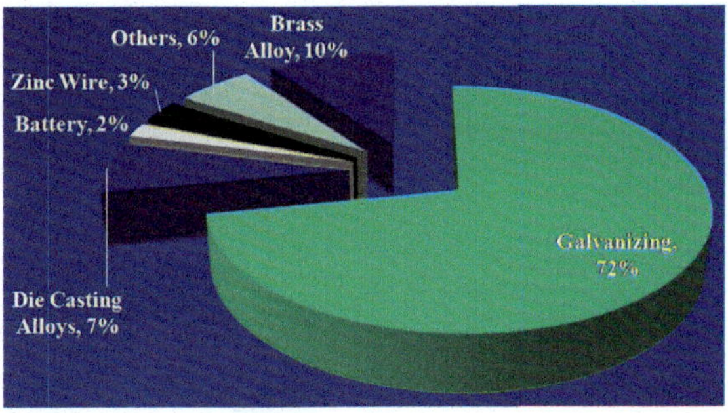

Fig. 13.2 End uses of Zinc—India

for its excellent corrosion resistance, is the most widely applied coating material of choice.

Zinc has the following inherent advantages:

- Excellent corrosion resistance
- Great natural affinity for steel
- Long maintenance free life
- Service life directly proportional to coating weight
- Metallurgicallly bonded coating
- High wear and abrasion resistance
- Good in pH range of 6 to 12.5
- Paintable, for aesthetic appeal
- Weldable, use recommended touch ups &
- Most economical, life cycle cost basis.

13.4 Hot Dip Galvanizing

Hot dip galvanizing is a process wherein well-cleaned steel products are dipped in molten zinc for a specified immersion time; the steel-zinc reaction leads to a metallurgically bonded coating. This coating gives both barrier as well as sacrificial protection, a value added property incidentally (Figs. 13.3, 13.4, 13.5).

13.5 Galvanizing—Past

Post-independent India obviously laid more emphasis on agriculture, irrigation, dams, power, transport, defence etc. The country had a very tight "licensing raj" with industrial licences as well as import licenses, till 1991. India also had a severe foreign exchange crunch; conservation of foreign exchange and import substitution were the priorities of the Government of India. India was totally import dependent for many commodities, including Zinc and Lead.

Fig. 13.3 A factory controlled metallurgical reaction of zinc and steel that provides "barrier" and "sacrificial protection" from corrosion

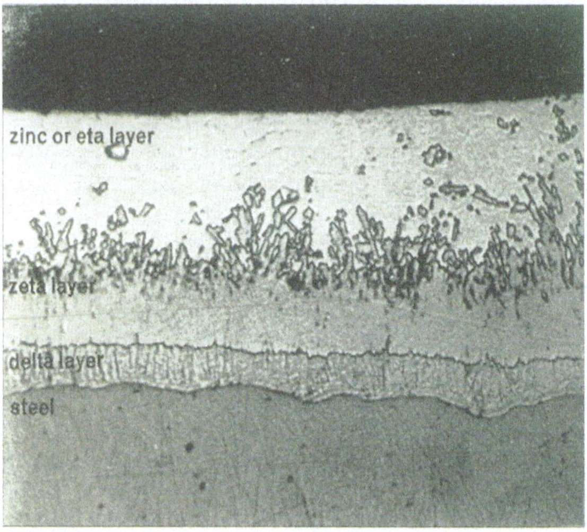

Fig. 13.4 Photomicrograph of a section through a typical hot dip galvanized coating

Galvanizing of steel sheets began in Jamshedpur in the then Tata Iron & Steel Company Ltd, the earlier avatar of Tata Steel; in the early sixties, they were cut sheets dipped in a molten zinc bath, extracted and passed through steel rollers for wiping away excess Zinc.

Tinplate Company of India, a subsidiary of TISCO also had similar cut sheet galvanizing lines. Indian Iron & Steel Co. Ltd. (IISCO) Burnpur also had a similar cut sheet galvanizing line.

In the late sixties the modern day high speed Sendzimir lines for continuous galvanizing of steel sheets were set up by Hindustan Steel Ltd (the predecessor of SAIL) at Rourkela and Bokaro (Fig. 13.6). In the mid-eighties many thin gauge sheet galvanizing lines were licensed and came up across the country; such sheets, being lighter, were preferred in the hilly regions of J & K, HP, North East etc. for roofing,

Fig. 13.5 In hot dip galvanizing, there are two types of processes: batch galvanizing and continuous galvanizing

paneling, sheds etc. Some of the players were Nippon Denro Ispat, Lloyds Steel, Sipta Coated, Usha Rectifiers, National Steel etc.

Many State Electricity Steel Boards (Punjab, Odisha, Kerala, Tamil Nadu) had their own captive galvanizing units for structural steels used for substation structures,

Fig. 13.6 Sheet galvanizing

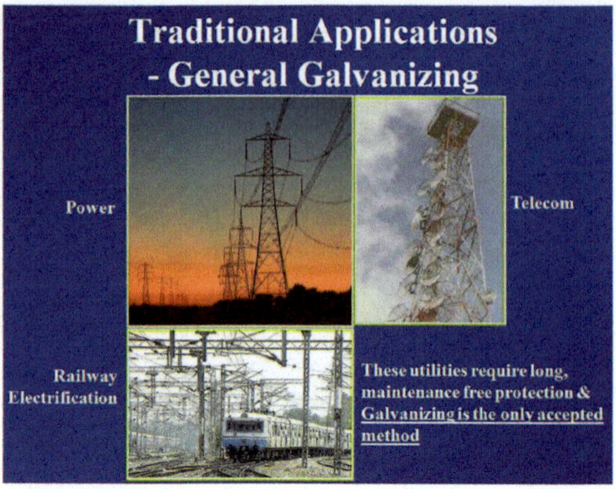

Fig. 13.7 Traditional applications of general galvanizing

power transmission etc. For galvanized nuts & bolts, Guest Keen Williams (GKW) had a centrifuge galvanizing plant in Howrah, West Bengal. Government of India set up a structural galvanizing unit at Triveni Structurals, Allahabad. Indian Railways also had a captive galvanizing plant at Raipur for galvanizing of railway electrification towers. Kamani Engg Corpn as well as Richardson & Cruddas setup multi-locational general galvanizing units in India for domestic as well as export markets.

In the seventies and eighties, there were a number of tube galvanizing units in India such as Indian Tube Company, Bharat Steel Tubes, Zenith Steel Pipes, Gujarat Steel Tubes, Ambica Steel Tubes etc. Their products had a ready market in India for drinking water pipes, irrigation, sprinkler irrigation etc. India was also a major exporter of galvanized ERW[1] tubes. Some of the above companies used to secure Export Performance Awards given by Engineering Export Promotion Council (EEPC) year after year.

India also had a number of continuous steel wire galvanizers such as Usha Martin Block (Wire Ropes), Special Steels Ltd, Industrial Cables India, Devidayal Wires, Hindustan Wires, Deccan Wires etc. These galvanized wires were mainly used for making wire ropes, barbed wire, cable armor etc., the above companies were also exporting their products to many overseas countries.

There were also a number of small units in India for galvanizing of steel buckets, pipe fittings, nuts & bolts with primitive technologies in Howrah, Ludhiana, Kanpur etc.

In general galvanizing, the galvanized steel structures were mainly used in traditional applications like power, railway electrification & telecom, giving a long maintenance free life (Fig. 13.7).

In 1985, the Lotus Temple (Fig. 13.8), using about 300 tonnes of galvanized rebars—India's first example—came up, taking the cue from the Opera House at Sydney. India Lead Zinc Development Association (ILZDA) was instrumental in

[1] Electric Resistance Welded.

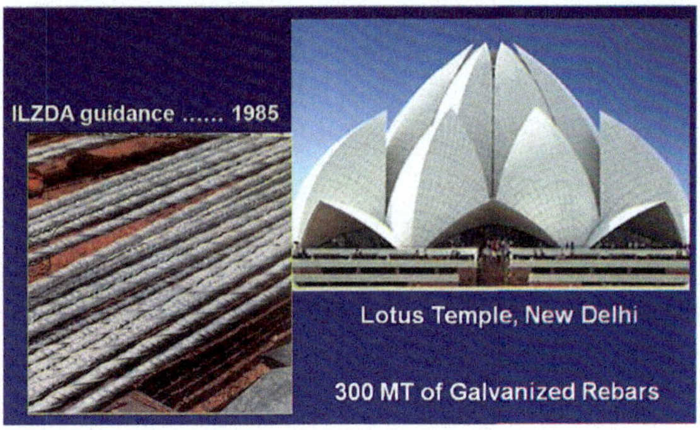

Fig. 13.8 Galvanized rebars for architecture

introducing this concept. Even today the temple looks bright and beautiful without any concrete cracking, rust stains etc.

India also started the process of standardization for the galvanizing industry, very early based on overseas standards like BS, JIS, ASTM etc., Indian Standards Institution (ISI), which has now become the Bureau of Indian Standards (BIS), played a catalytic role from the sixties and it is continuing its work now, aligning many Indian Standards with ISO (International Standards Organization) standards. ILZDA drafted many standards in the early years for the consumer sectors of zinc and lead.

13.6 Galvanizing—Present

Today, India has many galvanizing units which are competitive, energy efficient & eco-friendly in their operations. Improvements have taken place in heating & temperature control, ETP, metal economy, minimization of process wastes like Zinc ash, Zinc dross, flux regeneration etc. Shilpa Steel & Power is the first company in India to introduce Italian technology with enclosed covers over the galvanizing bath so that splashings are minimized, heat conserved and there is safety for the workers. (Fig. 13.9).

It is well known that, a couple of years ago, India was the second fastest growing economy in the world. As per the World Steel Association, India is also the second-largest steel producer in the world. Thanks to the impressive GDP growths and the increasing investments in infrastructure, (Fig. 13.10) galvanized steel consumption also has been on the rise.

Zinc is the main input for galvanizing of steel products and services. Hindustan Zinc Ltd, the only Zinc producer in the country, has been meeting 80% of India's demand. Table 13.1.

About 33% zinc goes in the sheet sector, 11% in pipes, 32% in general, 10% in wire and balance in miscellaneous applications. In the sheet sector, the capacity for GP/GC/Galvalume is approximately 10.00 million tones and for color coated steel sheets it is around 3.0 million tonnes (Table 13.2).

Fig. 13.9 Modern enclosed galvanizing plant

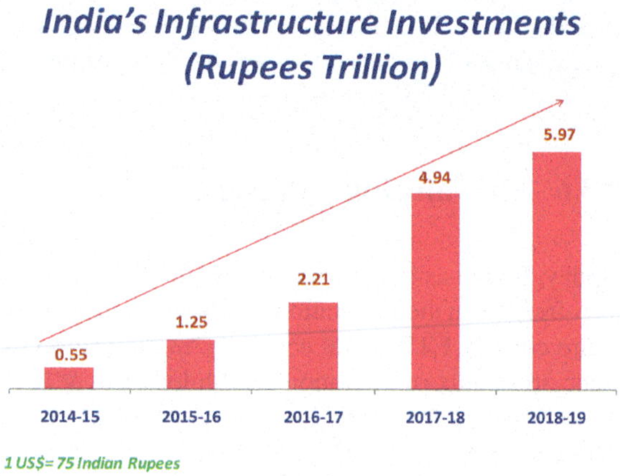

Fig. 13.10 Investents in infrastructure in India

Table 13.1 Zinc Production—India

Year	Production (tonnes)
2016–2017	671,988
2017–2018	791,461
2018–2019	695,321
2019–2020	688,000
2020–2021	716,000

Table 13.2 Production (in million tonnes) in the recent past

	2017–2018	2018–2019	2019–2020	2020–2021
GP/GC/galvalume	7.0	6.9	7.5	6.7
Color coated sheets	1.6	1.8	2.2	2.0

Concrete being porous, absorbs moisture, CO_2 and chloride ions due to capillary action, which attack the base steel leading to corrosion. Volume expansion of rust leads to concrete cracking, rust stains etc., Galvanized rebar is a proven solution for rebar corrosion particularly in coastal areas, corrosive locations, immersed structures, petrochemical complexes etc., During the last few years, more and more galvanized rebars were used in commercial and residential constructions, guest houses, railway coach washery etc. in Mumbai, Mangalore, Vizag, Ahmedabad etc.

After the launch of National Highway Development Program (NHDP), the usage of galvanized guardrails/ crash barriers (Fig. 13.11) picked up significantly across the country; the highway expansion program will continue in the future as well.

In recent years, the usage of galvanized high mast lighting columns has become popular in more cities and towns of India, airports, sea ports, railway yards, traffic junctions, bus terminals, stadiums etc. have been using such high mast lighting columns widely.

Due to the expansion of power & telecom sectors, the application of galvanized cable trays picked up momentum and this also will grow in the years to come.

Investment in solar energy and wind energy have been on the rise during the last few years; the steel structures supporting the solar panels are always hot dip galvanized, because they are totally maintenance free for several decades in remote areas.

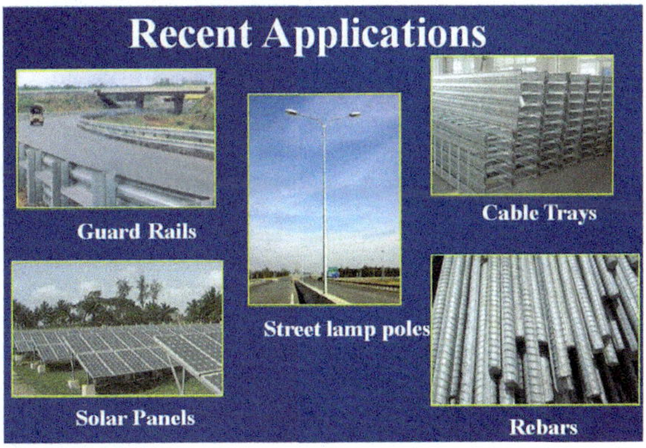

Fig. 13.11 Recent applications of galvanized steel structures

13.7 Galvanizing—Future

In order to ensure a long maintenance free life and to maintain its integrity, any structure exposed to the atmosphere, especially public infrastructure, should be galvanized. There are a number of potential applications where hot dip galvanizing should be adopted: steel railings, foot over bridges, traffic sign posts, bus terminals, platform structures etc. (Fig. 13.12).

Automobiles used in coastal zones and corrosive locations should use galvanized or galvannealed sheets for their bodies (Fig. 13.13), a practice widely used in many overseas countries. Galvanized steel sheets are already used in bus body building in India.

India has announced a number of infrastructure projects like Renewable energy mission, Sagarmala, Bharatmala, Power for All, Jal Jeewan, rural electrification, complete railway electrification, 100 smart cities, express-way and highway expansion, power capacity expansion, telecom growth, migration to 5G etc., where plenty of steel structures will be used and hence there is a huge opportunity for galvanizing.

India is planning to go in a big way for clean, renewable energy, massive investments and more FDIs are likely to flow into the country. Steel structures used in wind energy and solar energy are hot dip galvanized. By 2022 India aims at 175 GW of renewable energy (Fig. 13.14). This will be further increased in the coming years to

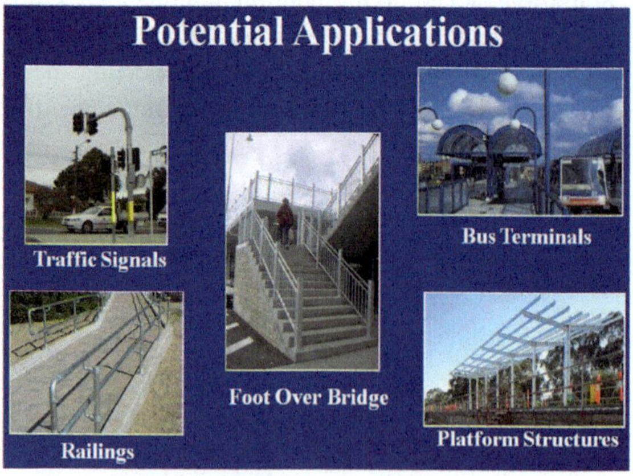

Fig. 13.12 Potential future applications for galvanized steel

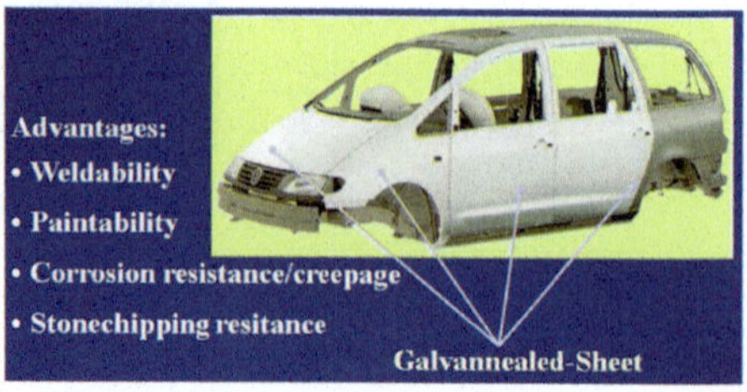

Fig. 13.13 Automotive applications for galvanized steel

Fig. 13.14 Growth of renewable energy in India

450GW by 2030. The priority is to shift from fossil fuels to natural resources like solar and wind bringing down the usage of thermal coal. Roof top solar panels are also being used widely in hospitals, hotels, colleges, schools, commercial buildings, malls, railways etc. Ultramega solar parks have also come across the country and more are likely.

The latest development in India is the launch of a Continuous Galvanized Rebar Plant, first in the country, setup by a mini steel plant in Punjab in 2020, in association with the International Zinc Association, with the support of Hindustan Zinc Ltd.

13.8 Conclusion

With increasing investments in infrastructure, construction and automobile sectors, along with more domestic steel and zinc production, higher economic growth, infrastructure investments etc., India is poised for a quantum jump in the application of hot dip galvanizing, thus minimizing corrosion losses. The huge financial savings made can be wisely used for more & more infrastructure or social projects.

With galvanizing, 'FIRST COST IS THE LAST COST'.

Chapter 14
Lead Recycling in India—Imperative Need for Affirmative Actions

L. Pugazhenthy

Abstract India's lead reserves are ~2.2 million tons which is ~2.5% of the world reserves. At present, India's total consumption of lead is ~1.2 million tons of which about 225,000 comes from primary lead, about 800,000 tons from the secondary lead industry, and the balance accounts for imports. An estimated 85% of lead in use today goes into batteries, mostly for automobiles. The current lead acid battery market size is estimated at US $10 Billion approximately and when the batteries run down, 99% of this lead is recycled to make new batteries, making it a perfect example of circular economy in India. This chapter presents the details of lead recycling in India.

Keywords Extended producer responsibility · Rotary · Collection · Eco-friendly

14.1 Introduction

Over the years, Lead batteries have become an integral part of our lives in India such that the numbers today are just mind-boggling. As a result, used lead battery recycling also has become so huge which is good considering the imperative need for Sustainable Development. At the same time, recycling of used lead battery also faces numerous challenges and opportunities.

L. Pugazhenthy
Past President, The Indian Institute of Metals (2008-09), New Delhi, India

L. Pugazhenthy (✉)
Executive Director, India Lead Zinc Development Association, New Delhi, India
e-mail: ilzda.info@gmail.com

R. Divakar et al. (eds.), *Indian Metallurgy*, Indian Institute of Metals Series,
https://doi.org/10.1007/978-981-99-5060-7_14

14.2 Evolution

In the 60 and 70s, Lead batteries were used in forklifts in shop floors & warehouses as well as in miner's cap lamps. During those days, a very small number of cars, two wheelers, trucks and buses were also manufactured in India. The 80s saw the entry of Suzuki of Japan joining hands with an Indian outfit Maruti, to roll out a very small compact car. A large number of Indians showed keen interest in these cars, with the result there was a long waiting time of several years before delivery. This enterprise, Maruti Suzuki Ltd brought about a churning and a revolution in India's automobile story.

Today India has particularly all the well known global players in the automobile segments such as passenger cars, commercial vehicles, two & three wheelers, SUVs, buses, trucks, tractors etc. A few years ago, India displaced Germany as the 4th largest automobile producing country in the world and is poised to become No.3 in the coming years. Severe power cuts in summer months across the country resulted in the rapid growth of Lead battery powered inverters as energy backups in India. This was followed by the arrival of the computer era which again led to the massive penetration of Lead battery powered UPS. Mobile phones brought along the necessity of telecom towers, at the base of which you have a bank of Lead batteries in air-conditioned cabins. Lead batteries are also used in railway coaches, defence communication, submarines etc., After the Paris Round, currently India is investing heavily in the renewable energy space, both solar as well as wind, where Lead batteries will be used for energy storage purposes. In order to meet the fresh challenges due to climate change as well as to bring down the imports of crude oil, India has recently launched the National Electric Mobility Mission, by which electrification of vehicles will be accelerated; the Lead battery powered e-rickshaws & electric two wheelers will see a rapid growth in the coming years. As a result of all the above developments, Lead acid battery practically touches our daily lives in several ways (Fig. 14.1).

Approximately 85% of the Lead consumed in India goes for manufacturing Lead acid batteries. The current Lead battery market size is estimated at US $10 Billion approximately. At present India's consumption of Lead, as per industry estimates, is put at 1.2 million tons of which about 225,000 comes from primary Lead produced by Hindustan Zinc Ltd. About 800, 000 tons of lead comes from the secondary Lead industry, both formal as well as informal. The balance accounts for imports by India.

From the sixties, India was recycling used Lead batteries, though very small quantities, in a primitive manner, with low recoveries and more emissions, due to a lack of appropriate technologies as well as environmental awareness. Because of the serious environmental and health risks, the Supreme Court of India banned imports of used Lead batteries and other hazardous wastes in 1996. As a result, the Lead battery sector as well as the Lead industry were in doldrums affecting transport, power, telecom, defence, railways etc.

In order to help the ailing lead industry, ILZDA organized an "International Conference on Lead & Zinc Recycling—Technology & Environment" at Delhi in

Fig. 14.1 Multiple applications of Lead batteries

1998. The conference deliberations decided to create an appropriate framework for ensuring a "close loop" arrangement for an effective collection and environment-friendly recycling of used Lead batteries. As a result, the Ministry of Environment, Forests & Climate Change (MoEF&CC), Govt. of India set up a "Core Group" of various stakeholders, including ILZDA to identify the required actions & strategies in India's interest. This ultimately resulted in the following steps:

Regulating Lead battery auctions

↓

Bring out Battery (Mgmt& Handling) Rules

↓

Registration of Eco-friendly Lead Recyclers

↓

Strict implementation of the norms & rules

↓

Monitoring/Fine tuning for improvements

In the earlier years, the auctions by bulk consumers were attended by middlemen, scrap merchants etc., and they were picking up the used Lead batteries and feeding informal or backyard recyclers. Therefore, the new regulation stipulated that only registered/authorized Lead recyclers could participate in such auctions (dissuading the middlemen, traders etc.,) so that the lead recyclables go to the eco-friendly recycling units only.

After a series of "CORE GROUP" meetings for two years, the MoEF&CC brought out "Battery (Management & Handling) Rules 2001" which included "Extended Produced Responsibility" and covered all stakeholders of lead batteries i.e., manufacturers, dealers, importers (of new batteries), battery assemblers, re-conditioners, auctioneers, individual consumers and bulk consumers; the main aim is to collect the old battery against the sale of the new battery on a "one-to-one" basis and to ensure that they are all processed by registered eco-friendly lead recyclers only. B(M&H)R also mandated that battery manufacturers should file returns with the State Regulatory Boards on the no. of old batteries collected as well as new batteries sold. State Regulatory Boards were requested to send these returns to the MoEF&CC so that there is a clear picture on the national inventory every year. The B(M&H)R also encouraged setting up collection centers across the country for used Lead batteries. The battery collection targets fixed in the rules were:

1 Year (2002): 50%

II year (2003): 75%

After III Year (2004): 90%

Ultimately, the aim was that India should collect back 100% used Lead batteries and send them for environment-friendly recycling only.

A registration or authorisation committee consisting of experts including ILZDA, which used to go through the applications of lead recyclers, visit the plants for effecting improvements, used to give registration/authorization to such eco-friendly units, on a case to case basis. Units using blast furnaces, the so called "Mandir Bhatis" which showed leakages, later on, shifted to close door operations like rotary furnaces (Fig. 14.2).

Even after the implementation of the B(M&H)R, the backyard lead recyclers were thriving and active because the battery dealers were diverting the collected batteries to

Fig. 14.2 Rotary furnace for Lead smelting

the traders and backyard recyclers. In order to check this trend, the battery dealers in the country were asked, through an amendment in 2010, to get themselves registered with the respective State Regulatory Boards and to file returns. The returns would indicate the number of batteries sold as well as collected batteries and also its pathway i.e., they are being sent to registered environment-friendly recyclers only.

Likewise, all the importers of new lead batteries were also mandated, through the same amendment in 2010, that they should also be registered with the State Regulatory Boards and file returns providing information on the numbers of batteries imported as well as collected and sending the collected old batteries to the registered lead recyclers only. In the same year 2010, the responsibilities of the Registration Committee for Hazardous Wastes as well as the implementation of B(M&H)R were shifted by MoEF / Central Pollution Control Board to the State Pollution Control Boards, for implementation at the state levels.

Thus as a policy, India has introduced an excellent package of initiatives for an organized collection and environment-friendly lead battery recycling. From now on, we should take those initiatives to their logical conclusions by taking the following affirmative actions:

- Voluntary Industry Initiative
- Strict enforcement/monitoring by State Regulatory Boards
- Focus dealers and importers
- Dissuade role of traders
- Tighten backyard smelting
- Encourage collection centers
- Stringent customs clearance (imports)
- Continue awareness programs
- Provide incentives for green recycling
- Introduce cleaner recycling technologies
- Implement occupational exposure precautions
- Recognize & motivate clean operators

The above measures should make India a country adopting green technologies for an organized collection, safe storage and transportation as well as eco-friendly Lead battery recycling, in the true spirit of Sustainable Development, thus proving that Lead is the best example for Circular Economy.

14.3 Summary

The demand for lead acid batteries in the automotive sector continues unabated, while the industrial battery segment adds further impetus. Electric vehicles and renewable energy are high thrust areas of the Government of India and plans for these are in fact already at the various stages. The demand for lead will therefore increase by leaps and bounds with an increase in the vehicle production, electric scooters, new telecoms networks, inverters and other sectors dependent on lead acid batteries. This trend is a clear indication of the fact that the country needs to start preparing to meet the increasing demand of lead in the coming years.

Chapter 15
Defence Metallurgical Research Laboratory: Relentless Journey Towards Materials Galore

G. Madhusudhan Reddy

15.1 Introduction

Defence Metallurgical Research Laboratory (DMRL) at Hyderabad had its origins as one of the earliest laboratories under the umbrella of the Defence Research & Development Organization (DRDO) during 1958. DMRL was initially carved out of the Technical Development Establishment (Metals), which was earlier called the Inspectorate of Metals and Steel, situated at Ichhapore (near Kolkata). Upon subsequent shifting to Hyderabad during 1963, DMRL made rapid strides in the development

G. Madhusudhan Reddy (✉)
Former Director, Defence Metallurgical Research Laboratory, Hyderabad 500058, India
e-mail: director.dmrl@gov.in; gmreddy_dmrl@yahoo.co.in

R. Divakar et al. (eds.), *Indian Metallurgy*, Indian Institute of Metals Series,
https://doi.org/10.1007/978-981-99-5060-7_15

of several strategic materials of particular interest to defence with applications in aircraft, ships, tanks, electronics, and missiles.

DMRL, tasked with the primary objective of design and development of advanced metallic, ceramic and composite materials, had also contributed significantly to the related processing technologies to meet the needs of the system laboratories of DRDO and armed forces as well. DMRL developed numerous strategic materials/technologies with specific applications of interest to defence forces during its nearly six decades of service to the nation—DMRL's R&D found direct applications in aircraft, ships, tanks, electronics, and missiles too. Simultaneously, DMRL's scope of functioning continued to grow encompassing friction materials, heavy alloys for armaments, steel projectiles and armour, ultra-high strength low-alloy steel, titanium and titanium-based alloys, nickel-based super alloys, investment casting of super alloys for aircraft applications, magnetic materials, etc. Over the years, the laboratory has acquired a special status as a premier centre for R&D in metals, alloys, ceramics and composites. Currently, the multidisciplinary core competence at DMRL includes diverse areas such as product engineering, production support and performance analyses on metallic/alloy/composite parts, process development & surface engineering, extractive metallurgy of Ti/Mg/Rare-earth metals, design & development of speciality alloys/intermetallics/ceramics/composites. The laboratory's efforts have resulted in significant contributions to DRDO systems, the armed forces, and civilian spin-offs as well. DMRL has also made a very prominent long-standing niche for itself in Failure Analysis support to the armed forces, apart from several civilian spin-offs such as the development of turbine blades for civilian energy applications, Cu-Ti anti spark tools, and many bio-medical devices.

The R&D contributions at DMRL have led to the formation of new technology and production centres in the country. They are: Mishra Dhatu Nigam (Midhani), Hyderabad, Heavy Alloy Penetration Plant (HAPP) an Ordinance Factory, Tiruchirapalli, Non-ferrous Technology Centre (NFTDC), Hyderabad, and International Advanced Research Centre for Powder Metallurgy Materials (ARCI), Hyderabad. Today, each of them is a nationally important organization in its own right.

15.2 Core Competence

Over the years, DMRL has established core competence in the following areas of metallurgical engineering and materials science:

- Knowledgebase in process-structure–property-performance relationships of advanced materials
- Design and development of speciality alloys, intermetallics, ceramics and composites, process development and surface engineering
- Extractive metallurgy of titanium, tungsten and magnesium
- Product engineering, production support and performance analyses of metals, alloys and composites

The laboratory's efforts are directed towards the development of capabilities and technologies that are crucial for high-end defence systems, but not readily available from foreign sources due to strategic or other reasons. Over the years, DMRL has developed and matured several technologies catering to the requirements of systems for various defence applications in crucial areas such as armour, ammunition, missile, aircraft, and naval systems.

The significant scientific and technological achievements of the laboratory in recent years are briefly summarized in the following pages.

15.3 Protective Armour Technologies

Over the past four decades, under a number of technologies driven projects, DMRL has indigenously developed and provided protective armour technologies for a wide range of combat platforms for all of our tri-service armed forces. The most notable among them include composite armour for main battle tanks (MBT), infantry combat vehicles (ICV), helicopters, body armour, upgrades for ICVs, and Wheeled Armour Platform (WhAP).

Protective armoured vehicles: (a) MBT, (b) Helicopter, and (c) WhAP

Several advanced metallic, ceramic and polymeric materials have been developed by DMRL, in close collaboration with various industries and academic institutions. Each of these materials technology applications has emerged as an outcome from years of extensive efforts on process development, scientific understanding of process-structure–property relations, and critical analyses of the ballistic & impact immunity against various threats.

With a constant increase in the demand for increased ballistic protection with a relatively lower weight penalty, DMRL has constantly been pursuing the development of advanced protective armour technologies for next generation weapons platforms. Futuristic developmental efforts have been in place all along through progress in realization of advanced lightweight materials, supported by modelling & simulation, optimally scaled ballistic & blast tests, and advanced imaging & diagnostic techniques-thereby contributing to India's complete self-reliance in this highly specialized domain.

15.4 Ammunition Technologies

Catering to the requirement of ammunition materials for the Indian defence, the army, in particular, is one of the major thrust areas of research at DMRL. The earliest success story in this domain was in the development of tungsten heavy alloy cores (also known as long rods) in Fin Stabilized Armour Piercing Discarding Sabot (FSAPDS) for MBT during the 1980s. A production agency for FSAPDS named Heavy Alloy Penetrator Project (HAPP) was established at Tiruchirapalli and successfully mentored by DMRL, as part of a joint initiative between DRDO and OFB. Significant quantities of KE penetrators, meeting the specified Depth of Penetration (DOP), have been supplied to army subsequent to the inception of HAPP.

The continual emergence of improved levels of armour protection in the Armoured Fighting Vehicles (AFV) has given rise to demands for enhanced ballistic performance in penetrators in the form of relatively higher DOPs. The cores to meet the requirement of higher DOP penetrators were developed at DMRL, using a W–Ni-Fe-Co alloy subjected to double swaging with an appropriate inter-pass heat treatment, realizing improved properties at desired lengths thereby resulting in successful firing trials.

DRDO has also initiated the development of an enhanced penetrator, in terms of consistent performance and higher DOP. The improved design relies heavily on cores of higher density and thus higher tungsten content. This has largely been achieved by optimizing the processing parameters. The development of cores being successful, a decision has been taken to validate their effectiveness through firing trials.

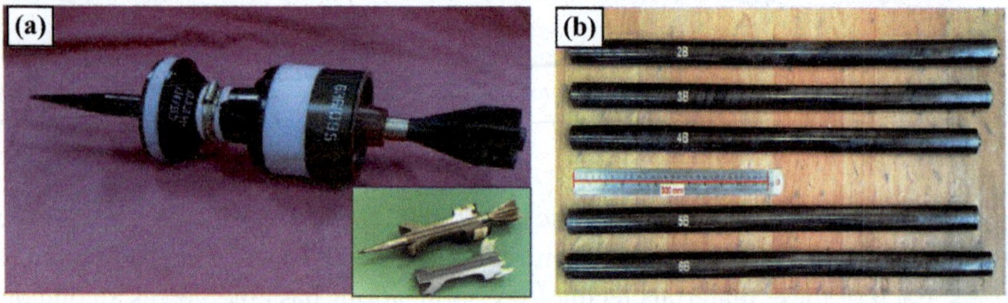

KE penetrators: Full Assembly: (a) FSAPDS and (b) WHA penetrator (long rod)

15.5 Shaped Charge Warhead Liner Technology

Shaped charge warhead liners are used in the weaponization of small & medium calibre armaments. DMRL has established the expertise in resolving metallurgical concerns through scientific determination of material chemistry, development of thermo-mechanical schedule to achieve the desired structure and properties, and development of suitable fabrication methods to realize the liner geometries. The tantalum explosively formed penetrator (EFP) liners developed by DMRL were successfully deployed in the technology demonstration of multi-layer penetration by the static missile systems developed by DRDO. Further, the EFP liners developed from the high purity Iron plates developed by DMRL were assessed for applications in Sensor Fused Munitions (SFM).

Shaped charge liners

Presently DMRL is working on the processing technologies for liner materials to facilitate the development of state-of-the-art warhead technologies, such as the Follow Through Warhead (FTW) technology and Low L/D shape charge warheads technology. Additionally, the processing route for certain advanced liner materials has also been successfully established.

15.6 Technologies for Naval Steels

DMRL has successfully developed technologies for the indigenous production of low-alloy speciality steels in the form of plates and bulb bars for naval applications, with the R&D efforts initiated during the year 2000. The initial lab-scale development was carried out in collaboration with NMRL, while the scale-up to industrial production was realized through SAIL and a few private firms as well. The Indian Navy strongly supported DMRL, both during the development and production efforts, finally culminating in the acceptance of the steels for naval use after due qualification. The steels are currently being used for the navy has also designated DMRL developed low-alloy steels as the standard default material for all new constructions and repairs of warships in future.

A large quantity of these steels, to the tune of thousands of tonnes, has already been supplied so far for fabrication at the industries. The indigenous plates and bulb bars are considerably cheaper than their imported equivalents. This has benefited the nation through substantial savings in fabrication cost and foreign exchange requirements.

INS Vikrant – the first Indigenously Aircraft Carrier developed by India at Cochin Shipyard

Type certification of the steel plates and bulb bars, which is mandatory for its underwater applications and simultaneously, efforts at establishing the indigenous production of high strength steels and their certification for underwater applications has been concluded, sucessfully achieving all major objecvtives., i.e., estabilshed indigenious production of high strenth steel semi-products under Russian consultancy using indigenous facilities.

15.7 Investment Casting Technologies for Superalloys

DMRL has been involved for the past several decades in the development of various technologies towards the production of Ni-base superalloy components, for advanced gas turbine engines, through vacuum investment casting route. These hot section components such as turbine blades, vanes, and rotor blisks are extremely complex in their geometry and characterized by close dimensional tolerances. During the initial stages (up to 1990s), equiaxed solid aerofoil components such as low and high pressure turbine blades were developed and supplied for aeroengine development efforts of DRDO. Equiaxed casting technology was also used recently for the development and supply of several critical components for a small turbo-fan engine (STFE) developed at GTRE. As the requirement for components with higher temperature capabilities increased, the technologies for the production of hollow components with complex internal air cooling channels were developed. Simultaneously, during the 1990s, the task of development of ceramic cores for investment casting of hollow aerofoil components was undertaken by a dedicated team at DMRL.

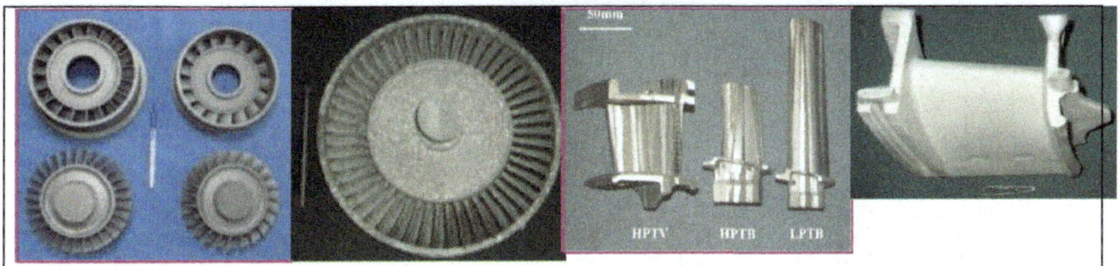

(a) **(b)** **(c)** **(d)**

Some of the gas turbine engine components developed at DMRL: (a) equiaxed components for jet fuel starter for Tejas aircraft, (b) low pressure turbine rotor blisk for small turbo-fan engine (STFE) used for powering unmanned air vehicles (UAVs), (c) DS blades and vanes for Kaveri engine and (d) SX vane casting currently under development for use in advanced aircraft engines.

During the development of Kaveri gas turbine engine (1990–2000), another technological milestone was surpassed in the form of development of a large number of directionally solidified hollow aerofoil components (such as the low & high pressure blades and vanes) for GTRE. Subsequently, the directional solidification (DS) casting technology was transferred to HAL, Koraput for industry-scale production of components. During 2015, the strategically important and unique single crystal (SX) component technology was also developed at DMRL. The demonstration of industrial viability of this technology is presently underway for the production of SX blades and vanes in large numbers for meeting GTRE's requirements. Additionally, DMRL is also working on futuristic technologies such as the production of double-wall SX aerofoil components, which are expected to possess significantly higher temperature capabilities in comparison to their single-wall counterparts. Further, environment friendly technologies such as recycling of pattern waxes and superalloy scrap for the production of aeroengine components are also being pursued at DMRL.

15.8 Life Prediction Technologies for Aero-Engine Components

Gas turbine aero engines accumulate damage in service as a result of their demanding operating conditions. This damage can manifest in several forms depending upon the component, engine type, and its operating environment. Replacement of service-damaged parts is expensive and is a significant factor to reckon in the life cycle cost of engines. Structural Integrity may broadly be quantified in terms of permissible usage in the service environment before crack nucleation (Safe-Life), crack propagation (Damage Tolerance (DT)), and evolving damage (Continuum Damage Mechanics (CDM)). The gains due to migration from safe-life to DT to CDM in the context of an aero engine lifting, with particular reference to fatigue loading and consequent failures, are quite significant.

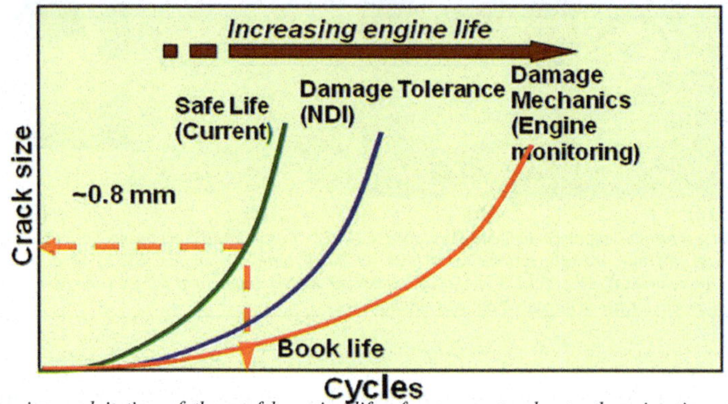

Increasing exploitation of the useful service life of components, due to the migration of design philosophy from Safe-life to DT to CDM.

DMRL has taken up evaluation of the potential for revision of Time Between Overhauls (TBO) of a unique Thrust Vector Control (TVC) jet nozzle of an advanced fighter aircraft engine and the Total Technical Life (TTL) of a transport aircraft engine—both systems being essential for IAF from an operational point of view. Both the tasks were successfully executed and the TBO life of TVC jet nozzle was doubled from OEM recommended limit and implemented in IAF fleet, after certification by CEMILAC. The upward TTL revision by about 15% for the transport aircraft engine was scientifically established and CEMILAC's approval is awaited for implementation in IAF. Simultaneously, the Science & Technology aspects of Remaining Life Assessment (RLA) of high temperature gas turbine components were also established.

A project on revision of TTL of an advanced gas turbine engine has been taken up by DMRL and is currently in progress. The project is aimed at developing a scientific understanding of assessing the remaining life of in-service jet engines powering advanced fighter aircraft. The methodology to be followed for RLA involves application of a variety of modelling techniques in combination, to ultimately quantify the level of degradation in materials of critical components as a function of service usage. This is a highly multidisciplinary and multiagency project, with DMRL functioning as the nodal lab for execution under aggressive PDC timelines-to cater to the immediate operational requirements of IAF.

15.9 Extraction of Reactive and Refractory Metals

15.9.1 Titanium Sponge

Titanium sponge is the nascent form of titanium metal, obtained through the high temperature reduction of $TiCl_4$ (Kroll process). DMRL has developed state-of-the-art Kroll technology in 3500 kg batches through extensive experimentation, equipment engineering, scientific understanding of the process technologies, advanced process instrumentation, and exhaustive data logging. The technology was demonstrated at DMRL on a lab scale and subsequently transferred to KMML, Kerala for establishing the titanium sponge plant for indigenous production. The plant has a capacity of 500 MT/year and was supported through funding by VSSC. The plant is now in regular operation and meeting the requirements of aerospace and defence programmes of the country. DMRL has also completed type certification of the titanium sponge produced at KMML for critical aero and naval applications, in close association with KMML and VSSC. DMRL has also filed an Indian patent on this process development.

High purity titanium sponge cake (3500 kg)

500 tpy titanium sponge plant at KMML established with DMRL technology

15.9.2 Magnesium

Fused salt electrolysis of anhydrous magnesium chloride ($MgCl_2$) a by-product generated during the production of titanium sponge, is developed to produce magnesium metal and chlorine gas, as an integral part of the commercially economical production of titanium sponge. The process essentially consists of the electrolysis of $MgCl_2$ in a pressure tight electrolytic cell operated at 690–700 °C. DMRL carried out extensive pilot plant operations of mono and multi-polar magnesium cells with a view to develop the technology for the production of magnesium metal at the rate of about 300 kg/day. Different types of cells were designed, developed and operated. The salient features of the process technology include molten salt feeding system, two module 5-bipole electrolyte zones, and a fool proof chlorine collection & neutralization system. Several hundreds of tonnes of $MgCl_2$ in each cell operation helped in the generation of valuable process engineering data. DMRL signed MoUs with NFC, DAE and VSSC (ISRO) for establishment and operation of Mg pilot projects with the technical knowhow developed at DMRL.

15.9.3 Tungsten

Tungsten metal is of significant strategic interest to defence and India has been importing the metal for the manufacture of various types of ammunition systems for the armed forces, such as FSAPDS at HAPP. Since China currently dominates global supplies of tungsten metal powder, it is felt essential to develop the indigenous tungsten supply chains.

DMRL in collaboration with two CSIR laboratories developed the processing technologies for the production of tungsten metal from heavy alloy scrap, Ammonium Paratugnstate (APT), and tailings material (process waste) accumulated at Hutti Gold Mines, Karnataka. Based on the pilot plant data of the extraction/recycling process technologies, it is proposed to establish and operate a technology demonstration plant of 100 MT/year capacities in association with an interested production agency.

15.10 Titanium Alloys for Aircraft Components

Over the past few decades, DMRL has been the nodal agency for the indigenous development of titanium alloys for defence applications. The primary alloy melting, processing, and characterization was routinely carried out at DMRL followed by industrial scale melting and processing at MIDHANI. DMRL has successfully undertaken alloy development and the realization of relevant components for the following alloys. (i) Ti-10V-2Fe-3Al is a high strength metastable beta Titanium alloy for closed-die forged components. Under the ADA funded project, melting and thermomechanical processing of cylindrical bars, to diameters of 60 and 100 mm, was completed together with type certification for their future programmes. The alloy will replace many steel components, resulting in considerable weight saving (ii) Titan 44 is a beta Titanium alloy used for applications in sheet form. It possesses excellent oxidation resistance because of higher Mo content. Despite being a beta Titanium alloy, its creep properties are comparable to those of Ti-6Al-4V. The technology for melting was established overcoming the challenge of non-availability of proper Al-Mo master alloy. A pilot melt has been processed to 2, 1.2 and 0.8 mm thick sheets to sizes 1100 × 1000 mm together with extensive property evaluation for aerospace applications. Two more melts and complete type certification are proposed to be undertaken in a new project.

Currently, DMRL is pursuing the Titanium alloy development and component manufacturing technology for AMCA structural application that involves: (a) Development of Ti-5Al-5Mo-5V-3Cr high strength deep hardenable beta Titanium alloy slabs of 200 mm thickness and higher for aircraft forging applications, (b) Development of Ti-6Al-4V slabs of 1300 × 1200 × 120 mm size, (c) Large size closed-die forgings of Ti-6Al-4V, (d) Development of investment cast components of Ti-6Al-4V, and (e) Manufacture of large size bulkhead frame from Ti-6Al-4V plates and forged sections by machining and welding. Presently DMRL is also providing expert consultancy for indigenous development of alpha Titanium alloys viz. BT1-0, PT-3B and PT 7M in various product forms for naval applications.

15.11 Powder Processing Technologies

DMRL has established unique Hot Isostatic Press (HIP) and Cold Isostatic Press (CIP) facilities during the mid-70s, for the manufacture of critical rotating components of defence systems. A broad-based expertise on several aspects of the isostatic pressing technology for different high temperature materials has been developed. DMRL has launched a hardware oriented indigenous R&D effort to establish a unique P/M (HIP) processing technology for the manufacture of Integral Turbine Rotors. The HIPed integral rotors were successfully produced out of stainless steel for one of the missiles and Ni-base superalloy rotors (both in monolithic/dual-materials) for space applications (GSLV, ISRO). DMRL has also established an advanced HIP technology for defect healing of investment cast components from Aluminium alloys, Ni-base superalloys and Titanium alloys for applications in military aircraft, missiles, and submarines respectively.

The HIP technology of DMRL was also adapted to the diffusion bonding of high strength special steels and fabrication of Test Blanket Modules (TBMs) for applications in fusion reactors. DMRL has recently established a sophisticated Inert Gas Atomisation (IGA) facility. Production of high-quality powders from different grades of superalloys for aeronautical applications (through HIP & 3D Metal Printing) has been demonstrated successfully. Simultaneously, indigenous development of P/M superalloys was also pursued through modelling and simulation resulting in the identification of a set of 5 new optimized alloy compositions. The development, production, and processing of these new alloy powders will be carried out by adopting the *Integrated Computational Materials Engineering* (ICME) approach to significantly save on the developmental time, funds, and efforts.

As part of one of the current projects, DMRL has taken up an exciting R&D effort towards the development of P/M Ni-base superalloy aeroengine discs with improved performance to meet the futuristic needs of military aircraft and helicopters. The sub-size superalloy N18 discs with targeted properties have been realized through innovative HIPing and near isothermal forging, duly followed by a novel heat treatment procedure. The technology established under this project will benefit immensely the aeroengine programmes of DRDO.

In the recent past, DMRL has developed a unique technology based on CIP and Sintering for the manufacture of fused silica radomes for high-speed target seeking missiles. This CIP technology has been transferred to major industries for industry-scale production. Further, DMRL has taken up an independent project for the development of ceramic radomes with high strength and high temperature capability for application in hypersonic missiles.

15.12 Materials for Hypersonic Vehicles

The objective of the project *HYPERMAT* was to develop materials and coatings for long duration hypersonic vehicles, apart from establishing specialized test facilities for their characterization. Owing to the severe aerodynamic heating during flight, the surface temperatures are expected to reach up to 1300 °C at the critical locations on control surfaces such as leading edges, nose tips, etc. Technology has been developed at DMRL to produce C_f-SiC panels as well as nose tip and leading edge shaped composites. A totally indigenous laboratory-scale CVI-CVD facility has also been established. Technology has been developed to synthesize the ZrB_2 powder on a 5 kg batch scale and to vacuum hot press it into 150 mm diameter discs, when temperatures encountered are expected to significantly exceed 1300 °C. Box type, truss type, and foam based metallic thermal protection system (MTPS) of cross-section 300 × 300 mm have been developed to protect the airframe. Technology has been developed using electron beam melting/vacuum arc melting to produce niobium based Cb 752 alloy sheets of 300 × 300 × 4 mm size possessing target composition and properties for the scramjet combustor. Protective coatings have been developed at coupon-level for all the materials developed in the project. All the important characterization facilities such as Gleeble, Laser Flash, IR Heating, and the Induction Plasma Erosion and Coating (IPEC) facility as envisaged in the project have been established and used extensively for the characterization of the above material. Since all the objectives of this project have been successfully realized, DMRL is currently working on developing an in depth understanding on manufacture and scale-up related issues in respect of the above materials. This is expected to be a significant step in the realization of the crucial hardware for the development of futuristic hypersonic vehicles.

15.13 Liquid Silicon Infiltration (LSI) Technologies

15.13.1 Reaction Bonded Silicon Carbide (RBSC)

Silicon Carbide is the most widely used structural ceramic material, for both ambient and high temperature applications. Among the two processing routes for consolidation of Silicon carbide, processing of Sintered Silicon Carbide (SSC) involves sintering temperature >2100 °C and >20% resultant volume shrinkage. On the other hand, processing of Reaction Bonding of Silicon Carbide (RBSC) is carried out at 1500–1600 °C by Liquid Siliconisation Infiltration (LSI) and the resultant volume shrinkage is marginal (~1–2%). DMRL has taken up the development of Reaction Bonded Silicon Carbide (RBSC) bearing bushes and thrust bearings to circumvent the restrictions imposed on the import of these critical components for the indigenous nuclear submarine program.

DMRL has established the processing facilities and optimized the plan for realization of RBSC with excellent high temperature strength in excess of 1000 °C, very high hardness (2500 VHN), and wear resistance. In the Reaction Bonding process, molten silicon infiltrated into the porous SiC/C preform, chemically reacting with the carbon to form secondary grains of SiC to bond the primary grains of SiC resulting in a near-net shaped dense RBSC product. However, the presence of some amount of free silicon limits its application temperature to below the melting point of Silicon (1410 °C). The RBSC Bearing bushes and Thrust Bearings developed by DMRL have been fully qualified by DMDE before their induction in the main feed water pumps for submarines. The RBSC technology was transferred to International Advanced Research Centre (ARCI) and industry through a MOU to fulfil the requirements of Navy on royalty sharing basis.

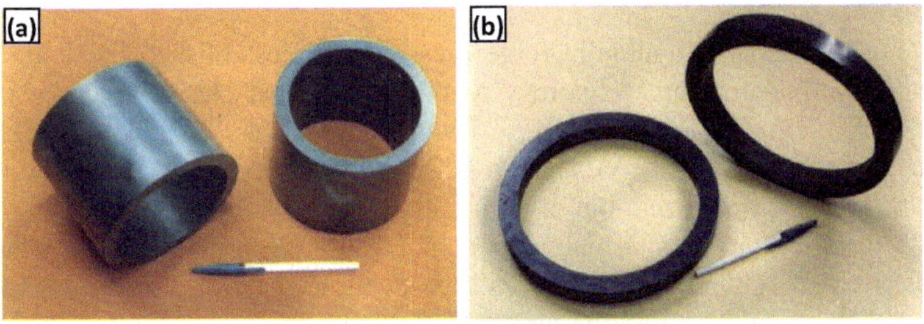

Reaction Bonded Silicon Carbide (RBSC) bearing parts: (a) RBSC Bearing Bushes and (b) RBSC Thrust Bearings

15.13.2 C_f-C-SiC Jet Vanes

Jet Vanes are the part of Thrust Vector Control (TVC) mechanism in interceptor missiles, where direction of flow of exhaust gases is controlled by a set of jet vanes to steer the missile in the required direction. The jet vane material should withstand highly erosive combustion gases with marginal levels of erosion at extreme temperatures and high speeds. DMRL, together with DRDL and ASL, developed a new composite material C_f-C-SiC. The expertise and facilities of DMRL in the area of Liquid Silicon Infiltration (LSI) were employed in the rapid development and validation of C_f-C-SiC composites for jet vane applications. These low density vanes could successfully withstand the high rate of erosion (due to the presence of SiC), high thermal shock (due to the presence of highly conductive carbon fibres), and also retain their high temperature strength.

The C_f-C-SiC vanes have been static tested for their performance at high temperatures using char motors and six component test beds. A protective collar was incorporated in jet vanes to overcome the jamming of the jet vane at its bearing due to the molten alumina present in the exhaust, enabling DRDO to successfully accomplish the first flight trial of interceptor missile. The effective team work by DMRL-DRDL-ASL resulted in achieving self-reliance in the area of C_f-C-SiC jet vane technology.

This material has become the lifeline for several other missile systems too. To take this development forward, ASL established the production facilities to exploit the LSI processed C_f-C-SiC composites to cater to the requirements of missile complex, thereby saving a significant amount of foreign exchange.

15.14 Near Net Shape Technologies

Using a unique 2000 MT hydraulic forge press, DMRL established the indigenous near isothermal forging technology to produce critical class-I components through a scientific methodology involving Dynamic Materials Modelling (DMM) and Finite Element Analysis (FEA). The rotor disc of 1–5 stages of the high-pressure compressor (HPC) was realized, out of a difficult-to-deform titanium alloy similar to IMI-685 for a military gas turbine aeroengine. The indigenous technology has been transferred to MIDHANI and the bulk production of these disc forgings is successfully being carried out by the firm using DMRL's forge press on cost basis. On similar lines, (a) HPC discs and compressor front shaft out of titanium alloy equivalent to IMI 834 for Kaveri Engine and (b) HPC components for aeroengine have also been forged, qualified, and supplied to GTRE. Indigenous technology to manufacture a mountaineering accessory Karabiner, out of an aluminium alloy has also been established and the technology was transferred to the Ordnance Factory Dehradun. Till date, 59,228 nos. of the Karabiner have been manufactured by the industry and supplied to the Indian Army.

15.15 Electronic Materials

DMRL has been working on electronic materials over the past two decades, a first of its kind effort in the country, with a unique core competence in the nano-level characterization of electronic and opto-electronic materials. Several projects were successfully completed in collaboration with SSPL/GAETEC on high-frequency devices, photonic device structures, and functional materials. Strategic significance of the single crystal SiC applications encouraged the initiation of a research activity during 2009, partnering DMRL and SSPL, leading subsequently to a joint S&T project in 2015—with the objective of developing SiC single crystal bulk growth and wafer fabrication processes. DMRL thereby successfully established the single crystal SiC technology, in association with SSPL and C-MET. Several issues with the technology were effectively addressed including the SiC single crystal growth, SiC wafer fabrication, SiC material's physical/electrical properties' characterization, process modelling, and deposition of the GaN epitaxy on indigenous SiC wafers. Presently, efforts are in place to develop the SiC devices for defence applications—especially sensors for harsh environments and high-power devices. A multidisciplinary effort is also being launched in association with numerous R&D labs/academia/foundries with relevant expertise in the domain.

15.16 Metal Foam Technologies

DMRL had successfully established the technology to produce closed-cell aluminium foams through the melt route, by way of mixing fine TiH_2 (Titanium Hydride) particles into the aluminium melt. Numerous fine cells are generated in the melt due to the release of hydrogen from the particles, which subsequently grow and form cells of 2–4 mm size. Aluminium foam ingots of 175 $\times$ 175 mm cross-sections of 400–700 mm lengths, plates of 300 $\times$ 300 $\times$ 100 mm sizes, and with a density of 0.3–0.7 gm/cc were successfully produced through this technology. Due to the unique hexagonal structure of the cells, foams could absorb high impact and blast energy at a low transmitted stress. Product shape, size, cell uniformity, and consistency in the foam quality are important factors from the application stand point. These specifications were successfully achieved through several process optimisation experiments and innovations. Modelling and simulation were also carried out simultaneously to explain the mechanisms of cell structure evolution, apart from understanding the deformation mechanics of cells in impact and blast situations to enable the eventual optimisation of component designs.

Al-foams produced through melt route: (a) Ingots, (b) Plates; Impact absorbers given to BARC (c) 285 x 285 x 400 mm size, (d) 300 dia. X 400 mm length size and (e) Energy absorbers designed by DMRL / DRDL

Various foam-based impact absorber components were fabricated and supplied to several DRDO laboratories and external agencies viz., OFB, BARC, TBRL, HEMRL, and DRDL for various applications' trials. DMRL also provided an aluminium foam impact absorber component to BARC, which could successfully absorb 90 kJ of energy at an average transmitted force of <100 gm force (as specified by the user). Similarly, for an application related to missile launches, an aluminium foam-based impact absorber with high L/D ratio was successfully designed and evaluated at high strain-rates, which displayed the desired energy absorption characteristics. The successful development of this technology and the increasing demands for it resulted in the initiation of ToT to industries, which is currently in progress. Two industries have come forward with interest in the technology and technical assessment of the industries is currently in place.

15.17 Advanced Magnetic Materials Technology

Over the past five decades, DMRL has developed technologies for the production of state-of-the-art permanent magnets, catering to the strategic needs of the country through AlNiCo magnets in the sixties and seventies. These achievements were followed by subsequent efforts on the development of more powerful Rare Earth based SmCo and NdFeB magnets. The contributions of DMRL were instrumental in meeting the requirements of defence R&D programmes in several navigational devices such as gyros, accelerometers & TWTs, and motor transmissions for torpedo launchers etc. The efforts at DMRL also resulted in the Transfer of Technology to M/s IREL (India) Ltd to establish first of its kind REPM production facility at Vishakhapatnam using indigenous resources.

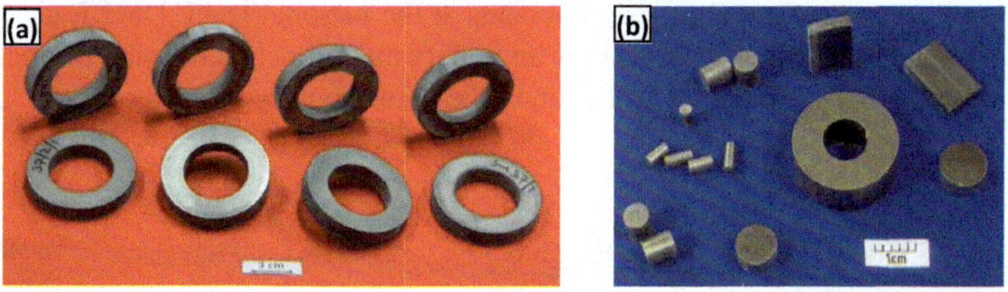

Typical permanent magnet components developed for applications in strategic technologies of defence, space and atomic energy sectors: (a) ring magnets and (b) various shapes of magnets

During recent years, DMRL has widened its focus on thin film magnetic materials for miniaturized sensor and actuator applications and a dedicated Class-1000 clean room facility has been established. DMRL has successfully developed different functional thin films and prototype devices based on spintronics, magnetostrictive, multiferroics, hard magnetic thin films etc. with desired functional properties for strategic applications. Typical devices have been successfully demonstrated for application in monitoring air pressure and controlling of wing movements in micro-aerial vehicle (MAV) which can be used for security surveillance and magnetic field sensors for naval applications. Stray magnetic fields of naval systems such as submarines and war ships need to be minimized in order to avoid detection by sea mines. These sensors will be useful in the detection and de-gaussing of stray magnetic fields of naval systems.

15.18 Advanced Metal Joining Technologies

DMRL has made significant contributions to resolving a wide range of fabrication issues by establishing welding technologies for similar & dissimilar welds of several advanced materials such as ultra-high strength steels, aluminium & titanium alloys, metal-matrix composites, ceramics and superalloys. Focussed research at the lab has promoted the engineering applications of these advanced materials in a big way through innovative fabrication techniques for realization of various defence systems such as combat vehicles, missile casings, base-mortar structures, nose-cap shell for an underwater launched missile, and compact heat exchangers. DMRL has also brought out the key influence of electron beam oscillation and pulsed laser beam techniques in reducing the Laves phase during the welding of superalloys, thereby improving the mechanical properties of aeroengine components. Further, a significant breakthrough was achieved by establishing ballistic capabilities in the welds, comparable to those of parent armour. DMRL's sustained initiatives, leadership, and dedicated efforts have resulted in the establishment of exclusive Friction-Stir Welding (FSW) &Processing Technologies, thus elevating DMRL to a unique position in the country in the manufacture of aerospace components at the national level. DMRL successfully fabricated the Nose-Cap shell for an underwater launched missile, through an innovative application of FSW for the first time in the country at the shop-floor level.

15.19 Tribological Technologies

DMRL has developed rigorous scientific understanding on the Tribological behaviour of advanced materials and associated coatings, in an effort to evolve state-of-the-art surfacing technologies for major systems for DRDO. Various types of coatings such as abradable, wear resistant, thermal barrier and dimensional restoration surfacing were developed.

A successful application of DMRL's research in tribology was towards the development of a solid lubricant for use in the canister of a long-range missile. The coefficient of friction between the Aluminium support blocks and canister of the missile had to be at a low value to meet the design requirements, that too at very high contact interface stresses. A suitable solid lubricant coating system has been developed on the Aluminium support blocks to reduce friction by 75%. The coated support blocks are now being used in all canisters of the missile successfully.

The interface frictional conditions undergo transition when the operating conditions such as stresses, temperature, and sliding speeds vary. DMRL has done innovative efforts in this regard on a variety of materials including the Carbon–Carbon composites and Metal Matrix Composites. Through meticulous and extensive research, it was discovered that C–C based materials undergo a transition in friction to higher levels when the interface temperatures reach a threshold value. This knowhow helped in scientific analysis of the premature failures of C–SiC based jet vanes in missile systems. Subsequently, a composite coating comprised of a high temperature solid lubricant was developed to successfully reduce the friction by half. This contribution also resulted in the enhancement of the factor of safety in the jet vane designs by over 200%. These efforts by DMRL played a key role in the qualification of the rocket motor of an advanced missile system.

The components operating under hypersonic conditions are exposed to very high velocity gases and consequent heating to extreme temperatures, thereby experiencing considerable heat flux in excess of 200 W/cm^2. DMRL has developed a state-of-the-art test facility known as Induction Plasma Based Erosion cum Coating (IPEC) facility. This test infrastructure can thus facilitate the simulation of very high Mach numbers, high enthalpy, and high heat flux. This experimental facility at DMRL is truly unique, going by the global standards too.

15.20 Additive Manufacturing

Additive Manufacturing (AM) or 3D printing (3DP) is the process of generating a three-dimensional solid object, of virtually of any shape, from a digital model. A distinguishing feature of the additive manufacturing processes is that the material is added layer-by-layer in successive patterns to produce a component to its final desired geometrical configuration, to replace the conventional material removal methods. Coupled with phenomenal developments in allied fields such as Laser & EBM technology, CNC, Modelling & Simulation, and Powder Metallurgy; 3D Printing offers enormous potential for extensive applications in the development of complex components for defence.

Realizing the technological significance and future potential for Additive manufacturing, DMRL has been working in this area since 2008—especially applications to metallic materials using the LENS technology. Over the years, DMRL developed expertise in the metal additive manufacturing of advanced materials such as superalloys, stainless steels, Ti-6Al-4V, etc. Tensile mechanical properties of the LENS deposited metallic materials are found to be on par/superior to their wrought counterparts. Functionally Graded Material (FGM) panels of YSZ-IN625 materials were developed using LENS for hypersonic combustor liner applications. Thermo-physical properties of these FGMs are found to be promising. The existing LENS-750 facility at DMRL is a 3-axis Directed Energy Deposition (DED) facility, where only relatively simpler shaped components may be generated. With the knowhow acquired in 3D Printing technologies, the development of a Fuel Nozzle component for Missile applications could be successfully accomplished by DMRL using the 3D Printing infrastructure at a collaborating agency. Innovative re-design of the component through appropriate modifications to its internal channels and manifolds enabled the development of 3D Printed unified assembly, without the need for support structures. Advantages realized through 3D Printing of the component include the assembly of 2 parts into 1, eliminating CNC machining & EB welding, and reducing material wastage.

The metal 3D Printing requirements of DRDO need to be addressed, offering end-to-end solutions together with complete process modelling, metallurgical, mechanical characterization, and qualification. Towards this objective, DMRL has embarked upon a project to develop different types of components, with widely varying sizes and complexities for defence applications, apart from the generation of standard alloy powders for 3D printing applications. Some of the major technological challenges identified to be overcome under this upcoming project include process development &standardization, process modelling of 3D printing for prediction of the properties, powder production technology for standard alloys, and qualification/certification methodologies for defence applications.

3D Printed Fuel Injector

15.21 Outlook for the Future

DMRL developed numerous strategic materials/technologies with specific applications of interest to defence forces over nearly six decades of its service to the nation—the R&D at DMRL had direct applications in aircraft, ships, tanks, electronics, and missiles as well. Simultaneously, DMRL's scope of functioning continued to grow encompassing friction materials, heavy alloys for armaments, steel projectiles and armour, ultra-high strength low-alloy steel, titanium and titanium-base alloys, nickel-base superalloys, investment casting of superalloys for aircraft applications, magnetic materials, etc. Over the years, the laboratory has acquired a special status as a premier centre for R&D in metals, alloys, ceramics and composites. Currently, the multidisciplinary core competence at DMRL includes diverse areas such as product engineering, production support and performance analyses on metallic/alloy/composite parts, process development & surface engineering, Extractive metallurgy of Ti/Mg/Rare-earth metals, design & development of speciality alloys/intermetallics/ceramics/composites. The laboratory's efforts have resulted in significant contributions to DRDO Systems, the Tri-Services, and Civilian Spin-Offs. DMRL has also made a very prominent long-standing niche for itself in Failure Analysis support to the armoured forces, apart from several civilian spin-offs such as the development of turbine blades for energy applications and many bio-medical devices. DRDO's mission on the design, development, and manufacture of advanced hardware equipment for our Defence Services often rests heavily on the timely availability of tailor-made optimized materials. Responding to the call for self-reliance in materials and allied technologies, DMRL has undertaken numerous flagship programmes making significant advances in the concurrent development of futuristic materials and related processing/manufacturing technologies for defence applications.

The development of new materials/technologies is considered successful only when the concept is converted into a realistic product, and the same is deployed into service with adequate confidence on its performance. However, the developmental times are very long due to delayed stage iterations in the developmental cycle and complexities inherent to material systems. In order to keep pace with changes in design, it is inevitable to accelerate the development of materials. This can be achieved through emerging novel approaches such as *Artificial Intelligence (AI)*, *Machine Learning (ML)*, and *Integrated Computational Materials Engineering (ICME)*. While the former two are essentially data driven, the latter can largely be based on physics-based models. More often, a judicious combination of all the above approaches is necessary for the accelerated development of optimized materials. ICME is a holistic way of seamlessly integrating materials development with the system's design. In other words, rather than being selected from the handbook/catalogue, material too becomes a design variable in the development of engineering systems. Therefore, materials are tailor-made for a specific application.

When the physical understanding on processing, material structure evolution and properties is at a mature level, several rigorous physics-based scientific models can be used to simulate the material behaviour. Each physics-based model can be visualized as a building block for ICME. Together with an appropriate integration platform, it is feasible to arrive at optimal materials solutions. Since ICME also includes performance models of the systems, it is in principle conceivable to solve the inverse problem of designer materials. Another complementary approach is based on application AI/ML techniques to operate upon databases. What is not comprehensible in terms of mathematical equations may be interpreted for inferences under the AI paradigm. With suitable structuring of data from multiple sources, an essential part of the material innovation infrastructure, digital data is created as illustrated in the figure below.

Since the curated data exists in the digital format, logical inferences may be drawn through ML using statistical algorithms and hence knowledge bases can be generated. Thus, mature AI systems can offer recipes to design newer materials in a very short time. A major breakthrough from adopting these approaches would be making materials' development keep pace with the design of engineering systems. Another aspect of AI is the design of intelligent experiments, based on prior data and autonomous reasoning of the cause-effect relationships.

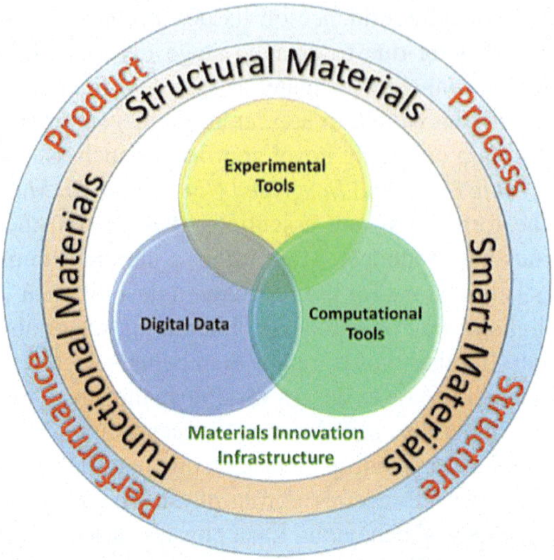

Materials Innovation Infrastructure covers process-structure-property-performance data

15.22 Summary

DRDO's mission on the design, development, and manufacture of state-of-the-art hardware equipment for our Defence Services hinges on the timely availability of strategic materials. Ensuring self-reliance in turn mandates indigenous development of all materials. Responding to the call for self-reliance in materials and allied technologies, DMRL has undertaken numerous flagship programmes making significant advances in the accelerated development of futuristic materials for defence/ engineering applications.

DMRL has traversed through nearly six decades in time, contributing effectively to scientific efforts in addressing the present and future requirements of materials for defence applications. The laboratory has done consistently good in basic science, its scale-up to the technology level, and realization of commercial production for self-reliance as well. The focus all along has been to evolve holistic materials solutions for strategic applications, through seamless multidisciplinary team work. DMRL's thorough research for defence has resulted in commendable civilian spin-offs too. With ever increasing requirements for tailor-made materials and allied technologies for critical applications, DMRL is embracing the application of emerging disruptive technologies such as ICME/AI/ML for accelerated development of optimized materials/products.

Chapter 16
Metallic Materials R&D in Vikram Sarabhai Space Centre: Past, Present and Future

S. Somanath, Roy M. Cherian, S. C. Sharma, and M. Mohan

16.1 Introduction

Achievements in aerospace science and technology are closely related to the successful development and production of engineering materials, meeting stringent quality requirements. Materials are the backbone of any engineering structure and aerospace materials technologies are closely guarded, in view of their dual use. Materials processing technologies are infrastructure intensive, unlike many other areas of research and are cost and time intensive as well. System designers specify the requirements for the development of materials to meet the project time schedules.

Indian Space Research Organization (ISRO)'s space projects such as the polar and geosynchronous launch vehicles, communication and remote sensing satellites, recoverable space capsule, interplanetary missions such as Chandrayaan and Mangalyaan, etc. are symbols of the country's pride and owe their success largely to the dedicated efforts of material scientists towards development and indigenisation of performance-critical aerospace materials and products.

The initiatives taken by ISRO in the early 1970s for the indigenous production of many of the structural and functional materials for the satellite launch vehicle program were the earliest seeds sown and these have yielded remarkable results in meeting the strategic material requirements of the present day, not only for the space sector, but also in atomic energy and defense programs of India. Technology denials have indirectly helped the country achieve self-reliance and in establishing large-scale facilities required for meeting the demands of the strategic sectors. Successfully meeting the critical requirements of all materials for Indian space programs indigenously has been made possible through the dedicated efforts of teams that overcame several hurdles and challenges. Figure 16.1 shows the evolution of expendable launch

S. Somanath (✉) · R. M. Cherian · S. C. Sharma · M. Mohan
Vikram Sarabhai Space Centre, Trivandrum 695022, India
e-mail: chairman@isro.gov.in

R. Divakar et al. (eds.), *Indian Metallurgy*, Indian Institute of Metals Series,
https://doi.org/10.1007/978-981-99-5060-7_16

Fig. 16.1 Evolution of expendable launch vehicles in ISRO. Materials have played a key role in the realization and success of these launch vehicles

vehicles in ISRO. It must be emphasized that materials have played a key role in the realization and success of these launch vehicles.

This article presents the research and development carried out in metallic materials in Vikram Sarabhai Space Centre, the lead centre of ISRO responsible for the design and development of launch vehicles.

16.2 Genesis of Metallic Materials Research at Thumba

Space research activities were initiated in India during the early 1960s by Dr. Vikram A. Sarabhai, the founding father of the Indian Space program. It was his vision to explore space and bring its benefits to national development and the driving force was his belief and conviction that India 'be second to none in the application of advanced technologies to solve the real problems of man and society'. Space activities in India started on a small scale with launching of a sounding rocket from Thiruvanantha-puram, the southern tip of India. The Indian space program took off as an organized activity with the setting up of the Space Commission and the Department of Space in 1972. Since then, India has made remarkable progress in establishing state-of-the-art space systems, which constitute an important element of national infrastructure. These necessitated major developments in the areas of materials, chemicals, processing technologies and related infrastructure.

In view of the non-availability of strategic materials due to their applications in dual-use technologies from international sources and with a vision to achieve self-reliance in the field of rocketry, indigenously developed materials were targeted from the initial stages of the space program. This demanded setting up of large-scale processing facilities and industrial participation in developing performance critical materials, chemicals and propellants. The last five decades of launch vehicle development in India have resulted in the development of several materials-related technologies and the progress has been truly phenomenal.

In the early 1970s, the facilities to produce space-quality materials were scarce and VSSC took the lead in the development of technologies for producing stringent space-quality materials (Suseelan Nair and Sinha 2009). Among them were structural, composite and special purpose materials including high strength low alloy steels, ultra-high strength maraging steel, aluminium alloys, titanium alloys, magnesium alloys, superalloys, powder metallurgy products, Beryllium, magnetic, electronic and optical materials.

Selection of materials for space applications involves simultaneous consideration of a number of conflicting factors such as high specific strength, high specific stiffness, ability to be fabricated into products through forming, welding, brazing, compatibility with corrosive propellants, etc., to name a few. ISRO made concerted efforts to indigenously develop (Shukla et al. 2014) and productionise these materials with the help of Indian industries and some of the major success stories of materials development are discussed here.

16.3 History of Development of Maraging Steel for Solid Rocket Boosters

Sounding rockets and early launch vehicle programs used AFNOR15CDV6 steels for the solid rocket motor casings. However, when the material for the large scale solid booster of PSLV was to be selected, the selection committee favoured a low-carbon iron-nickel martensitic steel that, in the presence of alloying elements, gets strengthened by precipitation hardening through the formation of a number of intermetallic compounds. The solid rocket motor casing design is based on fracture mechanics approach and the 18 wt% Ni (M250 grade) maraging steel has several distinct characteristics that make it an ideal choice such as high strength coupled with excellent toughness, good dimensional stability, ease of fabrication and simple heat treatment.

The development of maraging steels in VSSC started in the early 1970s through a button melting furnace. Since these are ultra-high-strength steels and need to be processed with controlled chemistry, with tolerable impurities and gas contents, facilities such as air induction melting, vacuum arc re-melting, forging press and rolling mill were installed and laboratory scale processing and characterization were completed by 1975. Towards the end of 1970s, melting of maraging steel at tonnage level could be started at M/s Mishra Dhatu Nigam Limited, Hyderabad and the material was realised for the successful fabrication and testing of a RH-300 rocket motor (300 mm diameter and 2.0 m in length). With materials processing technologies, such as melting, forging, rolling being established and fabrication technologies successfully demonstrated, it was proposed as a candidate material for the large PSLV solid rocket boosters. Considering the merit of the proposal, based on technical superiority over the other contenders, a national level committee under the chairmanship of Prof. Brahm Prakash selected M250 maraging steel for solid booster casings and this material is currently used in PSLV as well as GSLV and GSLV-MkIII. Maraging steel motor cases of 2.8 m diameter for PSLV and GSLV as well as 3.2 m diameter for GSLV MkIII are continuously been manufactured at M/s Larsen and Toubro, Mumbai and M/s Walchand Nagar Industries, Walchandnagar (Ghosh et al. 1987; Nageswara Rao and Narayana Murty 2019; Gupta et al. 2017).

Several technical problems cropped up during the developmental phase and subsequent production phases and were solved through experimental works at both VSSC and industries. Some of the problems were low fracture toughness in ring rolled forgings, appearance of white streaks in X-ray radiography of large-sized rings (Narayana Murty et al. 1997), HAZ II-parent material interface cracking (Ramesh Narayanan et al. 1990), need for repair welding and local aging for fully aged segments, observations of stress corrosion cracking (SCC) and hydrogen embrittlement (Narayana Murty et al. 2014; Manwatkar et al. 2016).

Presently, M250 grade maraging steel required for ISRO programs is primarily sourced from M/s MIDHANI in different forms. The steel is vacuum induction melted and vacuum arc remelted in dedicated facilities set up for the purpose and the billets are rolled at Rourkela or Bhilai steel plants of M/s SAIL, into the plates of about 8 mm thickness required for solid rocket motor casings. The large-sized rings for these pressure vessels are ring rolled at M/s Bay Forge, Chennai in a ring rolling mill set up for VSSC. Maraging steel strips for Mermen bands used in separation systems of launch vehicles are also rolled at MIDHANI. It is a matter of immense pride that the quality of M250 grade maraging steel produced by MIDHANI is far superior to that of any international supplier with respect to fracture toughness and India is fully self-sufficient in producing these performance-critical materials used not only in launch vehicles but also in several defense and nuclear systems.

Figure 16.2 shows maraging steel plate used for the fabrication of the solid motor booster cases, being rolled at Rourkela steel plant. As cobalt is one of the major alloying elements in maraging steels and as it is a strategic material with limited reserves concentrated in few countries, attempts have been made to replace cobalt with other elements and the result is the development of Co-free maraging steels. In these steels, cobalt is replaced by chromium along with an increase in titanium content. In order to further reduce the cost of solid booster case materials, 15CDV6, a high strength-low alloy steel (Beena et al. 1995) is modified by Niobium inoculation and increased carbon content to improve its strength and has been christened as ESR

Modified 15CDV6 or 0.3CrMoV steel (Manikandan et al. 2018; Ramkumar et al. 2021, 2017). Extensive development work has been carried out and technologies have been developed to realize solid motor case segments out of this novel grade steel which have been qualified for use in flight. Large diameter rocket motor casing made of ESR grade 0.3C-CrMoV steel being oil quenched is shown in Fig. 16.2c. The integration of the first stage of maraging steel motor case at SDSC-SHAR, Sriharikota is presented in Fig. 16.2d.

Fig. 16.2 Maraging steel used for the fabrication of the solid motor booster cases, being rolled **a** at Rourkela steel plant, Rourkela, **b** rings required for the fabrication of rocket motor cases being ring rolled, **c** large diameter rocket motor casing made of ESR grade 0.3C-CrMoV steel being oil quenched and **d** integration of the first stage of maraging steel motor case at SDSC-SHAR, Sriharikota

16.4 On-Demand Materials for Rocket Science

16.4.1 High Strength Aluminium Alloys

SLV-3 and ASLV used alloy steel/composites for different stages of solid rocket motors as they have 'all solid propellant' stages. When the PSLV was designed, the second stage (designated as PS2) consisted of a liquid stage with a hypergolic combination of unsymmetrical dimethyl hydrazine (UDMH) as fuel and nitrogen tetraoxide (N_2O_4) as oxidizer. The propellant tanks for these are made of an Al-4.5Zn-1.5 Mg alloy (AFNOR7020) and were realized by TIG welding of rolled AFNOR7020 sheets and end rings/domes at HAL, Aerospace Division, Bangalore. Subsequently, the same material was also used for the fabrication of L40 stages of GSLV, the only difference being in the size of the propellant tanks. Also, the PS2 tank has a common bulkhead, whereas L40 stage has separate tanks for oxidizer and propellant.

Aluminium alloy AFNOR7020 has been used for tankages from the time of Viking Engine technology transfer by Air Liquid, France and several PSLV and GSLV flights were successfully flown with these tanks. However, this high-strength aluminium alloy is prone to stress corrosion cracking and was phased out after a leak in the propellant tank on the launch pad in a GSLV flight (Kiran Kumar et al. 2015). The material for construction of earth-storable propellant tanks was changed to another aluminium alloy, AA2219, the same material that was being used in GSLV Mk-III propellant tanks (Narayana Murty et al. 2219; Manwatkar et al. 2019; Manikandan et al. 2021). Figure 16.3 shows the spectrum of aluminium alloys used in launch vehicles. The complexity and importance of different grades of aluminium alloys may be noted which are used in performance-critical applications.

The quest to indigenise aerospace quality aluminium alloys required for ISRO's programs started in mid-1980s with the participation of Ordnance factory, Ambajhari and several hurdles were faced such as degassing of the liquid melt to control hydrogen and on-line filtering of the liquid melt to remove, the unavoidable but undesirable, inclusions that can be deleterious in downstream operations like rolling and forging. These limitations were successfully overcome with the establishment of dedicated aerospace cast-houses within the country.

Other important aluminium alloys that are widely used in launch vehicles are AA2014 for interstage structures, AA6061 for water tanks and AA7075 for various control system components of liquid engines. Technology for melting and casting, meeting the critical aerospace specifications, is now available within the country and DC cast billets of these alloys are produced routinely at M/s Hindalco - Almex at Aurangabad. The large-sized ring rolled forgings are processed at M/s Bay Forge, Chennai. Figure 16.4a-d shows the processing of large-diameter aluminium alloy ring rolled rings essential for the fabrication of propellant tanks and inter-tank structures of launch vehicles. The open die forgings of aluminium alloys are made at several forging industries such as M/s. SIFL, Thrissur, Rachamallu Forgings, Deccan Smiths, Asha Forgings, Manjira and CHW Forge. Large number of facilities have been

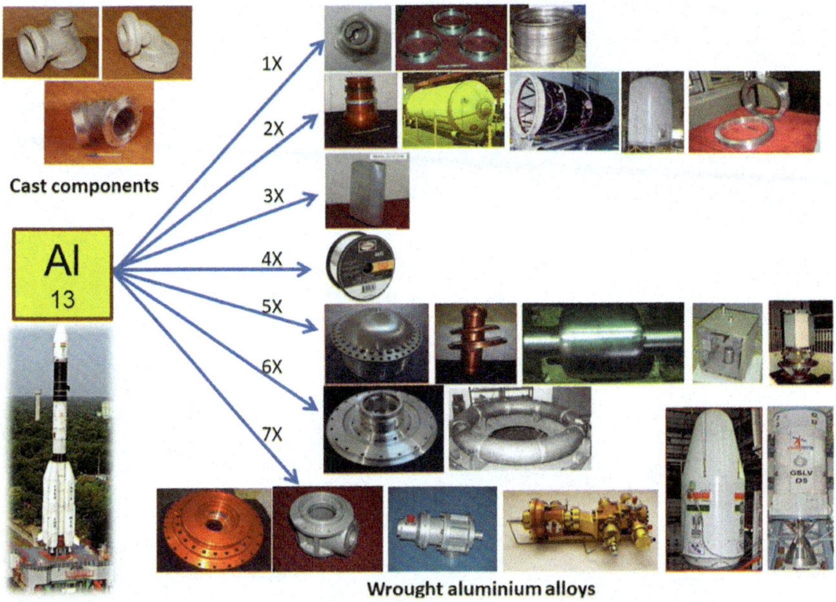

Fig. 16.3 Spectrum of aluminium alloys used in launch vehicles. The complexity and importance of different grades of aluminium alloys may be noted which are used in performance critical applications

established, including drop bottom quenching furnaces essential for heat treating aluminium alloys, at different aluminium alloy forging units. Presently, aluminium alloys are the most critical materials for launch vehicle structures and the requirements are mostly met through indigenous sources except for large-size extrusions, plates and sheets due to limitations in extrusion press and rolling mill in the country, respectively.

16.4.2 Titanium Alloys

Titanium alloys are the choice in launch vehicles for the upper stage structures, where every kilogram of the weight saved results in an equivalent payload gain. So, they are used in the satellite propellant tanks and high-pressure helium gas bottles for pressurization of fluids in upper stages. Figure 16.5 shows various titanium alloy products used in launch vehicles and satellites. The workhorse titanium alloy is Ti-6Al-4 V with the regular grade used for room temperature applications and the extra-low interstitial grade (ELI) used up to 77 K. On the other hand, Ti-5Al-2.5Sn is used for applications up to 20 K. There are other alloys like Russian grade VT14 and half alloy (Ti-3Al-2.5 V) for inter-tank structures (Anil Kumar et al. 2021).

Fig. 16.4 Processing of large diameter aluminium alloy rolled rings essential for the fabrication of propellant tanks and inter tank structures of launch vehicles. The figure shows the **a** ring rolling mill under operation, **b** solution treatment furnace with drop bottom quenching facility, **c** ring expander to impart the desired temper and **d** a large size rolled ring

Activities related to the development of titanium alloy products started in the 1990s with the establishment of in-house facilities in Materials Group at VSSC. Initial developments in early 1970s were through joining of Ti-6Al-4 V consumable electrodes suitable for melting in the vacuum arc melting furnace and thermomechanical processing of the billets to various shapes to establish the process parameters necessary for obtaining the desired microstructure and mechanical properties. This in-house experience was useful in drawing the specifications to procure billets and plates from abroad as well as developing them indigenously through MIDHANI.

One of the important applications of titanium alloy Ti-6Al-4 V is in the form of spherical gas bottles (200, 400 and 600 mm diameter) in annealed condition, for storing inert gases for pressurization systems (Gupta et al. 2016; Manikandan et al. 2019). These spherical gas bottles are realized by closed die hammer forging to hemispherical domes and electron beam welding them. Simultaneously, technology based on hot forming of plates was also developed to realize such hemi-spherical shapes and one such important requirement was a propellant tank for a spacecraft with a diameter of 1200 mm, which was successfully met in solution-treated and annealed condition, in view of its higher strength. Electron beam welding technology for joining these

Fig. 16.5 Titanium alloy products used in launch vehicles and satellites

hemispheres was established and these gas bottles are now routinely used in PSLV and GSLV/GSLV MkIII flights. While the production of the hemispherical dome forgings is carried out at M/s SIFL, Thrissur and M/s Bharat Forge Ltd., Pune, their welding is carried out at M/s BrahMos Aerospace Thiruvananthapuram Ltd., Trivandrum. Similarly, technology for production of LH2 submerged Ti-5Al-2.5Sn gas bottles has also been established and is being successfully used in GSLV MkIII flights.

Another critical application of titanium alloys is in the fabrication of large-sized propellant tanks for upper stages of launch vehicles and satellites. These are fabricated from large-sized rings in solution-treated and aged (STA) condition and are joined by electron beam welding. Satellite propellant tanks are made to stringent quality control specifications as they should serve up to 15 years in orbit. The rings required for the propellant tanks are processed at M/s MIDHANI, Hyderabad and subsequently welded (EB Welding) at LPSC, Bangalore. These propellant tanks are being routinely produced and flown.

It is a matter of immense satisfaction that VSSC, using technology developed by DMRL, Hyderabad has established a state-of-the-art Titanium sponge plant at KMML, Chavara, to produce 500 TPA titanium sponge. The sponge produced is used for melting titanium alloys of various grades at M/s MIDHANI. With the establishment of this plant, India joined select countries with end-to-end capability of producing titanium- from 'Sand to Space'.

16.4.3 Magnesium Alloys

Magnesium alloys are the choice where good strength and stiffness, good damping capacity, excellent thermal diffusivity are the criteria for selection. The difficulties in the processing of these highly reactive alloys are well known. In view of their importance to the space program, seeds for their development were sown in late 1960s itself. A 3-axis rate gyro mounting block was cast out of magnesium alloy AZ92 for SLV-3. Subsequently, ASLV used magnesium alloy components for inertial platform module of the Stabilized Platform Inertial Navigation System (SPINS) consisting of inner role gimbal body, inner role gimbal body cover, bottom cover and top plate. Magnesium alloy complex castings were developed for PSLV for cluster and housings for the Redundant Strap-down Inertial Navigation Systems (RESINS), which were subsequently outsourced to HAL, Koraput for production. One of the biggest achievements in magnesium alloy development in ISRO was the realization of a very large casting for the outer axis gimbal for an Angular Motion Simulator, to take advantage of the weight and damping capacity of magnesium. The weight of the casting was 400 kg and the weight of the melt was one ton and this was achieved in the 1990s, when the maximum capacity to melt magnesium was only 300 kg. Different components such as VHF transmitter and receiver body, top cover and bracket for transmitter, etc., made from magnesium alloy ZE41 (an alloy containing zinc, zirconium and rare earth) were supplied for SROSS satellite. Forged blocks of magnesium alloy AZ31 were melted in VSSC for S-band and C-band transponders meeting the stringent ultrasonic quality requirement as per AMS2630 Class A and the process parameters established were passed on to HAL (Foundry and Forge) for production in large quantities. Other significant developments in magnesium alloys include the Payload adapter in ZK30 ring rolled rings, etc. Large-sized magnesium alloy plates in AZ31 were processed through MIDHANI from the billets realized at M/s Exclusive Magnesium, Hyderabad. These very significant developments in the magnesium alloy development in VSSC could be achieved predominantly through in-house laboratory scale developments and subsequent scale-up through Indian industries. Today, we have self-sufficiency in all magnesium alloy products required for launch vehicle and satellite applications. Figure 16.6 shows the typical view of an equipment bay (center) and various types of magnesium castings used for mounting the electronic packages. The photograph at the bottom shows a large magnesium casting for angular motion simulator.

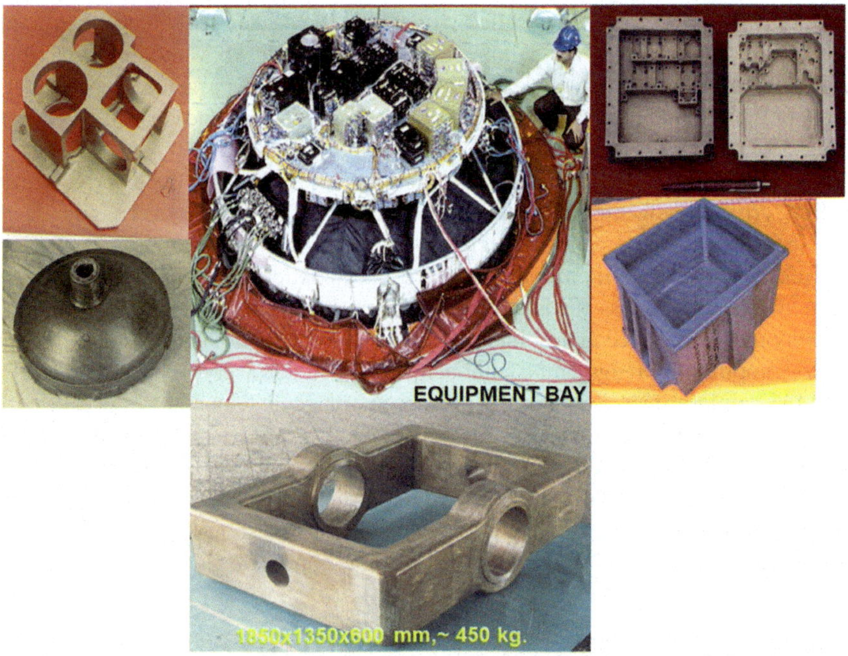

Fig. 16.6 Typical view of an equipment bay (centre) and various types of magnesium castings used for mounting the electronic packages. Bottom photograph shows a large magnesium casting for angular motion simulator

16.4.4 Steels for Cryogenic Applications

Cryogenic rocket engines of space launch vehicles use liquid hydrogen (boiling point 20 K) and liquid oxygen (boiling point 90 K) as fuel and oxidizer, respectively. Many of the parts of engine are exposed to these extreme temperatures. The retention of mechanical properties with decreasing temperature, in combination with other fabrication requirements like formability and weldability are the primary material selection criteria for structures operating at low temperatures. This was achieved by improving the existing austenitic stainless steels in terms of alloying additions and reducing the impurity levels through special melting and refining processes (ESR/VAR). Austenitic stainless steels, due to the absence of DBTT, excellent combination of mechanical and physical properties and adequate fabricability have long been used for a wide array of cryogenic applications. 07X16H6 (ICSS301) is the workhorse material used widely in cryogenic engine and stage components. Other lean austenitic grades like 304L, 316L, etc. are also recommended for cryogenic applications due to improved structural stability. No spontaneous martensitic transformation induced by quenching to low temperature is observed in this grade; though martensitic transformation through deformation persists. Hence, this grade is usually preferred for pre-formed thin-walled structures. The leaner grades are often cold-worked to higher extent for gaining improved strength while retaining ductility. AISI 202 grade austenitic stainless steel exhibits excellent cryogenic properties, for its impressive combination of toughness, corrosion resistance and cost-effectiveness. This steel is used for impeller applications operating at 20 K.

The low yield strength of austenitic stainless steels poses a major problem for weight-critical cryogenic applications, especially for high-thrust rocket engines. This has led to the development of precipitation hardenable (PH) stainless steels with martensitic and austenitic matrices for aerospace applications. The development of martensitic PH stainless steels, with a BCC matrix, requires control of interstitials by adopting special secondary melting practices (ESR/VAR) and alloy design that provides optimum strengthening without loss of toughness at cryogenic temperatures. The PH stainless steels used for applications at cryogenic temperatures in cryogenic and semi-cryogenic rocket engines and associated systems include 14-5PH, A286, 11Cr-9Ni and 12Cr-10Ni steels (Anoop et al. 2021, 2020, 2018). Several of these steels are used as closed die forgings. All these steel grades have been indigenously produced meeting the specified requirements. Figure 16.7 shows typical components fabricated from stainless steels for launch vehicle applications.

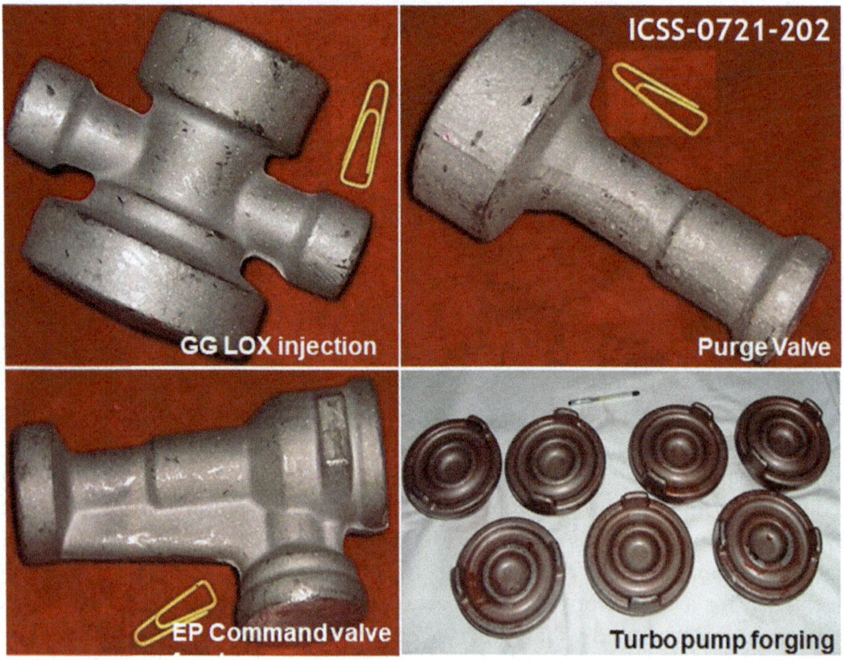

Fig. 16.7 Various components fabricated from stainless steels for launch vehicle applications

16.4.5 Materials for High Temperature Applications

High-temperature Co-20Cr-15W-10Ni alloy (KC20WN) and Nb-10Hf-1Ti alloy C103 are used for the fabrication of critical components for both launch vehicles and spacecraft control systems. VSSC has developed the processing technologies for these alloys in collaboration with M/s MIDHANI. During the development of Co-based Superalloy KC20WN, major challenges of achieving grain size and mechanical properties were resolved and process parameters were optimized to meet the regular requirement of liquid propulsion systems. The processing of refractory Nb alloy C103 is highly challenging due to its poor oxidation resistance. Problems faced in each of the processing stages were resolved to realize 2 mm thick Nb-Hf-Ti alloy sheets through innovative methods of warm rolling followed by cold rolling. These alloys are being used in various forms (sheets and rings in KC20WN and sheets and rods in C103 alloy) in liquid engines of both launch vehicles and satellites of ISRO.

16.4.6 Investment Castings for Cryo/semi-Cryogenic Engines

Investment casting process (also called lost-wax process) uses a ceramic shell mould. Wax pattern dipped in ceramic slurry is fired to obtain a shell to pour liquid metal to form very intricate shaped precision castings. Large number of investment castings is used in cryogenic and semi-cryogenic engines of launch vehicles. Figure 16.8 shows typical investment cast products used in the cryogenic and semi-cryogenic engines of launch vehicles. The GSLV used 14 types and GSLV-MkIII uses 4 types of investment castings, both of which have been successfully flown proving the indigenously developed complex technologies and inspection procedures.

Semi-cryogenic engine is being developed to meet the future launch vehicle requirements which work on an oxidizer rich staged combustion cycle and deliver a vacuum thrust of 2000kN. The propellants are liquid oxygen (LOX) and Isrosene. This engine is to operate for a nominal duration of around 355 s. As many as 18 types of investment castings are required for the realization of this engine including 11 types for realization of turbopump subsystems and 7 types for control components. All these intricate castings are realized through investment casting route to achieve close dimensional tolerance and very good surface finish (Sharma et al. 2012; Nimbalkar et al. 2015; Dineshraj et al. 2015). These castings are mainly made up of stainless steels and superalloys.

The challenges involved in the development of these investment castings include design of the gating systems, prediction of the location of possible defect formation, meeting the quality control requirements with respect to radiography, optimization of heat treatment schedules, repair of defects and establishment of procedures. Extensive modelling and simulation studies using software like Procast™ are routinely used for design of the casting processes. Indian industries like Mishra

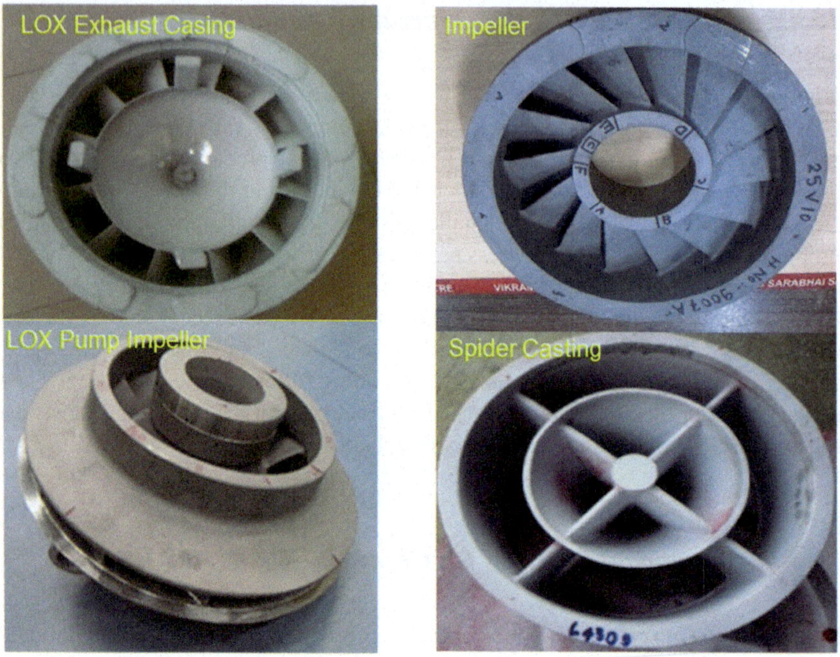

Fig. 16.8 Different types of investment cast products used in launch vehicles

Dhatu Nigam Limited, Hyderabad, PTC castings Lucknow, HAL Koraput, UDL Nasik, IPCL Bhavnagar are involved in the production of these castings.

16.4.7 Special Purpose Materials

A large number of special-purpose materials are used in ISRO launch vehicles and many of these are functional materials unlike the structural materials described above. These include magnetic materials, electronic ceramics, composite materials including carbon cloth, pyrolytic graphite and thermal protection materials like silica tiles, metal-ceramic coatings, powder metallurgy products like nozzle throat inserts for bi-propellant reaction control thrusters and beryllium for space craft mirrors. Infrastructure facilities for processing of critical powder metallurgy products like nozzle throat inserts, double layered bronze seal rings for cryogenic engine, etc., have been established in-house at VSSC. The small volume of these critical products evoked low interest among industries for productionisation. Efforts are being put in place to transfer the technology to outsource the items continuously and few of the critical powder metallurgy products are being realized through these sources presently. Figure 16.9 shows various powder metallurgy products used in satellite launch vehicles. These include W-Cu nozzle throat inserts for SLV-3, Mo-Cu

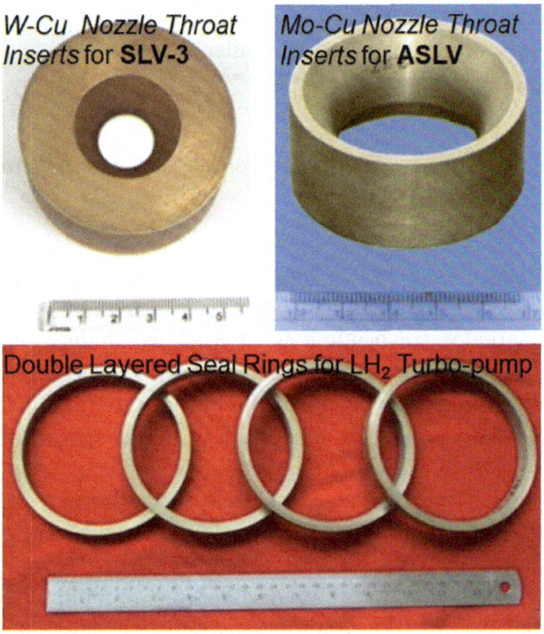

Fig. 16.9 Powder metallurgical products used in satellite launch vehicles. These include W-Cu nozzle throat inserts for SLV-3, Mo-Cu nozzle throat inserts for ASLV and double layered bronze seal rings for LH2 turbo-pumps of C25 engine of GSLV MkIII

nozzle throat inserts for ASLV (Mahesh Paidpilli et al. 2018; Mishra et al. 2010) and double-layered bronze seal rings for LH2 turbo-pumps of C25 engine of GSLV MkIII.

ISRO's requirements of high-energy permanent magnetic materials were met by import till 1990 and subsequent geopolitical developments resulted in their non-availability. This prompted VSSC to indigenously develop them. VSSC in collaboration with DMRL has successfully developed volume-efficient, high-energy density SmCo5 magnets for ISRO's missions for use in packages for navigation, guidance and control of satellites.

16.5 Welding Metallurgy

Welding is the most important fabrication process in the realization of space structures. Since the weld joints must meet stringent aerospace quality control norms, the processes adopted should result in joints that pass the non-destructive test procedures. The launch vehicle structures like solid rocket motors processed out of maraging steel are multi-pass GTAW welded, the earth-storable and cryogenic propellant tanks fabricated out of aluminium alloy AA 2219 are GTAW welded and the upper-stage/ satellite propellant tanks of Ti-6Al-4 V are electron beam welded. The liquid and cryogenic engines have a number of complex similar and dissimilar weld joints. Welding metallurgy is one of the important areas of research in VSSC. The past five decades of metallurgical research have paved the way for the establishment of WPS for several similar and dissimilar materials.

16.5.1 Friction Stir Welding of Aluminium Alloy Propellant Tanks

Conventional TIG welding process is being used for the fabrication of earth-storable and cryogenic propellant tanks made of aluminium alloy AA 2219. With the evolution of friction stir welding as an attractive alternate fabrication process for increased productivity of aerospace hardware, ISRO has taken up this activity to demonstrate the FSW technology on the flight hardware (Manikandan et al. 2021). Since FSW is a solid-state welding process, heavy fixturing is involved. Extensive weld trials were conducted and technology is demonstrated on long-seam weld joints of propellant tanks. The FSW welded tanks have been flown since PSLV-C39. Figure 16.10 shows the conventional and friction stir weld methodologies used in the construction of propellant tanks. Figure 16.10a shows the photograph of (a) tungsten inert gas (TIG) welding station used for welding the circumferential welding of propellant tanks and Fig. 16.10b shows the friction stir welding station. Figure 16.10c shows the strap-on L40 stages consisting of aluminium alloy propellant tanks under integration at SDSC-SHAR and Fig. 16.10d shows the optical microstructure of aluminium alloy AA2219-T87 welded using different welding processes viz. TIG, FSW and EBW show distinctly different macrostructural features. It can be clearly seen from Fig. 16.10d that the partially melted zone responsible for the lower strength of weld joint in TIG welding can be avoided by solid-state welding through friction stir welding.

16.6 Developments in Non-destructive Testing of Materials

Non-destructive testing is one of the most important areas of research in VSSC as it may be noted that malfunctioning of a single system can lead to catastrophic failures. Aerospace design demands materials with high specific strength and designers wish to exploit the full potential of material with minimum margins of safety, unlike ground-based systems. This calls for extensive testing and qualification of flight hardware. The material being procured is melted with controlled chemistry, impurities, gas contents and is thermo-mechanically processed to the desired shapes/sizes and qualified through various NDT techniques like Ultrasonic Testing, Dye Penetrant Test, Fluorescent penetrant inspection, Liquid Penetrant Test and Radiographic method as per the specification requirements. Subsequently, the hardware realized is also subjected to proof tests that are essential for flight certification and non-destructive testing is an in-built exercise post every test protocol. Periodic inspection

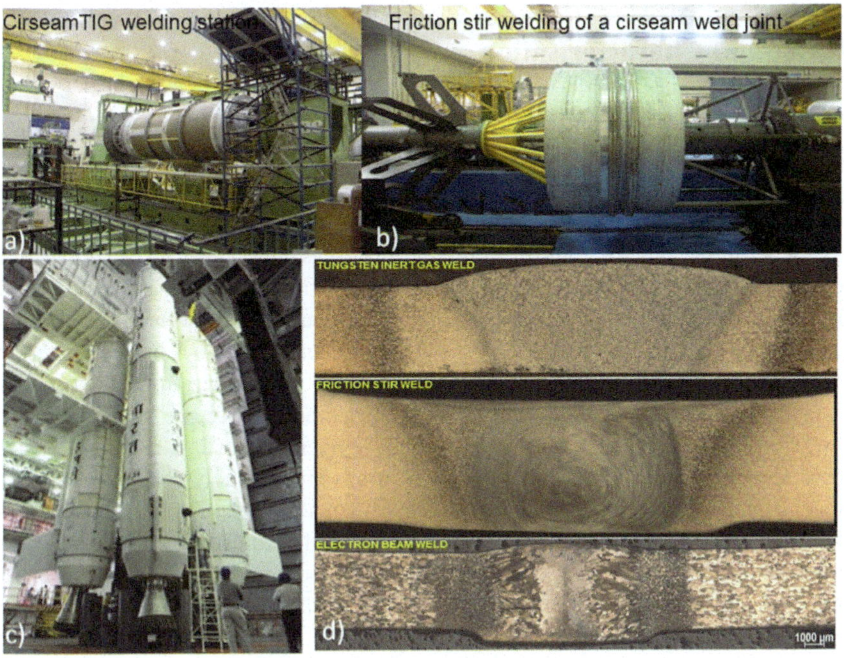

Fig. 16.10 **a** tungsten inert gas (TIG) welding station used for welding the circumferential welding of propellant tanks; **b** Friction stir welding (FSW) of a circumferential weld joint of aluminium alloy propellant tanks structure; **c** strap-on L40 stages consisting of aluminium alloys propellant tanks under integration at SDSC-SHAR and **d** optical microstructures of aluminium alloy AA2219-T87 welded using different welding processes viz. TIG, FSW and EBW showing distinctly different macrostructural features

and effect of long-term storage on the quality are ascertained through non-destructive techniques.

Since the solid rocket motor design is based on the principles of fracture mechanics, detectability of defects exceeding the critical flaw size is one of the most important aspects for certifying the pressure vessels. Starting from the formative years of VSSC, a strong team of non-destructive testing and evaluation has been formed. Phased array ultrasonic testing and computer tomography for the detection of defects have been introduced as advanced non-destructive test techniques for the routine inspection of flight hardware.

16.7 Facilities for Material Processing and Characterisation

During the initial years of satellite launch vehicle development, the approach was to develop these technologies in-house and wherever possible, transfer the technologies to industries This was basically due to the low volumes of material requirement which the industries were reluctant to take up during the formative years as well as the stringent quality control norms the materials must meet for acceptance. In order to meet the developmental requirements, several state-of-the-art, custom-built, infrastructure facilities have been established in-house. These were used for development of maraging steels, titanium alloys and magnesium alloys before attempting industrial processing. Similarly, extensive material characterization facilities were established that include mechanical testing machines for fatigue and fracture, bi-axial testing, thermo-mechanical fatigue, high-temperature testing. Hundreds of standard test procedures for a variety of applications are routinely conducted on these machines. Further, proof loading of the components is a routine activity using these facilities. Extensive metallurgical characterization facilities like scanning electron microscopy and transmission electron microscopy, both with energy dispersive spectroscopy and X-ray diffraction have helped the materials development for production, quality control and failure analysis investigations. An exclusive corrosion testing laboratory helps in studying the corrosion, stress corrosion cracking susceptibility of launch vehicle materials. All these facilities are used for the clearance of launch vehicle materials routinely. Figure 16.11 shows some of the state-of-the-art facilities established in VSSC for the processing and testing of materials.

16.8 Materials Research Activities at Liquid Propulsion Systems Centre

While the material development activities have largely been concentrated in VSSC right from the beginning of ISRO's launch vehicle programs, the necessity of establishing certain materials development facilities related to the liquid propulsion systems arose later. With the establishment of a dedicated centre for liquid propulsion systems, materials development activities related to liquid and cryogenic stages were also initiated. Since this centre deals with the development and production of complex liquid stages consisting of large-sized propellant tanks and engines (Vikas engine of different categories L37.5, L40, L110), several material fabrication technologies have been developed and established at different work centres. For example, the conventional TIG welding of propellant tanks has been replaced by friction stir welding (FSW) in PSLV and launch vehicles carrying FSW tanks have been flown. Technologies required for vacuum brazing of inner copper alloy chamber with outer steel chamber in regeneratively cooled nozzles have been successfully developed and were successfully flown. The Cu-Cr-Ti-Zr alloy used in the inner liners of cryogenic

Fig. 16.11 State-of-the art facilities established in VSSC for processing and testing of materials

and semi-cryogenic engines has been successfully indigenised with the support of NFTDC, Hyderabad (Sarkar et al. 2020a, b; Sudarsana Rao et al. 2017). Similarly, many different grades of steel are used in cryogenic and semi-cryogenic engine systems which are performance critical. These include steels like AISI 202 which is used for impellers in cryogenic engine and braze joints of this steel must perform at 20 K operating at more than 36,000 rpm during flight. For GSLV-MKIII, the impellers are made of Ti-5Al-2.5Sn-ELI which is diffusion bonded and technologies for the same have been successfully established. Further, thermal barrier coatings based on Calcia stabilized Zirconia coatings on cobalt-based superalloy KC20WN, silicide coatings on columbium alloy C103, metal-ceramic ignition resistant coatings on nickel-based superalloy and a number of metallic coatings for braze joint interfaces are being routinely used in launch vehicle programs. As a part of the indigenisation program, many of the metallic materials used in liquid and cryogenic propulsion systems have been indigenised through several Indian industries which are regularly supplying them meeting the stringent quality control norms.

As the work involves working with highly hazardous, corrosive, earth-storable propellants (UDMH/UH-25, N_2O_4 for launch vehicles, MON-3 for upper-stage propellant tanks), establishing compatibility of materials with these liquid propellants is a challenge. Similarly, for cryogenic and semi-cryogenic engines, materials are exposed to extremely low temperatures and one of the primary requirements is the ability of materials to retain the toughness at the operating temperatures. Therefore, several facilities were established for testing and characterization of these materials. These include facilities for testing materials for the operating environment of liquid oxygen (90 K) and liquid hydrogen (20 K). This is one of the unique facilities in the country for mechanical testing of materials down to 20 K.

16.9 Fifty Years of Materials Development Activities in ISRO and Future Goals

The last fifty years of materials development activities in VSSC have successfully resulted in technologies that have propelled ISRO launch vehicle and satellite programs. Over the past five decades, large numbers of sophisticated, state-of-the-art experimental facilities have been established that cater to material development activities. During the early years of the launch vehicle program, industries were in a nascent stage to take up the challenges of materials requirements meeting the stringent quality control requirements. This necessitated VSSC taking up both the laboratory scale development work and subsequent scaling-up activities to realize the systems. However, the last three decades saw enormous growth in the infrastructure facilities in the country for processing various types of metallic materials and fabrication of launch vehicle structures. Even though the qualification of materials and inspection of flight hardware is still in the ambit of VSSC, slowly the Indian industry has acquired the know-how based on the extensive experience that has been gained and has matured to deliver materials on demand, meeting aerospace quality requirements.

The present demands of space include low-cost and fast realization of rocket systems that have prompted the reusability of vehicle structures as well as disruptive technologies such as additive manufacturing. Research and development in this direction is underway currently and it is intended to induct these technologies in the coming years. Simultaneously, type testing of materials meeting the human rating of launch vehicle structures for India's manned space program, Gaganyaan, is underway.

16.9.1 Additive Manufacturing

The realization of components and structures in rocketry is through the conventional manufacturing techniques involving casting, forming, forging, extrusion, machining, riveting, brazing and welding, etc. The high buy-to-fly ratio in conventional manufacturing processes leads to both cost and lead time disadvantages. While there have been attempts in the past to promote near-net shape forming/forging technologies like closed die forgings, shaped forgings/rings, etc., still the manufacturing processes involve considerable material removal by machining. Further, large volume of welding leads to increased requirements of inspection to identify defects. These limitations of the conventional manufacturing processes have been of late addressed by the advent of various additive manufacturing processes wherein a bottom-up approach is used for component fabrication. The materials for fabrication could either be in powder or wire form and the heating source could be laser, electron beam or plasma arc. Using processes such as laser powder bed fusion or directed energy deposition, performance-critical components have been successfully realized and their performance is being evaluated in hostile environments. VSSC has started implementing additive manufacturing technologies through extensive technology development studies on several aerospace materials such as Inconel-718, Ti6Al4V, Ti6Al4V-ELI, AISI 316L, AlSi10Mg and 17–4 PH and has now matured to realize launch vehicle components from these materials successfully. Figure 16.12 shows an example of velocity trimming module (VTM) brackets realized through laser powder bed fusion process in AISI 316L for launch vehicle applications (Pradeep et al. 2020). Monolithic components as large as 350 × 1500 × 2100 mm in Ti6Al4V are also envisaged to be processed through electron beam additive manufacturing route. Many technology development projects are in progress at VSSC to implement different technologies for realizing structural and functional materials (Pradeep et al. 2020, 2021a, b).

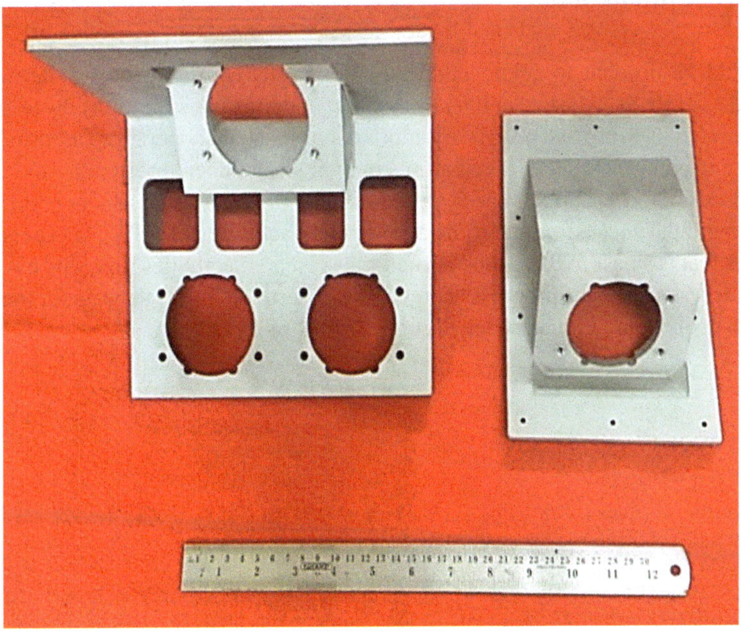

Fig. 16.12 Two types of VTM brackets processed through direct metal laser sintering of AISI 316L stainless steel for Small Satellite Launch Vehicle (SSLV)

16.9.2 Indigenisation of Flow Forming Alloys

Large diameter, ~3 m, solid rocket motors of several meters length is realized through roll bending of maraging plates of about 8 mm thickness, welded longitudinally to form shells and are joined circumferentially to different shells/rings/domes. Several meters of welding are involved for each segment of the motor case and usually, these are multi-pass welded, up to 4 passes. Non-destructive inspection is done in an extensive manner and defects if any are observed, and approved repair schemes are implemented. All this is cumbersome and time taking leading to cost disadvantages. VSSC has been attempting to implement flow-forming technology for the fabrication of solid rocket motor segments made from ESR15CDV6 steel (0.3C-CrMoV steel) and shells have been successfully formed and are under metallurgical characterization. It is expected that this technology would soon be implemented in Indian launch vehicles. Similarly, LPSC is developing technology for flow forming of 15-5PH steels for secondary injection thrust vector control (SITVC) tanks and it is expected that these tanks will be inducted soon into the launch vehicle program.

16.10 Conclusion

It would be prudent to acknowledge the enormous contributions by the material scientists of VSSC who had neither prior experience nor know-how on how to develop these materials in the first place. Self-motivation, spirit of cooperation, intuitive skills, value judgement, ability to take failures in their stride, open discussions and respect for other's opinions, all being the hallmarks of VSSC family of researchers, are the epitome of the success of materials development activities in VSSC. It is heartening to note that the past five decades of materials research activities in VSSC have been very successful and the progress achieved so far certainly paves the way for the future launch vehicle and human space flight programs of India.

References

Anil Kumar, V., Gupta, R.K., Prasad, M.J.N.V., Narayana Murty, S.V.S.: Recent advances in processing of titanium alloys and titanium aluminides for space applications: a review. J. Mater. Res. **36**, 689–716 (2021)

Anoop, C.R., Prakash, A., Narayana Murty, S.V.S., Samajdar, I.: Origin of low temperature toughness in a 12Cr-10Ni martensitic precipitation hardenable stainless steel. Mater. Sci. Eng. A **709**, 1–8 (2018)

Anoop, C.R., Singh, R.K., Kumar, R.R., Jayalakshmi, M., Narayana Murty, S.V.S., Tharian, K.T.: Development and Validation of Processing maps for hot deformation of modified AISI 321 austenitic stainless steel. Mater. Perform. Charact. **9**, 150–169 (2020)

Anoop, C.R., Singh, R.K., Ravi Ranjan Kumar, Jayalakshmi, M., Antony Prabhu, T., Thomas Tharian, K., Narayana Murty, S.V.S.: A review on steels for cryogenic applications. Mater. Perform. Charact. **10**, 16–88 (2021)

Beena, A.P., Sundaresan, M.K., Rao, B.N.: Destructive tests of 15CDV6 steel rocket motor cases and their application to lightweight design. Int. J. Press. Vessels Pip. **62–3**, 313–320 (1995)

Dineshraj, S., Mayukh Acharya, Alok Agarwal, Girikumar, S., Govind, S.C. Sharma, Koshy M. George: Development of hot isostatic pressing technology for investment cast products. Mater. Sci. Forum, **830–831,** 9–22 (2015)

Ghosh, B.R., Murthy, M.S.P., Sharma, S.C., Mittal, M.C., Nagarajan, K.V.: TIG Welding technology of 7mm thick 18 Ni 250 grade maraging steel plates for space applications. Trans. Indian Inst. Metals **40–3**, 227–234 (1987)

Gupta, R.K., Anil Kumar, V., Sumit Chhangani: Study on variants of solution treatment and aging cycle of titanium alloy Ti6Al4V. J. Mater. Eng. Perform., **25**, 1492–1501(2016)

Kiran Kumar, B., Yezhil Arasu, A., Thomas Kurian, Jayaprakash, J., Somanath, S.: Assessment on the failure observed in a liquid fuel tank in an aborted mission of a geosynchronous vehicle, 66th International Astronautical Congress 2015, Jerusalem IAC-15-C2.1.2,28717

Mahesh Paidpilli, A. Upadhyay, Khanra, G.P., Sharma, S.C.: Investigation of sintered properties on infiltrated tungsten-copper composite along the infiltration direction. Canadian Metall. Quart. **57**(1), 120–128 (2018)

Manikandan, S., Mishra, A.K., Nayak, G., Narayana Murty, S.V.S., Rao, G.S., Narayanan, P.R., Ram Kumar, P., Venkitakrishnan, P.V.: Residual strength and fracture toughness of 0.3 C-CrMoV (ESR) ultrahigh-strength steel used for solid booster motor cases. Mater. Perfor. Charact. **7** (2018)

Manikandan, P., Rao, G.S., Muneshwar, P., Narayana Murty, S.V.S., Narayanan, P.R.: Tensile and fracture properties of commercially pure titanium (CP-70) Hemispherical Forgings. Trans. Indian Inst. Metals **72,** 1469–1473 (2019)

Manikandan, P., Prabhu, T.A., Manwatkar, S.K., Rao, G.S., Narayana Murty, S.V.S., Sivakumar, D., Pant, B., Mohan, M.: Tensile and fracture properties of aluminium alloy AA2219-T87 friction stir weld joints for aerospace applications. Metall. Mater. Trans. A., 1–18 (2021)

Manwatkar, S.K., Narayana Murty, S.V.S., Narayanan, P.R., Sharma, S.C., Venkitakrishnan, P.V.: Stress corrosion cracking of a maraging steel shear bolt used in the interstage structure of a satellite launch vehicle. Metall. Microstruc. Analy. **5**:411–420 (2016)

Manwatkar, S.K., Sunil, M., Prabhu, A., Narayana Murty, S.V.S., Joseph, R., Narayanan, P.R.: Effect of grain size on the mechanical properties of aluminum alloy AA2219 parent and weldments at ambient and cryogenic temperature. Trans. Indian Inst. Metals **72,** 1515–1519 (2019)

Mishra, D.K., Saravanan, T.T., Khanra, G.P., Giri Kumar, S., Sharma, S.C., Sreekumar, K., Sinha, P.P.: Studies on the processing of nickel base porous wicks for capillary pumped loop for thermal management of space crafts. Adv. Powder Technol. **21**, 658–662 (2010)

Nageswara Rao, M., Narayana Murty, S.V.S.: Hot deformation of 18% Ni maraging steels: a review. Mater. Perf. Charact. **8–5**, 742–765 (2019)

Narayana Murty, S.V.S., Sreekumar, K., Natarajan, A., Sinha, P.P.: Metallographic investigations on the appearance of streaks in radiographic inspection of maraging steel forged ring-plate weldments. Proceedings of the International Conference on Advanced Metallurgical Processes, New Age International Pvt. Limited, pp. 1399–1404 (1997)

Narayana Murty, S.V.S., Sudarsana Rao, G., Venugopal, A., Ramesh Narayanan, P., Sharma, S.C., Koshy M. George: Metallurgical analysis of defects in the weld joints of large-sized maraging steel rocket motor casing. Metall. Microstru. Analy. **3**, 433–447 (2014)

Narayana Murty, S.V.S., Aditya Sarkar, P. Ramesh Narayanan, P.V.: Venkitakrishnan, Microstructure and micro-texture evolution during large strain deformation of aluminium alloy AA 2219. Mater. Sci. Eng. A **677**, 41–49 (2016)

Nimbalkar, V.M., Bhanushali, B., Mohape, M., Pandav, S.G., Deshmukh, V.P., Dineshraj, S., Mayukh Acharya, Alok Agarwal, Govind, Sharma, S.C.: Development of thin walled A-356 components by new rheocasting semi-solid metal processing technology (NRC). Mater. Sci. Forum **830–831**, 27–29 (2015)

Pradeep, P.I., Anil Kumar, V., Sriranganath, A., Satish Kumar Singh, Ankit Sahu, Sasi Kumar, T., Ramesh Narayanan, P., Arumugam, M., Mohan, M.: Characterization and qualification of LPBF additively manufactured AISI-316L stainless steel Brackets for Aerospace Application. Trans. INAE (2020). https://doi.org/10.1007/s41403-020-00159-x

Pradeep, P.I., Akhilesh, A.M., Anil Kumar, V., Manikandan, P., Naresh Kumar, K., Venkateswaran, T., Sahu, A., Singh, S.K.: Characterization of titanium alloy Ti6Al4V-ELI components made by laser powder bed fusion route for space applications. Trans. INAE (2021a). https://doi.org/10.1007/s41403-021-00254-7

Pradeep, P.I., Anil Kumar, V., Venkateswaran, T., Aswin, S., Vishnu S. Nair, Asvin Krishnan, Akhil, M. Agilan, Satish Kumar Singh, Ramesh Narayanan, P.: Processing and characterization of 3D printed Inconel-718 component through laser powder bed fusion route for high temperature space application, Trans. INAE (2021b). https://doi.org/10.1007/s41403-020-00196-6

Ramesh Narayanan, P., Sreekumar, K., Natarajan, A., Sinha, P.P.: Metallographic investigations of the heat-affected zone II/parent metal interface cracking in 18Ni maraging steel welded structures. J. Mater. Sci. **25,** 4587–4591 (1990)

Ramkumar, P., Anil Kumar, V., Gupta, R.K. Karthikeyan, M.K., Narahari Prasad, S., Gino Prakash, F., Chakravarthi, K.V.A., Maruthi Prasad, Y., Venkitakrishnan, P.V.: Melting and thermomechanical processing of high strength 0.3%C–CrMoV (ESR) steel. Trans. Ind. Inst. Met. **71,** 1475–1485 (2017). https://doi.org/10.1007/s12666-018-1283-2

Ramkumar, P., Gupta, R.K., Anil Kumar, V., Muthupandi, V.: Effect of pre- and post weld heat treatment on microstructure development and mechanical properties of 0.3%C-CrMoV (ESR) high-strength low-alloy steel. J. Mater. Eng. Perform (2021). https://doi.org/10.1007/s11665-021-05929-4

Renu N. Gupta, Raja, V.S., Mukherjee, M.K., Narayana Murty, S.V.S.: On improving the quality of gas tungsten arc welded 18Ni 250 maraging steel rocket motor casings. Metall. Materi. Trans. A **48**, 4655–4666 (2017)

Sarkar, A., Narayana Murty, S.V.S., Prasad, M.J.N.V.: Dynamic recrystallization in Cu-Cr-Zr-Ti alloy under large plane strain conditions. Metall. Mater. Transact. A. **51**, 4742–4752 (2020a)

Sarkar, A., Prasad, M.J.N.V., Narayana Murty, S.V.S.: Effect of initial grain size on hot deformation behaviour of Cu-Cr-Zr-Ti alloy. Mater. Charact. **160**, 110112 (2020b)

Sharma, M. Acharya, Agarwal, A., Govind, S.C. Sharma, Sreekumar, K., Sinha, P.P.: Study of shell cracking behavior and its remedies in investment casting process using quick cast rapid prototype polymer patterns. Mater. Sci. Forum **710**, 214–219 (2012)

Shukla, A.K., Sharma, V.M.J., Narayana Murty, S.V.S., Ramesh Narayanan, P., Sharma, S.C.: Integrity of structural and thermo-structural materials for Indian space programme. Proc. Eng. **86**, 8–17 (2014)

Sudarsana Rao, G., Srinath, J., Raman, S.G.S., Sharma, V.M.J., Narayana Murty, S.V.S., Ramesh Narayanan, P., Venkitakrishnan, P.V.: Effect of temperature on low cycle fatigue behavior of annealed Cu-Cr-Zr-Ti alloy in argon atmosphere. Mater. Sci. Eng. A. **692**, 156–167 (2017)

Suseelan Nair, K., Sinha, P.P.: Evolution of materials research for Indian space programmes, ISRO Head Quarters, Bangalore (2009), ISBN 81-908956-1-3

Chapter 17
Nuclear Fuel Complex—A Five Decade Success Story of Indian Nuclear Power Program

Komal Kapoor and Dinesh Srivastava

Abstract Nuclear Fuel Complex (NFC), Hyderabad was conceived as a pivotal industrial arm of the Department of Atomic Energy (DAE) in the late 60s with a mandate to fuel the nuclear power program of India and came into existence in 1972 as major industrial arm of DAE. This article traces its inception, growth, achievements and ambitious plans to support all strategic sectors.

Keywords NFC · Nuclear materials · Zircaloys · Steam generator tubes · Quality Assurance · Fuel production

17.1 Relevance of Nuclear Power

Owing to ever-growing demand for electricity, due to the increase in population and industrial requirements, several forms of electricity generation have been explored. They include coal-based thermal power, hydroelectric, gas-based power, renewable sources (solar and wind), nuclear power and their contributions are shown in Fig. 17.1.

However, the availability and accessibility of quality coal and long term consequences in terms of environmental effects and ecological changes are important points to be considered in running these plants. In view of the recent global concerns about environmental effects of coal-based power production, there is a renewed effort to find suitable alternatives to coal-based power production. Even though hydroelectric power stations account for about 12% of power production, they are restricted to certain geographical locations adjacent to rivers or waterfalls. Similarly,

The original version of the chapter has been revised: The author name "Komal Kapoor" has been updated as corresponding author. A correction to this chapter can be found at
https://doi.org/10.1007/978-981-99-5060-7_33

K. Kapoor (✉) · D. Srivastava
Nuclear Fuel Complex, Hyderabad, India

D. Srivastava
e-mail: dsrivastavabarc@yahoo.co.in; dsrivas@nfc.gov.in

R. Divakar et al. (eds.), *Indian Metallurgy*, Indian Institute of Metals Series,
https://doi.org/10.1007/978-981-99-5060-7_17

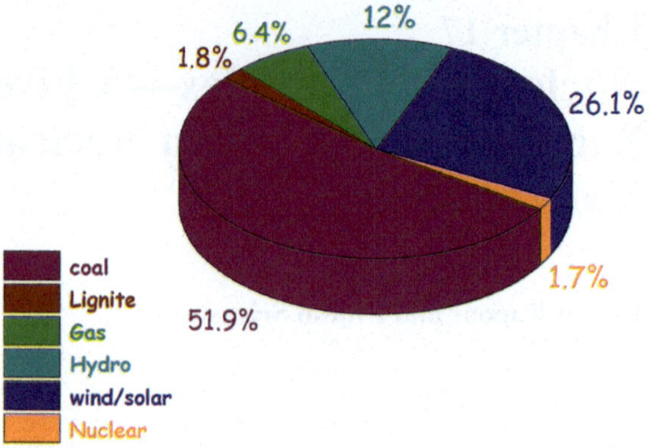

Fig. 17.1 Distribution of the installed power generation capacity of 393.389 GW in India as of Sept 30, 2021, as per Ministry of Power, Government of India

power stations using renewable sources (wind and solar) are also restricted to certain geographical locations and thus not able to cater to the requirements across the length and breadth of the country.

In view of these points and due to several advantages such as providing clean energy, sustainable power generation and availability of reliable power in all climatic conditions and geographical locations, nuclear power has become a promising and competitive source for generation of power. Given the huge resources of Thorium, India can rely on long term power sustainability through nuclear power, along with other energy sources.

17.2 Dr. Homi J. Bhabha—Father of the Three-Stage Nuclear Power Program in India

In view of the advantages as cited above, Dr. Homi Jehangir Bhabha sowed the seeds for nuclear energy in India in the year 1944 and conceived the idea of a Three Stage Indian Nuclear Program. Subsequently the Tata Institute of Fundamental Research (TIFR) was founded in 1945. TIFR was the backbone for the formation of Atomic Energy Commission (AEC) in 1948, followed by the formation of Department of Atomic Energy (DAE) in 1954. The pace of establishment of nuclear power had been hastened in the country after setting up of Nuclear Power Corporation of India Limited (NPCIL).

Nuclear fission of an unstable heavy atom is the genesis for nuclear power generation. U235, Pu 239 and U233 are the various isotopes used in nuclear fission reactions and therefore, one of these isotopes-based fuel with an appropriate coolant and moderator is used in nuclear power reactors. The three-stage Nuclear Power Program as envisaged by Dr. Bhabha was adopted by India and is shown in Fig. 17.2. The Three stages Indian Nuclear Power Program incorporates closed fuel cycle and thorium utilization as the main stay for sustained growth.

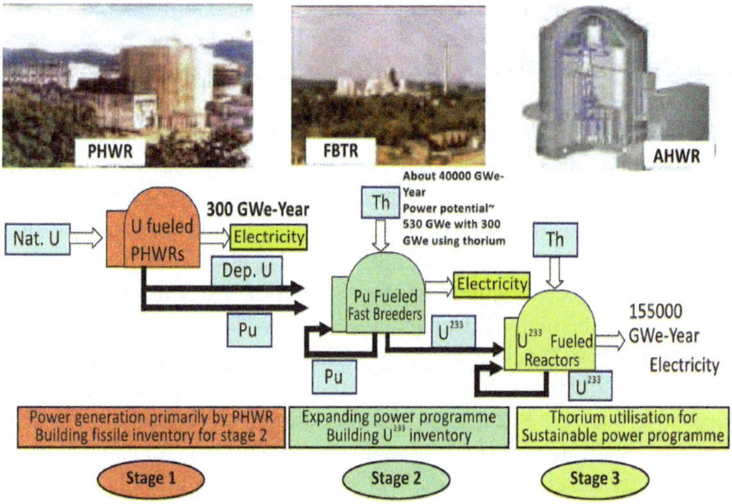

Fig. 17.2 Three stage Indian nuclear power programme adopted by India

This 3-stage Indian Nuclear Power Program involves the use of Pressurized Heavy Water Reactors (PHWRs) with natural Uranium containing small amounts of U235 as fuel and D_2O as coolant and also as a moderator in the 1st stage. Pressurized heavy water exchanges its heat with normal water and the steam thus generated is sent to a turbine to produce electricity. Plutonium-239 is extracted by reprocessing the spent fuel from 1st stage. In 2nd stage, this Pu239 will be used as input fuel for Fast Breeder Reactors (FBRs). Fertile isotope Th232 surrounds the fast reactor fuel core in FBR and the fertile thorium gets converted into fissionable U233 upon neutron absorption. And, as envisaged in 3rd stage Advanced Heavy Water Reactors (AHWRs) with U233 as fuel is to be adopted for power generation. The vast resources of Thorium in India can establish the sustainability of the three-stage nuclear power program adopted for energy security of the country for a long time. Pressurized heavy water reactors (PHWRs) are the major contributors to the nuclear power generation in India.

17.3 Birth of Nuclear Fuel Complex

Nuclear Fuel Complex (NFC), Hyderabad was conceived by Dr Bhabha as a pivotal industrial arm of Department of Atomic Energy (DAE) in the late 60s with a mandate to fuel the nuclear power program of Govt. of India and came into existence in 1972 as major industrial arm of DAE. The survey of land selection for establishing NFC site was personally supervised by Dr. H. J. Bhabha himself in 1965.

Being a premier industrial unit of DAE, Nuclear Fuel Complex (NFC) is engaged in manufacturing and supply of fuel sub-assemblies and other reactor core components with very stringent specifications to India's Pressurized Heavy Water Reactors (PHWR), Boiling Water Reactors (BWR) and Fast Breeder Reactors (FBR) operated by NPCIL. As an ISO 9000, 14,000 & 45,000 certified organization, NFC adheres to all three ISO standards during manufacturing & supply processes. Natural UO_2 and enriched UO_2 in the form of pellets are being used as fuel in PHWR & BWR respectively and MOX/MC as fuel for FBR. Whereas core components include zirconium-based alloys (Zircaloys) such as Zr-4 alloy and Zr-2 alloy as fuel-clad materials for PHWR & BWR respectively and Stainless Steel for FBR. The Zr–Nb alloy is being used for pressure tube manufacturing.

The production of nuclear-grade materials with acceptable quality involves mastery of many diverse technologies in the fields of mechanical, chemical and metallurgical processes. Successful and cost-effective production is a team effort involving, mechanical, chemical, metallurgical, fabricator engineers and quality control personnel. Over these years, NFC has evolved as a bright example for the successful transfer of technology from lab scale to large scale production. This would have not been realized without the contribution of strong work force that is there behind every achievement of NFC.

With comprehensive nuclear fuel manufacturing cycle under its belt, NFC is the only organization in the world to have the capabilities to process uranium and zirconium streams from ore to core, all under one roof.

17.4 Timeline of Milestones

After conceptualization of NFC during 1960s, a pilot plant was established in 1961 to produce nuclear-grade natural UO_2 powder and pellets and the half core of first PHWR (RAPS-I) was fabricated at AFD, Trombay. The rich experience gained during this time has enabled DAE to embark upon commercial production of nuclear fuel and Nuclear Fuel Complex (NFC) has come into existence with a production capacity of 100 Te of PHWR fuel and 24 Te of BWR fuel in 1972 at Hyderabad. The first 19-element PHWR fuel bundle with wire-wrapped spacers was produced in NFC on 8th June 1973 with the available indigenous technologies and assistance from Russia & Canada.

It is appropriate to recollect the past when it took almost 21 years for NFC to produce the first one lakh fuel bundles and next four lakh fuel bundles could be produced in the next 20 years time. Thus it took 40 years to manufacture the first 5 lakh fuel bundles. However, NFC is now able to manufacture one lakh bundles in a year. This has been possible only with the adaptation & implementation of innovative and break-through technologies into manufacturing and inspection & testing methodologies. Every production plant in NFC has progressively increased their plant capacities by incorporating state-of-the-art automation into their process and quality control systems.

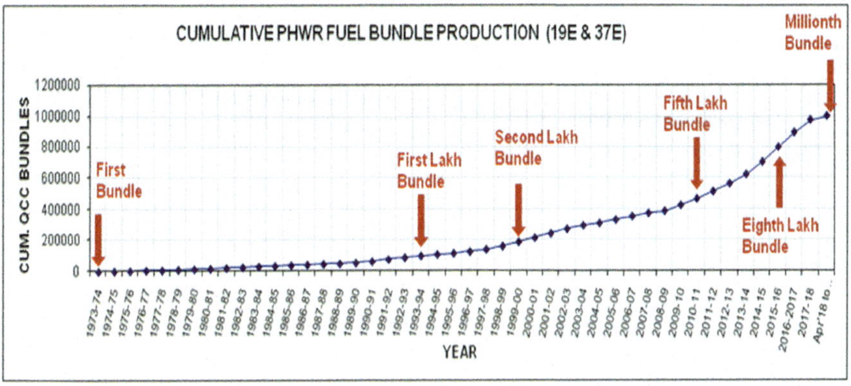

Fig. 17.3 Cumulative PHWR fuel bundle production

Capacity expansion of zirconium sponge production was already achieved in the year 2009 at Zirconium Complex (ZC), Pazhayakayal, Tamil Nadu. A new green field project called as Nuclear Fuel Complex-Kota (NFC-K) is being set-up at Rawthbhatta, Rajasthan as a fuel production facility to augment the additional requirements of India's Nuclear Power Program. Today the emphasis is being made on production with consistent quality and safety with the inclusion of concept of clean & green environment.

During its 50 years of journey, NFC has attained complete self-reliance in the manufacture & supply of fuel and core components for PHWRs and BWRs operating in the country. In this process, it has achieved the following important milestones during PHWR fuel bundle production and the cumulative PHWR fuel bundle production is shown in Fig. 17.3.

To sum up, the successful journey that NFC made during the last 50 years of its existence towards technological excellence from technology denial regime can be described in decade wise as below:

1970–1980: Acquiring Knowledge and 1st mile stone

- India forayed into the domain of PHWRs and NFC was established at Hyderabad with a capacity to manufacture 100 TPY of PHWR fuel.
- Plants were constructed and critical machinery was imported from Canada & Russia.
- It was all new for scientists and engineers to produce Uranium and Zirconium on industrial scale.
- Acclimatization of the processes and machinery of U & Zr production technologies.
- First 19-element PHWR fuel bundle with wire-wrapped spacers in 1973.

Inception of NFC	June, 1972
First Bundle Production	June, 1973
First Lakh Bundle Production	January, 1994
Fifth Lakh Bundle Production	December, 2011
Millionth Bundle Production	September, 2018

1980–1990: Understanding the existing processes and introducing new ones

- The equipment and processes were either replaced or modified to make them suitable to Indian raw materials
- Indigenization of processes has helped in scaling up the manufacture of Nuclear Fuel to 180 TPY
- Welded split spacer design of PHWR fuel sub-assembly
- Graphite coating of clad tubes
- The processes still remained manual and laborious, where Uranium is seen moving from one process step to next process step.

1990–2000: Developing Applications and 2nd milestone

- The decade was significant for NFC as it had to cater to the new PHWR fuel requirements for Kaiga and Kakrapara plants besides meeting the fuel requirement of operating reactors at Kalpakkam and Rawatbhatta.
- This extra demand for fuel had thrown a serious challenge due to huge gap between demand and supply.
- Processes were modified suitably and scaled-up equipment was procured, installed & commissioned successfully to accommodate the newer demands.
- The accumulated raw materials and rejects to the tune of 500MT were successfully processed and converted into finished fuel.
- Manufacturing of First one lakh fuel bundle was completed in 1994.

2000–2010: Decade of Innovations

- During this decade, NFC implemented many process innovations and indigenized process equipment across all operations.
- Manufacturing technology for extremely thin-walled tubing was established and Seamless Calandria tubes were produced for two 540 MWe PHWRs at Tarapur.
- Chamfered, double dish pellets and admixed lubrication.
- Empty tube welding, curved bearing pad.
- 37-element fuel bundle.
- Triple melting of end-plug ingot.
- NFC concentrated on the development of the engineering industry for its futuristic innovative requirements with respect to various reactor components, accessories for fuel assemblies and advanced automatic equipment.

2010–Till Date: Decade of Technological Revolution and significant milestones

- NFC has significantly exceeded its designed capacities by implementing Automation in fabrication and inspection.
- Flexible manufacturing systems have been developed for simultaneous processing of different input materials like Magnesium diuranate (MDU), HTUP, uranium ore concentrate (UOC) and sodium diuranate (SDU).
- An innovative process of direct reduction of ammonium diuranate (ADU) to UO_2.
- Introduction of new route of double radial forging and single pass pilgering for manufacturing Pressure Tubes.
- Technology was developed to manufacture U-bend Alloy 800 tubing in 30-m straight length for Steam Generators of 700 MWe PHWRs.
- Established new manufacturing processes for seamless tubes in circular, square, hexagonal cross sections, Double clad tubes, multi-clad varieties for different types of power reactors.
- Standardization of UO_2 powder characteristics for short sintering cycles.
- Manufacturing of 300MT of NG Zirconium sponge at ZC, Pazhayakayal.
- NFC has achieved the world's highest production of 1512 MT of PHWR fuel during 2016–17.
- Manufacturing of the first one millionth PHWR fuel bundle in 2018.

17.5 Production Activities in NFC

Nuclear grade natural Uranium Dioxide (UO_2) pellets are produced at NFC by converting a variety of raw materials using a well-established conversion process comprising different stages like dissolution, solvent extraction, precipitation, calcination & reduction to get UO_2 powder. Subsequently, pellet fabrication is accomplished through granulation, pre-compaction and high temperature sintering of UO_2 powder. The resultant UO_2 pellets are then qualified for physical integrity and chemical purity for encapsulating them into zirconium alloy fuel tubes for fabricating the fuel sub-assemblies. The three types of fuel sub-assemblies (bundles) fabricated in NFC are 19 Element bundle for 220MWe PHWRs (16.5kg weight), 37 Element bundle for 540/700 MWe PHWRs (23.8 kg weight) and 6×6 BWR bundle for 160 MWe BWRs (230 kg weight) and are shown in Fig. 17.4.

The desired specifications are achieved by having strict control on material manufacturing process. Control on manufacturing process can be achieved through monitoring the quality of materials taking part in the process and also with a strict vigil on process conditions. Chemical Quality Control (CQC) of raw materials, process intermediates and final products will ensure the desired quality of final products and thus, it becomes an integral part of QA/QC program. As an independent department, Quality Assurance (QA) group ensures the quality at intermediate stages and final stage of production to ensure the compliance to the requirements of the customer. Safety and Environment Protection are also ensured with the designated departments during entire manufacturing process.

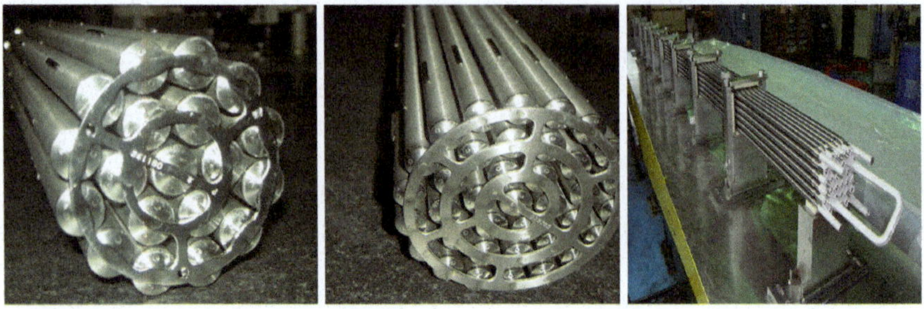

Fig. 17.4 The three types of fuel sub-assemblies (bundles) fabricated in NFC **a** 19-element bundle for 220 MWe PHWR, **b** 37-element bundle for 540 and 700 MWe PHWR, and **c** 6 × 6 BWR bundle for 160 MWe BW

Besides this, NFC also manufactures different reactor components using special alloys including special steels for strategic application and also produces high purity materials of 5N & 6N purity for electronic & tool applications. Apart from principle customer NPCIL, the list also includes DRDO, ISRO, HAL, BHAVINI, IGCAR, BARC, RRCAT, etc.

The diversified production activities of NFC for nuclear and non-nuclear applications are summarized in Fig. 17.5.

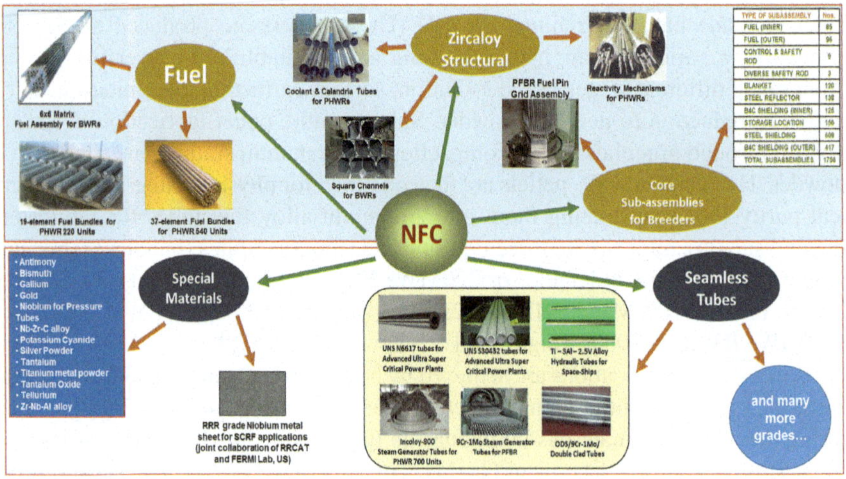

Fig. 17.5 Diversified production activities of NFC

17.6 Contributors to the Success of NFC

The success of NFC is the success of its production facilities and the technically strong & skilled team of workforce. They are engaged in manufacturing of nuclear fuel, in-core and out-of-core structural, special tubes and special materials backed up by a strong quality monitoring group to ensure the quality of the materials produced. Safety group is functioning to take care of the safety of all operations. The activities of individual production plants at NFC are described below.

17.6.1 Nuclear Fuel Manufacturing Plants

Nuclear Fuel manufacturing is very critical and plays a vital role in nuclear fuel cycle. The stringent chemical, physical and metallurgical requirements of nuclear fuel need to be engineered into the nuclear fuel during manufacturing. The physical characteristics also need to be engineered based on the operation conditions of the nuclear fuels which are quite harsh and also vary from reactor to reactor. The nuclear fuels during their operation are subjected to high-temperature operating conditions, high neutron flux environment which leads to physical and metallurgical changes.

Nuclear fuel is made for three types of nuclear power reactors, namely Pressurised Heavy Water Reactor (PHWR), Boiling Water Reactor (BWR) and Fast Breeder Reactor (FBR).

PHWR Fuel Assemblies: Natural uranium dioxide (Nat.UO_2) is the fuel for PHWRs and is obtained from different raw materials like Magnesium Di-Uranate (MDU), Sodium Di-Uranate (SDU) or Uranium Ore Concentrate (UOC). MDU/SDU concentrate is obtained from the indigenously milled uranium mines at Jaduguda, Jharkhand/Tummalapalli, AP and supplied by Uranium Corporation of India Limited (UCIL) and UOC is imported from different countries. MDU/SDU/UOC is subjected to dissolution followed by solvent extraction and subsequently precipitation of uranyl nitrate with ammonia to get Ammonium Di-Uranate (ADU). A further step of controlled heat treatment of calcination of ADU gives Uranium dioxide (UO_2) powder. The UO_2 powder is then processed to high-density cylindrical pellets by various operations like pre-compaction, final compaction and sintering at high temperature (1700 °C) in reducing atmosphere. The sintered UO_2 pellets are then centre-less ground to desired dimensions.

The finished UO_2 pellets are encapsulated in thin-walled Zircaloy tubes, both ends of which are sealed by resistance welding. Appendages such as spacers and bearing pads are resistance welded on these elements and 19 or 37 such elements of specified configuration are assembled together by welding them on to end plates at either end to form 19-element fuel assembly designed for 220 MWe reactors and 37-element fuel assembly designed for 540 and 700 MWe reactors.

BWR Fuel Assemblies: Cylindrical UO_2 pellets of varying enrichments and chemical compositions imported from other countries are encapsulated in thin-walled tubes of zirconium alloy, both ends of which are sealed by TIG welding. Elements with varying compositions are placed in a specified configuration such as 6 × 6 array along with spacer grids, stainless steel tie plates, zirconium alloy spacers and flow nozzles to form 6 × 6 nuclear fuel assemblies for BWRs.

Fast Breeder Reactor (FBR) Fuels: NFC is also responsible for fabrication of core sub-assemblies for Indian Fast Breeder Reactors deployed under 2nd stage of Indian Nuclear Power Program. The facility at NFC presently caters to the requirements of core sub-assemblies for two reactors namely 13 MW(e) Fast Breeder Test Reactor (FBTR) and 500 MW(e) Prototype Fast Breeder Reactor (PFBR).

NFC fabricated all the core sub-assemblies such as Fuel, Blanket, Nickel Reflector, Carrier and Special Assemblies for its initial core of FBTR in the beginning. Since then it is also engaged in continuously supplying annual requirements of Fuel and special sub-assemblies of FBTR. A typical FBTR fuel sub-assembly consists of 511 intricately machined components of 35 different types.

The Core sub-assemblies are hexagonal in shape with very thin wall special grade Stainless Steel (SS) tubes (circular and hexagonal) and precision SS components. These were fabricated with in-house developed know-how and equipment/fixtures built with indigenous capabilities. Pelletisation of the thorium oxide (ThO_2) has been carried out for the first time on a large scale that involved considerable ingenuity and effort.

The fabrication of various FBTR core sub-assemblies involves unique operations like button forming and chrome plating on hexagonal tubes, crimping on clad tubes, bead forming on spacer wires, helical wire wrapping on pins, etc. These operations are executed on special-purpose machines which were also developed indigenously. The fabrication of FBTR fuel involves substantially high radioactivity, which in turn increases complications in manufacturing. Deploying radiation protection lead shields at various process workstation help to overcome these challenges.

NFC fabricated and supplied core sub-assemblies like Fuel, Blanket, Control & Safety Rod, Diverse Safety Rod, Reflector, Inner Boron Carbide Shielding, Diluent, Purger, Source and Instrumented Central Sub-assemblies for initial core of PFBR. A typical PFBR Fuel sub-assembly consists of 1541 intricately machined components of more than 35 different types. Also, it involved the supply of about 60,000 crimped tubes and 2.5 Lakhs precision fuel pin components fabricated in-house/ outsourced meeting the stringent specifications involving Quality Assurance team of NFC and Quality Surveillance team of NPCIL/BHAVINI. The Core sub-assemblies are hexagonal in shape and are 4.5 m long. The fabrication processes of most of the machining components were developed and outsourced due to bulk quantity requirements. The various equipment, fixtures and radiation shields were developed indigenously for unique operations like Spacer Wire Bead Forming, Wire Wrapping, Automatic TIG Welding, Button Forming, Robotic fuel pin assembly, etc. NFC has also developed Automatic autogenous TIG welding technique for welding of very thin-walled Fuel clad tubes and Hexagonal wrappers. The complexity of fuel sub-assemblies involves handling and manufacturing of substantially high radioactive fuel pins, which were overcome by deploying special automation techniques and (composite) radiation shields for protection against gamma and neutron radiation as per regulatory requirements and safety standards. In addition to this, a special fuel assembly shop/facility has been established close to reactor site at Kalpakkam.

NFC is also involved in planning and setting up of nuclear fuel fabrication facilities for fabrication of various types of pins and sub-assemblies required for annual reloads of PFBR and FBRs. Fabrication of Fuel for FBTR & PFBR using In-house/indigenous made equipment and fixture is an example of India's self-reliance and Make in India policy of GoI.

17.6.2 Reactor Grade Zirconium Metal Production Plant

Reactor Grade (RG) Zirconium metal (RG Zr metal) is produced by limiting critical impurity element Hafnium (Hf) making it suitable for nuclear applications.

Zircon sand is the raw material for production of Zirconium metal. It comprises of Zirconium Silicate containing 67% Zirconium with about 2% Hafnium. Hafnium being a neutron absorber element (due to its high neutron absorption cross-section), its removal is an essential step in the nuclear metallurgy of Zirconium. The permissible limit for Hafnium is very less in Zirconium. The first step of removal of Silica involves the treatment of zircon with Caustic Soda. The resulting frit is washed with water to remove water-soluble Sodium Silicate. Then Zirconium and Hafnium are brought into solution with Nitric Acid. The separation of Zirconium from Hafnium and other impurities is achieved by solvent extraction using Tri-n-Butyl Phosphate (TBP) in Kerosene as the solvent. Zirconium is preferentially extracted into the organic phase leaving Hafnium and other impurities in aqueous phase. The Zirconium-laden organic is then stripped to recover pure Zirconium in aqueous solution. Zirconium Hydroxide is precipitated from the pure solution using Ammonia and the Hydroxide is filtered, washed, dried and calcined in a rotary kiln to obtain pure Zirconium Oxide. The oxide is hammer milled to fine powder.

Hafnium-free Zirconium Oxide is converted to Zirconium Tetrachloride ($ZrCl_4$) intermediate through chlorination operation in static bed reactors. Zirconium Tetrachloride is then reduced to Zirconium through magnesio-thermic reduction (Kroll's reduction) operation. The reduced mass of Zirconium after mechanical separation of $MgCl_2$ by-product is subjected to pyro-vacuum distillation operation for removal of entrapped Mg and $MgCl_2$ to get pure Zirconium metal.

17.6.3 Fuel Cladding and Assembly Components Production Plants

Chemically qualified Zirconium sponge is converted into different types of zirconium alloys after addition of required quantity of alloying elements and melting. Zirconium metal and the alloying elements are compacted in hydraulic presses to

Table 17.1 Composition of various Zircoloys

Zirconium alloy type	Alloying elements (Weight %)				
	Sn	Fe	Cr	Ni	Zr
Zircaloy-1	2.5	–	–	–	Balance
Zircaloy-2	1.5	0.12	0.1	0.05	Balance
Zircaloy-3	0.25	0.25	–	–	Balance
Zircaloy-4	1.5	0.22	0.1	–	Balance

obtain compacts/briquettes. These compacts are welded to each other by electron beam welding under vacuum to obtain a long cylindrical electrode. These electrodes are melted multiple times by consumable electrode in vacuum arc re-melting furnace in water-cooled copper crucibles, with intermediate stage machining for obtaining final ingots. The composition of different Zircaloys is given in Table 17.1.

Zircaloy ingots are subjected to 1st stage of extrusion, machining and cutting. After making a hollow billet, it is subjected to beta-quenching, machining and 2nd stage extrusion in order to obtain a hollow blank. This hollow blank is stress relieved in vacuum and passed on for multi-stage pilgering with intermediate vacuum annealing. Tube finishing operations like straightening, grinding, cutting, etc. are also performed to obtain the requisite stringent quality.

For this purpose, NFC possesses state-of-the-art fabrication facilities such as extrusion & piercing press, cold rolling mills, vacuum annealing furnaces, special surface finishing and treatment equipment are available to achieve the desired mechanical and metallurgical properties of cladding tubes.

Fuel tubes are assembled together with components, like End plates, End Caps, Bearing Pads & Spacer Pads, to form a fuel bundle. Bars required for manufacturing end plugs are produced by cold rotary swaging operation of the extruded bars in different passes with intermediate vacuum annealing, final straightening and center-less grinding to required sizes. As the end plug acts as the first sealing of fission products, the produced bars undergo rigorous quality checks. Similarly, Zircaloy Ingot is extruded to slabs which are then rolled to sheets. Bearing pads and spacer pads are punched out of these sheets. These components are made out of either Zircaloy-2 (for BWRs) or Zircaloy-4(for PHWRs). For these, NFC is equipped with mechanical presses, Automatic Vibratory finishing system, centre lathe machines, automated machining centres, etc.

17.6.4 Nuclear Reactor Core Component Production Plants

Seamless tubes of different sizes are being manufactured using alloys of Zirconium, titanium and special-grade stainless steels. Pressure Tubes (Zr-2.5wt% Nb alloy), Calandria Tubes (Zircaloy-4) and Garter Spring (Zr-2.5wt% Nb-0.5wt% Cu alloy) are the critical core structural of Pressurised Heavy Water Reactors (PHWRs). Square

Channels (Zircaloy-4) are used in Boiling Water Reactor (BWRs) and Hexcans (SS316/ D9 alloy) are used in Fast Breeder Reactors (FBR).

Manufacturing process routes for these critical cores structural are successfully developed and continue to be supplied to all the PHWRs, BWRs and FBRs.

Also, the manufacturing process route for reactor control assemblies required for PHWRs is successfully developed and continues to be supplied to all the PHWRs. These assemblies are designed for reactor power monitoring, control mechanisms and shut down. These are made of zirconium alloys and require high precision, reliable components and high quality tubes before welding. The manufacturing processes use hot extrusion, forging, pilgering, drawing of various sizes of tubes and punching, machining of components. These assemblies use a combination of electron beam welding, TIG welding and resistance welding and have stringent quality specifications for soundness of welds and accurate dimensional control.

For this purpose, advanced welding and machining facilities such as electron beam welding equipment, specialized TIG welding machines, CNC machines, pilger mills, precision roll joint machines, vacuum heat treatment furnaces, special surface finishing and treatment equipment are available in NFC.

In addition, NFC has contributed to various developmental programs of DAE such as Compact High Temperature Reactor, Upgraded APSARA Reactor through advanced machining and welding of exotic materials.

17.6.5 Stainless Steel and Special Alloy Tubes Production Plants

These are exclusive facilities for development & manufacturing the seamless tubes using various advanced grades of Stainless Steels & Special alloys, Nickel-based super alloys, Iron-based super alloy, Titanium alloys, Maraging steels for Nuclear, Space and Defence strategic applications. They house state-of-the-art manufacturing facilities like Cold rolling mills (Pilger mills), Tube straightening mills of different capacities & sizes, Draw-bench, Heat treatment facilities like Bright Annealing furnace, Vacuum Annealing furnace, Roller Hearth (LPG fired) Annealing furnace, Chemical operations like De-glassing, Pickling, Passivation, Alkaline degreasing, Solvent degreasing and Inspection facilities like Ultrasonic, Eddy Current & Hydrostatic Pressure Testing.

Manufacturing starts with forged & machined cylindrical billets which are hot extruded to hollow blanks. Extruded blanks undergo series of thermo mechanical processing to achieve desired properties. Finished tubes are subjected to various Inspection & Tests as per customer requirements. NFC has played a pivotal role in indigenous development & manufacturing of these products as import substitute and is an excellent example for Make in India policy.

17.6.6 Special Materials Plant

Nuclear Fuel Complex is also the country's premier facility engaged in the manufacture of a variety of high purity materials (5N/6N) and they find numerous applications in Electronics, Defence, Nuclear Industries, Scientific & industrial research organizations, institutions of higher learning and even in general engineering industry.

High purity materials such as antimony, bismuth, cadmium, selenium, tellurium, gallium, phosphorous oxy chloride, antimony trioxide, gold, gold potassium cyanide (GPC) and silver are produced. In addition, tantalum pentoxide, tantalum metal and reactor grade niobium metal in the forms of rod, sheet and crucibles are also produced. The high purity materials are used in semiconductor technology for the synthesis of compound semiconductors, and as dopants, diffusants, solders, etc. Tantalum is used in a variety of high temperature and corrosive atmosphere applications. Tool grade tantalum pentoxide finds its application in tool industry. Reactor grade niobium is used in nuclear industry for alloying of zirconium to produce special ZrNb alloys. NFC has produced Residual Resistance Ratio (RRR) grade niobium metal for use in super conductivity (SC) applications. Advanced alloys such as NbTi, NiTi, NbZrC, have been developed by electron beam melting route.

The production of these materials involves a variety of highly sophisticated equipment, advanced techniques, clean working environment and specialized technical skills. The gamut of operations include Hydrometallurgy, Pyro-metallurgy, Electrolytic processes, Solvent extraction, special distillations, zone refining, Electron beam refining, etc. Rigorous quality control is exercised at all stages of production to achieve quality and reliability of products. The availability of wealth of talent, advanced equipment and state-of-art of technology, backed up by excellent quality assurance processes ensures the quality of products. The characterization of high purity materials and intermediates using specialized analytical techniques such as AAS, ICP-OES, XRFS, UV–Vis Spectrophotmeter, IGF-based gas analyzers and Classical chemical methods are being done.

Further, NFC has transferred developed technologies to prospective entrepreneurs and they include technology for the production of Magnesium granules, Zirconium metal power, production of high purity materials such as Phosphorous Oxychloride, Indium, Sodium Iodide, Gallium, Gold, Silver, Capacitor grade tantalum powder and Tantalum anodes, etc.

A new production plant is being set-up in collaboration with ISRO for the production of niobium for exclusive usage in Indian Space programme.

17.6.7 Quality Assurance Group

Quality is important in any field of human endeavour, more so in a critical, high-tech area such as nuclear power plants where the costs of failure are extremely high with respect to material loss and also from the societal angle. The demands thus made on the Quality Assurance (QA) programs in DAE in general and NFC in particular, are on an altogether different level compared to those in other industries. NFC is unique in its integrated approach to the manufacture of a variety of finished products from ore to core through employing rigorous Inspection, QA and Non-destructive Evaluation (NDE) on an industrial scale. Because of these demands, NDE in NFC has, through years of experience, attained a high level of maturity. The high standards of the products manufactured at NFC are highlighted by their excellent real-time performance. NFC is striving to achieve Six Sigma strategy as a part of continual improvement in its operations, products and services through technological excellence and well structured QA program.

The exacting performance required by the nuclear fuel and hardware in the Power Reactor demands fulfilling stringent quality requirements of each product's specifications. NFC adopts well-structured quality assurance program. The goal is realized by employing a set of sophisticated equipment and techniques with qualified manpower. The State-of-Art NDE facilities, Centralized Analytical Quality Control Laboratory, Metallurgical & Mechanical testing and Advanced Material Characterization Laboratory carry out the testing, analysis and measurement in order to ensure process control and product compliance to specifications. Periodic reviews and brain storming are conducted to improve recoveries.

The sophisticated non-destructive evaluation facilities developed and techniques established which include ultrasonic test, eddy current test, X-ray radiography, dye-penetrate test, mass spectrometric leak detection and automated machine vision systems. Precision dimensional measurement systems and automatic physical measurement facilities cater to the in-house needs for measurement and qualification of products. An array of analytical techniques like Inductively Coupled Plasma Atomic Emission Spectrometry, Atomic Absorption Spectrometry, X-ray Fluorescence Spectrometry, Mass Spectrometry, Gas and Ion Chromatography, Laser Photometry are employed for analysis of raw materials, intermediates and final products.

A number of online measurement and control equipment have been introduced all along the processes and production lines. The Advanced Material Characterization Lab is equipped with several sophisticated characterization instruments viz. TEM, SEM-EBSD, XRD, Dilatometer, X-ray residual stress analyzer for studying metallurgical characteristics like microstructure, texture, dislocation density, re-crystallization behaviour, crystallography, nano-feature characterization, residual stress analysis etc. These facilities help in developing a variety of products and advanced alloys.

The emergence of several Quality Circles bears testimony to the fact that each and every employee working in their area is concerned and dedicated towards improving quality. Every year, the month of November is observed as Quality Month. As part of Quality Month Celebration, several lectures by eminent speakers and competitions like quiz, slogan and essay writing are conducted to create Quality Awareness among the employees.

NFC has over the past five decades earned a justifiable pride place in DAE as a reputed, reliable, quality-conscious supplier of critical inputs to the nuclear power program in India.

17.6.8 Safety Engineering Division

NFC gives utmost importance to the safety of workers, environmental protection and prevention of accidents/incidents at field level. It is regulated by Atomic Energy Regulatory Board (AERB). For taking care of overall safety aspects, to achieve the objectives of safety and co-ordinate with various plants, Safety Engineering Division (SED) was formed in the initial years of NFC. SED coordinates with AERB on regular basis for implementation of Factories Act, 1948 and Atomic Energy Factories Rules, 1996 and other statutes.

17.6.9 Environment Protection

While fulfilling the mandate of supplying the nuclear fuels and nuclear reactor components to Nuclear Power Program and support to strategic program of India, NFC gives utmost importance to protect environment by way of proper handling and disposal of by-products and effluents. A dedicated Effluent Management section comprising of expert chemical engineers is working towards this goal. The section is responsible for safe & prompt disposal of various process effluents generated from the various activities at NFC to firms authorized by Telangana State Pollution Control Board (TSPCB). Also, various reports are sent to statutory authorities viz. TSPCB, Atomic Energy Regulatory Board, Ministry of Environment and Forest & Climate Change. Items viz. Used oil, Spent resin, Boiler soot, Used polyurethane sponge plugs, Thermocol, etc. which were posing a significant safety/fire hazard are also disposed properly.

17.6.10 Zirconium Complex, Tamil Nadu (a Unit of NFC, Hyderabad)

In order to meet the additional requirements of production of zirconium metal, Zirconium Complex (ZC) was conceived in 2001 as a green-field project at Pazhayakayal, Tuticorin, Tamil Nadu. The complex was commissioned in November 2009 to produce Zirconium Oxide and Zirconium Sponge to meet the enhanced demand. During the last 13 years, the production capacity of the Plant has gradually increased and the rated capacity was achieved in FY2014–15.

Table 17.2 Expansion of nuclear power in India

Reactor type	No. of units	Capacity (MWe)
PHWRs	40	19,860
LWRs	34	39,820
FBRs	9	4,500
Total	83	64,180

Use of zirconium and its alloys for nuclear applications requires stringent monitoring of quality of raw materials, chemicals and intermediate process streams/products. A Materials Testing Laboratory is set up for analysis of samples using state-of-the-art analytical instruments for process control and quality control of products.

Zirconium Complex was certified for Quality Management System as per ISO-9001:2015 by Bureau of Indian Standards on 15th May 2018. Continuous efforts towards improvement/development of production processes are being made for higher productivity, better recovery, lesser consumption of chemicals and reduction of effluents. With these efforts, ZC is achieving newer heights and excelling at fulfilling the mandate of the Department of Atomic Energy.

Zirconium Complex has future plans of setting up of Magnesium Recycling Technology Development & Demonstration Facility and capacity augmentation of zirconium sponge production to meet the future demand commensurate with the nuclear power programme and the ground works in this direction have been already initiated. Further to this, a plan of action has been worked-out to enhance Zr production to 1300 TPY through capacity expansion at Zirconium Complex, Pazhayakayal in two phases.

Based on the ambitious plan of generating 60 GWe of nuclear energy by the year 2050, about 20 GWe is proposed to be generated through PHWRs and the remaining through LWRs and FBRs as shown in Table 17.2.

Accordingly, PHWR design has been successfully scaled up to higher capacities of 540 MWe (TAPS 3 & 4) and 16 numbers of 700 MWe. Towards this, presently 4 nos. of 700 MWe PHWRs are under advanced stages of completion at Rajasthan (RAPS 7 & 8) and Kakrapara (KAPP 3 & 4). Further, Government has accorded administrative approval and financial sanction for construction of 12,700 MWe PHWRs. This expansion in nuclear power program requires additional annual supply of 2000 MT of Fuel, Structural for 2–3 Reactors, Reactivity Mechanisms for 2–3 Reactors and supply of 40 Sets of steam generator (SG) Tubes. In view of this, NFC has geared up to meet these additional requirements and accordingly set up the following facilities.

17.6.11 NFC-Kota, Rawatbhata, Rajasthan

A new fuel fabrication facility is being set up (NFC-Kota) at Rawatbhata near Kota, Rajasthan. This green field project is established with plant capacity of 500 TPY

PHWR fuel fabrication and 65 TPY fuel cladding fabrication. The capacity of fuel cladding fabrication will be further augmented by 100 TPY. NFC-K project envisaged to establish 37 element fuel bundle manufacturing facility to cater to fuel requirement of up-coming 4 Nos: of 700 MWe PHWRs. The project is in an advanced stage of completion and is expected to take-up the production shortly.

17.7 Experiences in Indigenous PHWR Fuel Manufacturing

With its expertise in multi-faceted technical areas, NFC has become mature, self-sufficient in manufacturing technology of nuclear fuel fabrication and structural materials. The lean grade ores from mines with varying geology, leaching methodology & concentration and ores with diverse composition with wide variety of impurities pose several challenges in achieving consistent UO_2 powder quality & UO_2 pellet sinterability. This has thrown a challenge to NFC in meeting the enhanced fuel fabrication requirements to meet energy demands. However, after conducting several lab scale/pilot scale experimental studies and with thorough understanding of the results from these studies, NFC has overcome the technological challenges and successfully demonstrated large scale production of consistent quality fuel pellets from the raw materials of various chemical forms received from indigenous sources.

Introduction of many innovative state-of-art technologies for various processes, equipment mechanization and automation led to ergonomic advantages for the plants. Continual analysis and optimization of the process parameters resulted in improvement of overall process recovery with higher productivity, environmental and radiological safety.

The primary mandate entrusted to NFC could be successfully met with the adaptation of technology up-gradation, introducing automation and capacity augmentation. The following paragraphs highlight some of those achievements.

17.7.1 Innovative Technologies/Technological Breakthroughs in Fuel Manufacturing

17.7.1.1 Automation in NFC

Fuel fabrication involves large no. of chemical, metallurgical & mechanical processes and involves handling of very large no. of intricate components with intensive inspection at every stage of manufacturing/fabrication. Also, 'Zero Failure' target requires reliability, reproducibility and precision across all stages of production & inspection involving raw materials, intermediate stages and final products.

Automation is the key and essential to achieving these goals. Apart from this, automation benefits in safe handling of radioactive & hazardous materials on large scale and thereby reducing the radiation exposure to working personnel and air born activity levels. Elimination of manual operations with automation will enhance the safety in operations through remote/centralized controls & provision of alarms/ safety inter-locks. Additionally, it economizes the process by minimizing the failure rates & manpower requirements, especially to meet the high demands of production. Automation further improves the digitization and documentation processes.

17.7.1.2 Developments in UO2 Powder Production

Sinterability and physical integrity of UO_2 pellet is largely dependent on the physical characteristics of the UO_2 powder, which in turn is closely related to powder process parameters, such as dissolution temperature, organic/aqueous (O/A) ratio, uranium concentration, free acidity of uranyl nitrate solution, precipitation temperature, flow rate of the precipitant and temperature profile during heat treatment of the intermediates. It has been observed qualitatively that the effective filtration of uranyl nitrate extract before precipitation, plays a crucial role in achieving defect-free UO_2 pellets by filtering out various solid insoluble impurities.

For production of UO_2 powder with required characteristics that are acceptable in fuel pellet production coping up with varying impurity levels in raw materials, various process improvements and parameter optimizations have been carried out at various stages of UO_2 powder production process as detailed below.

Increase in dissolution temperature of raw material: A raise of 10°C in raw material dissolution temperature from the existing condition has ensured the complete conversion of dissolved organics and soluble silica present in the raw material.

Change of Uranyl Nitrate feed in Slurry Extractor: Introduction of Uranyl Nitrate Feed (UNF) to the Slurry Extractor at 3rd stage in place of 1st stage has reduced the aqueous entrainment in Uranyl Nitrate extract, which also decreased the tendency of formation of emulsions in subsequent stripping operation.

Increase in Uranyl Nitrate Feed concentration. The feed concentration has been increased by 75 gpl from the existing concentration to decrease the aqueous entrainment in the Uranyl Nitrate Pure Solution (UNPS) extract which is required to have better phase separation during solvent extraction process.

Introduction of Vapour Ammonia during precipitation The vapour ammonia precipitation of UNPS was introduced in place of liquid ammonium hydroxide solution to get ammonium diuranate (ADU) powder with desired size & morphology. Also, time of precipitation was reduced and minimized the liquid effluent generation as well.

Increase in % TBP in solvent extraction process: It is necessary to increase extract saturation to reduce the impurities in the extract and it was achieved by increasing the percentage of Tri Butyl Phosphate (TBP) by 3% from the existing condition during extraction. This has resulted in increase of U concentration in the extract by 10 gpl which is equivalent to an increase of extract saturation from 90 to 95%.

Introduction of modified baffle in Calcination furnace: For homogeneous mixing and complete conversion of ADU to U_3O_8, heat transfer simulation studies were carried out by modeling the calcination and reduction operations involved in the powder production. Accordingly, the baffle plates of calcination & reduction furnaces were suitably modified and a unique design of baffle cage as shown in Fig. 17.6 was introduced in the rotary furnaces for calcination, reduction and stabilization operations. This has resulted in increased powder lifting and heat transfer to the powder. After the modification, there was a remarkable decrease in calcination temperature (by about 20°C) with the required powder quality consistently due to the cascading effect realised with the provision of lifters.

Modification in AU Precipitation: The powder morphology has been customized by optimization of critical parameters like Ammonia Flow rate, UNPS concentration and Precipitation Temperature, terminal pH. Using the optimized parameter, the batch precipitation time has been reduced and resulting in more chemically active AU powder. Furthermore, the UO_2 powder produced from this AU powder has resulted in reduction in the temperature of subsequent calcination operation and played a pivotal role in reduction of sintering cycle.

Fig. 17.6 Unique design of baffle cage for rotary furnaces and the picture of calcination furnace

Modifications in Stripping Operation. The Port and the Mechanical Coalescers were modified to increase the stripping efficiency by 100%. The port modifications (before and after) and Mechanical Coalescers are shown below in Fig. 17.7a–d respectively.

The several process modifications carried-out in UO_2 powder production along with benefits obtained so in the plant are summarized in the following Table 17.3.

Automation in UO_2 Powder Production In order to have reliability in process with safety interlocks and reproducible quality in UO_2 power production, automation in powder production has been introduced. They include

(i) Automated operations through PLC-Based SCADA
(ii) Automated Radioactive Slurry Transfer System

The automation has resulted in an increase in recovery to around 99% (one of the highest recovery worldwide), 20% reduction in process variability, 40% reduction

Fig. 17.7 **a**, **b** Port modifications, **c** mechanical coalescer (MS), **d** MS placed in mixer settler

Table 17.3 Process modifications in UO_2 powder production and capacity increase

Sl. no	Process	Capacity increase (%)	Technological advancements
1	Dissolution	100	(a) Oxidative dissolution (b) Automated process control
2	Slurry extraction	100	(a) High Conc (b) High flow rate feed (c) Increase in TBP Conc (d) Hydrodynamic modifications
3	Stripping	100	(a) Mechanical coalescers (b) Port modifications
4	Precipitation	50	Change of AU morphology
5	Filtration	85	(a) Automated vacuum receivers (b) Powder chute modification
6	Drying	70	(a) Additional heaters (b) Thermal insulation
7	Calcination & reduction	60	Furnace internal modification

in man shifts, 55% reduction in average air activity with the complete elimination of manual operations.

There has been more than tenfold increase in the production of UO_2 powder using the existing infrastructure after the process modifications.

17.7.1.3 Developments in UO_2 Pellet and Fuel Bundle Fabrication

Development of Advanced 3-D blending machine Powder blending is one of the important process steps where UO_2 granules of various size distribution and powder lubricant are mixed homogeneously before undergoing pellet fabrication in final compaction press. An advanced 3-D blending machine was designed and developed indigenously to ensure homogeneity in mixing of UO_2 powder with lubricant. This has reduced the powder blending time drastically from 30 min to about 4 min and helped in lowering the variations in green pellet density by five times. The 3D Blending Machine as shown in Fig. 17.8 has replaced the existing carbouy rolling machine.

Integrated Blending & Granule Transfer System and Rotary Compaction Press. A newly designed Integrated Blending & Granule Transfer System (Fig. 17.9a) and Rotary Press (Fig. 17.9b) for Final Compaction of UO_2 powder was introduced to reduce density variation in green pellets by 50% and for two-fold increase in the productivity. Also, consistent Sintered Density and reduction in handling defects like chips (reduced by 3%) was achieved by using these systems.

Development of automatic pellet stacking machine. An industrial scale image processing-based automatic pellet stacking machine has been developed successfully for the first time in NFC. The machine is based on vision technique where the image of un-stacked pellets row is captured and is processed to find individual pellet length and its position. An advanced algorithm is developed for selection of pellets from row being stacked and buffer to obtain required stack length ($480.68 + 0.00/-1.60$ mm). A robot is used for interchange of pellets as decided by the algorithm. After stacking, stacked row is verified for stack length by Linear Variable Differential Transformer

Fig. 17.8 3D blending machine

Fig. 17.9 **a** Integrated blending & granule transfer system. **b** Rotary compaction press

(LVDT). Machine speed is equivalent to an operator speed. This auto pellet stacking machine as shown in Fig. 17.10 provides great ergonomic advantages by eliminating dependence on skilled operator for stacking and it also minimizes radiation exposure to the working personnel.

Fig. 17.10 Automatic pellet stacking machine

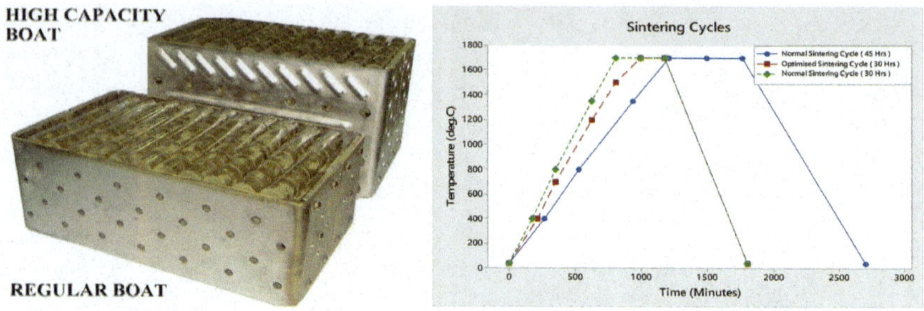

Fig. 17.11 High capacity molybdenum boat with MLR material and modified sintering profile

Design & Development of High Capacity Molybdenum Boat with MLR Material. Existing Molybdenum sintering boat design was modified with MLR (Molybdenum-lanthanum oxide recrystallized) grade material for improved sintering quality, thermal life and capacity. The innovative corrugation design at free surfaces has resulted in better heat transfer and sintering. With modified boat, both capacity per boat and thermal fatigue life of the boat has increased by 50%. The modified boat is shown in Fig. 17.11.

Modified Sintering Profile. A multistep heating profile has been developed after in-depth analysis of the UO_2 powder morphology. The pushing interval has been reduced, resulting in reduction of UO_2 sintering cycle time by almost 50%. Furnace throughput also increased by 40%. This temperature profile is being used in regular production activities. The modified sintering profile is shown in Fig. 17.11.

With these modifications, the overall sintering output has doubled during regular production.

Automated Green Pellet Charging System. The system is being used for automatic transfer of UO_2 green pellets into Mo boats and the developed PLC-based Pellet Charging System has been integrated with Compaction press. It works on vacuum pick and place concept. Elimination of manpower for boat preparation and minimizing radiation exposure to operator, elimination of chipping defects at green stage due to zero manual handling, increase in visual recovery by 3% are among the advantages. It is shown in Fig. 17.12a.

Fig. 17.12 **a** Automated Green Pellet charging system. **b** Sintered Pellet discharging system

Automated Sintered Pellet Discharging System. The system is being used for automatic discharge of sintered UO2 pellets from Mo boats into SS rod trays for inspection and grinding. Vision-based SCARA Robot processes the image from Area Scan Camera and analyses the position of the pellets to place on SS trays. Elimination of manpower and minimizing radiation exposure to operator, minimizing of chipping defects are among the advantages. It is shown in Fig. 17.12b.

17.7.1.4 Automated Appendage Welding

Integrated Spacer & Bearing Pad Welding Unit. This automatic PLC-based control system as shown in Fig. 17.13 is involved in handling, loading & un-loading of fuel-clad tubes and feeding of spacer & bearing pad along with pick & place for positioning of appendages. It can do 1800 appendages welding per shift per machine.

Integrated End Cap Welding & Machining (IECWM) Unit. Introduction of IECWM unit has eliminated manual handling of fuel element/tube and there-by reducing Radiation Exposure to operating personnel and improving overall safety. It can do 120 elements per hour of production which almost an increase of 150% from exiting production and with a recovery of 99%. The unit operations are shown in Fig. 17.14a–d.

Robotic End Plate Welding Machine for 37 Element PHWR Fuel Bundle. This facility has been established for fabrication of 37 Element PHWR Fuel bundles meant for forthcoming 700 MWe PHWRs. Annually 65,000 PHWR Fuel Bundles are being produced in this facility. Consistent Bundle Recovery of around 99% was achieved with a reduction of 15% in mandays. It has given an improvement of 20%

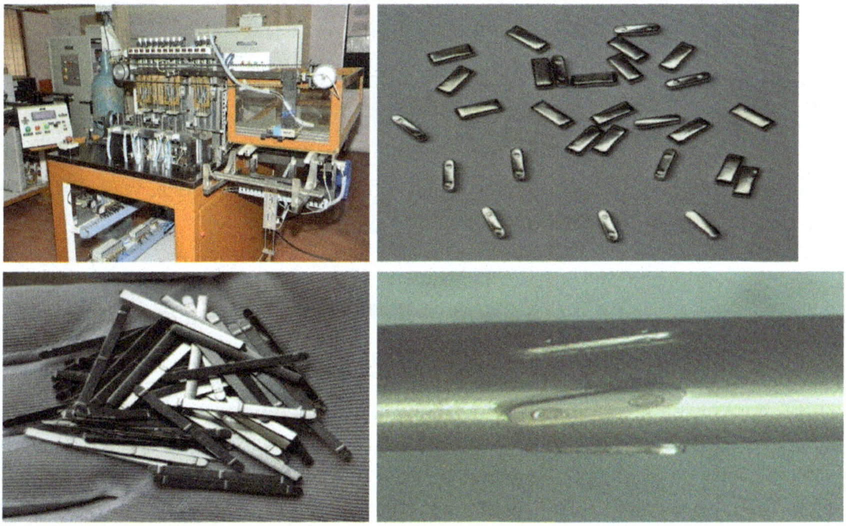

Fig. 17.13 Integrated spacer & bearing pad welding unit

Fig. 17.14 Integrated end cap welding and machining unit. **a** Tube machining, **b** End cap feeding, **c** End cap welding, **d** Element machining

productivity/person, 10% improvement in machine availability. There was significant reduction in operator fatigue. It is shown in Fig. 17.15.

17.7.1.5 Automated Radioactive Material Handling Machines

In order to improve Radiological Safety for occupational personnel, several state-of-the-art machines are introduced into regular plant operations involving radioactive materials. The introduction of such automated machines has resulted in reduction of Occupational Dose by 40%, Average Air Activity was reduced by 5% and samples above DAC were reduced by a factor of 2. They include Automatic Boat Transfer

Fig. 17.15 Robotic end plate welding machine for 37 element PHWR fuel bundle

using an Autonomous Guided Vehicle (AGV), Automatic Pellet Feeding to centre less grinding (CG) machine and Automatic Pellet Loading System.

17.7.1.6 Automation in Fuel Bundle Inspection

With an aim to increase productivity along with reduction in radiation exposure and also to reduce operator fatigue, the following automation systems have been introduced in fuel bundle inspection. They include Automatic Sintered Pellet Density Measurement System, Automated Ultrasonic testing (UT) for End Cap Welded Elements of PHWR Fuel Element, Integrated Helium Leak Test (HLT) and UT System for Fuel Elements and Automated Fuel Bundle Inspection System.

Automatic Sintered Pellet Density Measurement System. It works on Archimedes' principle for measuring the density of sintered UO_2 pellets (immersion technique) and uses SCADA with data logger system. The advantages include density measurement being independent of shape—dishing accounted (error in measurement due to discontinuity in pellet geometry is eliminated), elimination of manpower there-by reduction in radiation exposure and significant increase in through-put.

Automated UT for End Cap Welded Elements of PHWR Fuel Element. Fuel bundles are manufactured to high quality standards following stringent quality control program fulfilling the customer requirements. Manufacturing of PHWR fuel assemblies involves fabrication of clad tubes by pilgering process, resistance welding of different appendages on to the clad tubes, graphite coating on ID & baking of clad tubes, encapsulating of fuel within clad tubes by resistance butt welding of end caps and final assembly of fuel elements into fuel bundles by end plate welding.

Out of all the processes involved in the fabrication of PHWR fuel assemblies, end cap welding process is considered to be the most critical, as it hermetically seals the nuclear fuel inside the clad tube and plays a major role in controlling the fuel failures. Hence the quality of end cap weld joints is being monitored to high level of in-house developed standards of metallography and ultrasonic testing for evaluation of integrity of weld joints.

An automatic Ultrasonic Testing (UT) system was developed at NFC for 100% UT of end cap weld joints for ensuring their quality, replacing the random-based manual UT system. Automation of UT process eliminates human error, reducing the operator effort and time. The system is completely automated to meet the mass scale production requirements.

Automatic UT system as shown in Fig. 17.16 uses an immersion type pulse echo shear wave technique with a 20 MHz spherically focused probes. Individual fuel elements are picked from the element loading tray and immersed in a water bath. Probes move automatically near to the end cap weld regions and are indexed automatically with a pitch of 75μ over a sampling length of 900μ. No operator intervention is required as the UT probes are automatically indexed and moved near to end cap weld joints. The fuel element is rotated slowly for full circumferential scanning of the weld joints simultaneously. PC-based defect detection system is used for recording the signals and defective elements are identified automatically. The system has a repeatability of ±3% of FSH, three times faster than manual scanning and takes 30–40s per element for scanning and analysis.

This in-house developed integrated system has eliminated separate Helium Leak Test (HLT) operation and made it possible to have 100% ultrasonic testing (UT) (from earlier 25% UT sample check) along with HLT operation. This has resulted in an increase of manpower by 100%. Using this system already more than 30 Lakh welds are tested. The system is shown in Fig. 17.17.

Development of Automatic Fuel Bundle Inspection system. For ease and quality in manufacturing of PHWR fuel bundles, a Robot aided end plate welding station and a Supervisory Control And Data Acquisition (SCADA)-based automated bundle inspection system were developed in-house. After the end plate welding operation is complete, the Robot handles the fuel assembly and places on a conveyor for inspection

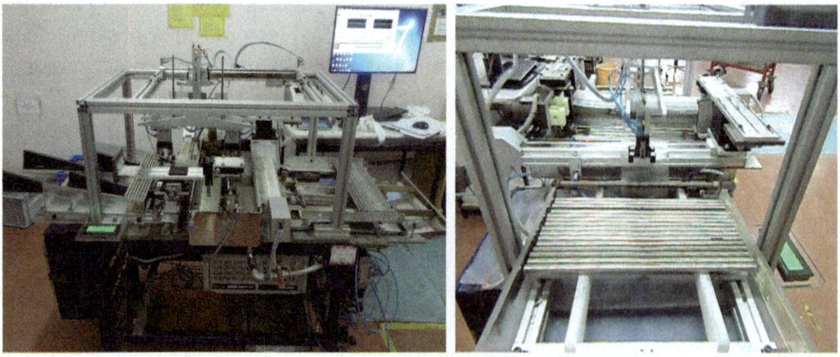

Fig. 17.16 Automated ultrasonic inspection system for end cap weld joints

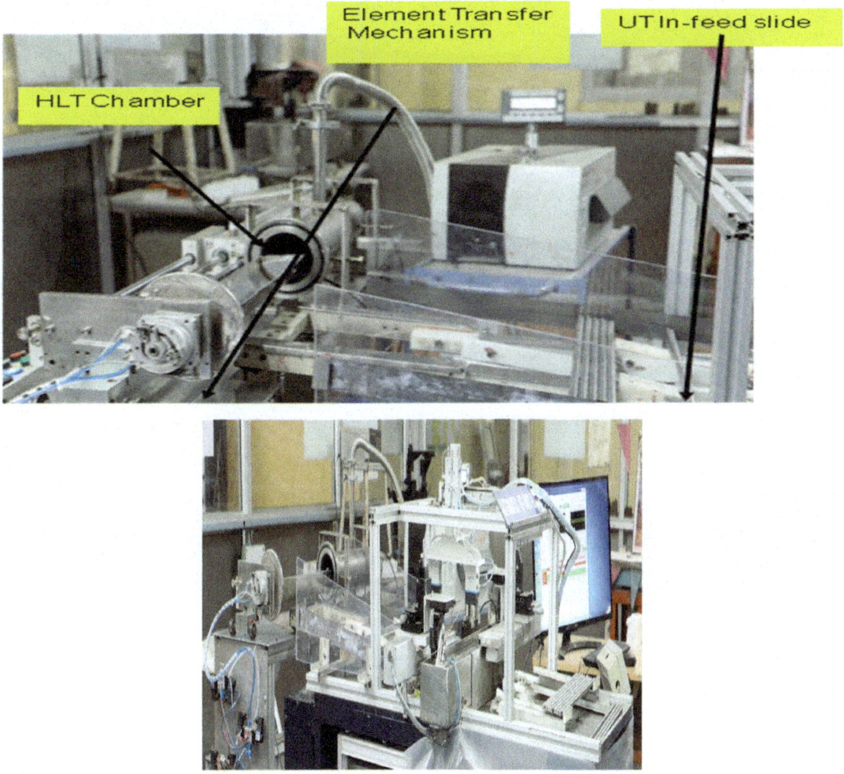

Fig. 17.17 Integrated HLT and UT system for fuel elements

activities. The inspection system performs a series of inspection activities such as back filling, Helium leak testing, bundle diameter gauging, weight measurement, etc., reducing human fatigue and thereby increasing productivity by 50%. In Automated Fuel Bundle Inspection system as shown in Fig. 17.18, the fuel bundles are placed in Helium chambers kept at 6 kg/cm^2 for about 30 min, then excess helium is air washed and passed through Helium leak detection chamber. Subsequent to HLT (Helium Leak Test), the bundles are sent through gauging and dimensional testing stations followed by final visual inspection. All the operations are controlled by a computer-based SCADA system and inspection data is logged on to a computer.

Automated System for UO2 Pellet Visual Inspection. Prior to loading them into empty fuel tubes, UO2 pellets are subjected to 100% visual inspection for defects like cracks, end capping, end chips, pits as shown in Fig. 17.19. Introduction of automated visual inspection system has reduced the subjectivity-related errors due to manual operations. Further, it has minimized the exposure to radioactivity during inspection.

Fig. 17.18 Automatic fuel bundle inspection system

Fig. 17.19 Automated visual inspection of UO2 pellets

17.7.1.7 Advanced QA Systems for Improving PHWR Bundle Quality

Vision-Based Tube Inspection System (VBTIS). An automatic vision-based tube and end cap inspection system has been introduced in the production line. The system is designed to identify and separate tubes having variations beyond the specified limits with respect to Tube diameter, Scratches, Dents, Pits on the outer surface, End chamfer, etc. This is shown in Fig. 17.20a.

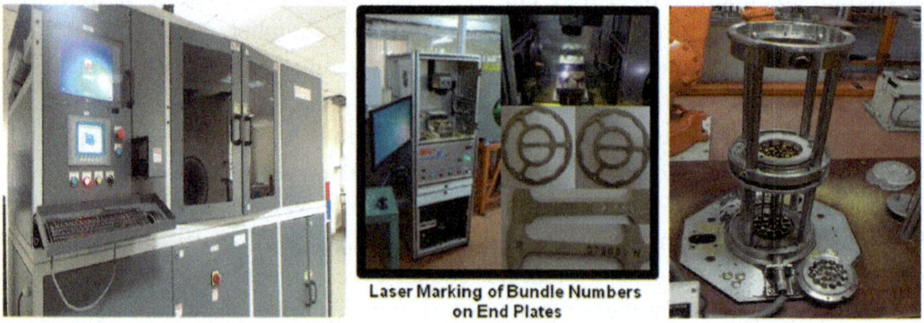

Fig. 17.20 **a** Vision-based tube inspection system, **b** Laser marking of bundle numbers on end plates, **c** Metallic Jig for PHWR bundle fabrication

Fig. 17.21 1500kg Zr-metal produced and Facility for its production

Laser Marking of Bundle Numbers on End Plates (LMBNEP). Laser marking machine introduced for marking identification number on PHWR end plate has drastically improved readability, repeatability, accuracy. Marking cycle takes <7 s. This is shown in Fig. 17.20b.

Metallic Jig for PHWR Bundle Fabrication (MJPHWRBF). Implementation of metallic jigs instead of Perspex jigs at end plate welding of both 19 and 37 Element bundles has increased life of jigs (3000 Bundles/pair against 150 Bundles/pair). This is shown in Fig. 17.20c.

Using these automated systems in UO2 pellet production, inspection and fuel bundle fabrication, there is a tenfold jump in production activities during the last ten years.

17.7.1.8 Developments in Zirconium Sponge Metal Production

Technology Demonstration Unit has been successfully commissioned at Zirconium Complex, Pazhayakayal, Tuticorin, Tamil Nadu for the production of 1500 kg batch against regular batch size of 950 kg for the first time. Several batches are produced through this process. Efforts have been made to reduce the production cycle time and to meet the additional requirements of sponge metal. The chemical analysis of the 1500 kg Zr sponge batches produced is meeting the technical specification. Higher productivity, improved purity, energy savings and improved recovery by 2% are the advantages. Figure 17.21 shows the production facility with a picture of typical batch produced.

17.7.1.9 Developments in Structural Components Production

Development of Novel Route for Production of Pressure Tubes. After successful completion of trials for fabrication of Zr-2.5%Nb Pressure Tubes (PT), a novel modified route based on double radial forging, single stage extrusion and single pass pilgering route was taken up for their production of industrial scale. During the extrusion, high extrusion ratio of >10 is used to obtain favourable crystallographic texture. This crystallographic texture has primarily basal pole orientation (0002) in the transverse (circumferential) direction. During further single-stage pilgering, there is no modification of this transverse texture. Such a texture is favourable for creep properties of the tube, which is one of the main life limitation properties for this structural material.

17.7.1.10 Quality Improvement in PT Through Modified Route

Metallurgical Improvements

- Introduction of two stage radial forging has led to coarser overall grain size and more uniformity in properties along the length of the tubes
- Higher extrusion ratio results in more favorable aspect ratio of alpha grains and higher fT i.e. stronger transverse basal pole texture compared to older route (0.65–0.70 in new route versus 0.5–0.6 in old route)
- Single pass pilgering has eliminated the requirement of intermediate annealing, thus ensuring more continuous beta phase network along the grain boundaries of alpha grains Increasing the Q factor from ~2 to 4 has changed the hydride orientation from mixed to predominantly transverse.

Visual Improvements

- Modification of autoclave carriage has resulted in significant reduction in jerks and vibration encountered by the tubes during loading, unloading and autoclave operation thus marks created by autoclave carriage on tube surface have been practically eliminated
- The customer has given a note of approval about these steps taken to improve the visual quality of the tubes.

Dimensional Improvements. Modification in pilgering pass schedule and fine-tuning of tooling profiles has resulted in significantly better dimensional control with the DCRs for dimensional deviations getting reduced from ~84% (341 tubes out of 403 tubes) for RAPP 8 to 20% (67 tubes out of 326 tubes) for KAPS II.

Advancements in Grinding & Honing. Modification in Honing machine: Increase in stroke length of honing machine to 7.5 m to carry out honing in single stroke. Advantages of OD grinding are confirmation of Blank Wt., UT Status before Pilgering, minimizing of reworks in Final Tube stage. Where-as minimized reworks in final pilgered tubes, better compliance towards stringent dimensional requirements in final tube, improved efficiency & recovery are the advantages of ID conditioning. The operations are shown in Fig. 17.22.

Advancements in Packing of Pressure Tubes. In view of its criticality and cost of production, it is equally important to dispatch the pressure tubes to reactor sites in the most secured manner so that they can be introduced into reactor without any re-works. Figure 17.23 shows the modified packing for dispatch of Pressure Tubes

Fig. 17.22 Grinding & honing operations in PT production

to Reactor sites. This simple modification has resulted in reduction of re-works and thereby savings to NFC.

QA improvements in PT/CT. Laser-based OD Measurement for Pressure Tubes was introduced for digital data acquisition in place of Thermal Strip Chart prints, for digital output in place of manual measurements and Automatic Logging of WT values at every 400 mm interval.

Development of new design for Garter Spring Manufacturing. A new design of Corrugated Girdle wire has been introduced in place of Plain Wire for manufacturing the Garter Springs. The new design has the ability to accommodate up to 6% creep expansion of pressure tubes and good sensitivity in detection of garter spring location. Figure 17.24 shows typical pictures of garter springs.

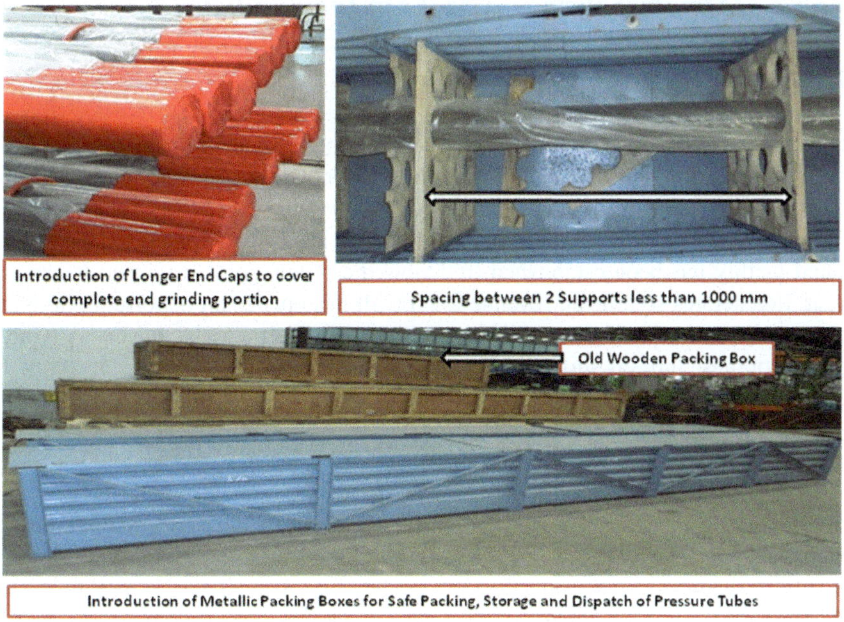

Fig. 17.23 Packing for dispatch of pressure tubes to reactor sites

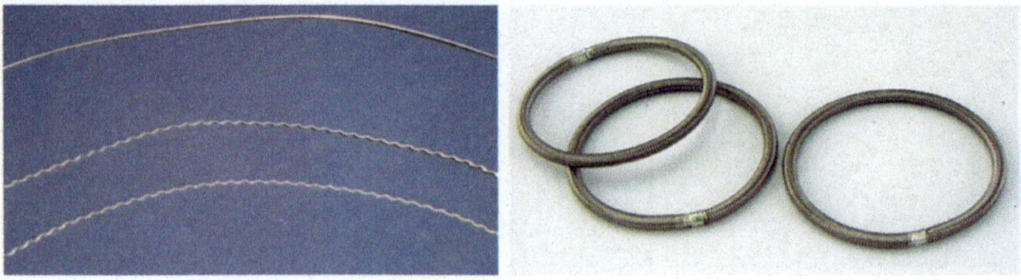

Fig. 17.24 Garter Springs made with corrugated girdle wire

Fig. 17.25 Incoloy-800 U bend SG tubes

17.7.2 'Atmanirbharta'—A Make in India Initiative

17.7.2.1 Steam Generator Tubes for PHWR

NFC successfully manufactured Incoloy-800 U bend steam generator (SG) tubes first time in India. The production of 30-m-long Incoloy-800 U bend tubes is a technological challenge to any engineer and NFC could do it successfully and delivered 8 sets for RAPS 7 & 8, KAPS 3 & 4 reactors. In view of very high demand, a dedicated facility has been established to double the production capacity to 6 sets per year. It generates significant revenue to NFC. It has opened an opportunity to NFC to become potential supplier of U bend SG tubes in International market. A typical picture of these tubes is shown in Fig. 17.25.

17.7.2.2 Indigenous EB Melting Furnace

In view of restricted technology, a successful effort was made in NFC to built indigenous electron beam melting furnace. The indigenously built EB Melting Furnace is used in melting and purification of refractory & reactive metals and alloys for strategic applications in Nuclear, Space, Defence fields. The efforts have resulted in huge revenue savings for the department and eliminated the dependency on external agencies for the maintenance of the furnace. The EB Furnace was built for the first time in India using indigenously available resources and in the process, India has become 4th country in the world to have such a facility. The facility was inaugurated by the President of India in 2018 as shown in Fig. 17.26 and since then being used for different purpose.

Fig. 17.26 Indigenously built EB melting furnace and inauguration by President of India

17.7.2.3 Manufacturing of Special Tubes

Several special grade steel tubes have been manufactured for special & strategic applications for first time in India as import substitutes such as Incoloy-800 U bend SG tubes, Zr-1% Nb tubes, Titan-24/11 tubes, SuperNi 42, Inconel 690/600, Alloy 617 etc. Some of them are shown in the following paragraphs.

Titanium Tubes Manufacturing: Zirconium and Titanium have similar metallurgical characteristics and processing routes (Extrusion, Pilgering, Heat Treatment, etc.). NFC with its vast experience of manufacturing Zircaloy tubes is well equipped to take up bulk manufacturing Ti-alloy tubes as import substitutes. Over the years, various Titanium alloy products like Titan-24 Tubes for Strategic Nuclear application, Ti-half alloy Truss rod Tubes for PSLV & GSLV, Ti half alloy Hydraulic Tubing for Light Combat Aircraft (LCA) have been developed. It is planned to augment capacity through an exclusive facility.

SuperNi-42 Tubes: These tubes as shown in Fig. 17.27 are of small diameter, extremely thin-walled and have stringent specifications with respect to dimensional tolerances and metallurgical properties. After initial trials for indigenous development in collaboration with BARC Mumbai, the manufacturing route was established for bulk production. The process consists of production through 10 stages of thermo mechanical processing followed by final finishing operations & stringent quality checks of mechanical testing, ultrasonic testing, dimensional & visual inspection. Optimization of the process was done using the experience gained on thermo mechanical processing on this special alloy, to achieve significantly improved recovery and productivity. It is planned to augment capacity through an exclusive facility.

Fig. 17.27 Typical picture of SuperNi 42 tubes

17.7.2.4 Other Developmental Works Carried Out in NFC

With its structured and well-established manufacturing process under its belt, NFC continuously made technological improvements to refine the processes. Some of them are as follows.

- NFC demonstrated the production of 1500 MTe of natural uranium fuel (world's highest-ever production).
- Manufactured and supplied 19/37 element Natural Uranium Fuel bundles for all the PHWRs and Enriched Uranium Fuel Assemblies of 6 × 6 types to BWRs. For the first time, 37 element fuel bundles with modified bearing pad design was made for the initial core requirement of India's 1st 700MWe PHWR at KAPS-3.
- In-house development of Auto Ring gauging for PHWR fuel bundles, a mandatory requirement prior to loading of bundles in to reactor core.
- Development of automated vision-based inspection system for surface examination and dimensional measurement of fuel bundle appendages viz. Bearing and Spacer pads to increase through-put with reliability.
- Development of High corrosion resistant SS-304L pipes for Fast Reactor Fuel Cycle Facility (FRFCF).
- Development and manufacturing of D9 Fuel Clad Tubes and Pure Nickel Tubes for Prototype Fast Breeder Reactor (PFBR).
- Development of Alloy 617 tubes for Advance Ultra Super Critical (AUSC) power plant.
- Development of and supply of Zr-1%Nb alloy tubes for strategic applications.
- Development and manufacturing of RRR Grade Niobium sheets for fabrication of Super Conducting cavities.

17.8 Association with NPCIL

As NFC enters golden year of its existence and it is a matter of privilege to recollect the fruitful association with Nuclear Power Corporation of India Limited (NPCIL), its principal customer who are responsible to operate nuclear power reactors in India. The association dates back to 1974 soon after Canada's withdrawal from cooperating India in PHWR program due to Pokhran-1 restrictions. Since then NFC has come long way in supporting NPCIL starting from 220 MWe at NAPS to 540 MWe at TAPS to present indigenous variant of 700 MWe at KAPS.

17.8.1 Role of NFC in First Criticality of KAPP

It is a happy and momentous occasion to DAE, especially NPCIL Team on achieving the 'First Criticality' of 3rd Unit of Kakarpar Atomic Power Project (KAPP-3) on 22–07–2020 at 09.36 AM. Indeed, it is a proud moment to all of us to see a congratulatory message from Shri Narendra Modi, The Prime Minister of India saying 'development of indigenous reactor as shining example of Make in India and a trailblazer for many such achievements'.

It is interesting to note that KAPP-3 finds a special place in Indian Nuclear Power Program as India's First 700 MWe and biggest indigenously developed variant of Pressurized Heavy Water Reactor (PHWR). Operationalisation of India's first 700 MWe reactor marks significant scale-up in design and economies of existing 540 MWe reactor without significant design changes and also addressing the important issue of excess thermal margin. It is also important to note that the milestone moment has come ahead of its original schedule. Thanks to coordinated efforts of many related agencies including NFC. This has given a tremendous boost and confidence to DAE to handle & ensure the successful completion of such mega projects. Thus, KAPP-3 has become the backbone to new fleet of 10 such reactors sanctioned by government in 2017. This will help DAE in ramp-up the nuclear power capacity to 22,480 MWe from existing 6780 MWe by 2031, where 700 MWe reactors find a lion share.

Nuclear Fuel Complex (NFC) has played an important & substantial role in this achievement through the timely supply of several reactor-core components to this indigenous variant and directly contributing to the government stand of self-reliance. The core components include 125T of natural uranium fuel, 392 channels of pressure tubes, 1568 numbers of garter springs, 30 m length of 'U-bend' Incoloy800 steam generator tubes, 96 numbers of reactivity mechanism assemblies. Therefore were challenges in executing this time bound assignment and but, were successfully met by NFC with untiring and commendable efforts of its employees. The technological solutions to these challenges were developed and seamlessly implemented. For example, development of two pass forging route with single pass pillgering in place of extrusion route was carried out to produce 6 m long pressure tubes with better stability and uniform mechanical properties under reactor conditions. Similarly, blank machining, ID honing and OD turning operations were introduced during fuel clad manufacturing process to reduce wall variations in clad thickness and also to obtain better recovery in UT pass. The design of garter springs was also changed to corrugated & welded spring to improve integrity of griddle wire under dynamic flow of coolant and improved testing capability during ISI.

17.9 Summary

One can confidently say that NFC always sets a new benchmark in *N*ever *F*ails in its *C*ommitments attitude and continue to play a significant role in all NPCIL's ambitious future expansion programs and delivering the requirements of other departments like DRDO, ISRO as well.

Acknowledgements The works reported in the article have been a collective work of many colleagues at Nuclear Fuel Complex (NFC), Bhabha Atomic Research Centre (BARC), Indira Gandhi Centre for Atomic Research (IGCAR), Mishra Dhatu Nigan Ltd (MIDHANI) and other collaborating institutions. I gratefully acknowledge them. I also thank Sri Y. Balaji Rao, GM (C.Lab, NUMAC & SIRD), NFC for his painstaking efforts in the preparation of this article.

Chapter 18
RDCIS-SAIL: A Perspective of Five Decades of Relentless Excellence in Ferrous Research and Innovation

Debasis Mukherjee and Sanak Mishra

Abstract This article traces the history of the Research & Development Centre for Iron and Steel (RDCIS) in India, from its inception in 1972 to its present status as a world-class research centre. The centre was conceived as the first corporate R&D in the public sector steel industry with the objectives of achieving strategic improvement in critical performance parameters of the steel industry, developing niche steel products for market expansion and import substitution. The initial task was to formulate a detailed 'Industrial R&D Plan for Iron and Steel' that included preparing a document highlighting the technological gaps in the Indian Steel Industry and using that as a baseline, to formulate programmes and underlying projects for implementation. The article also details the creation of an appropriate human resource and a world-class laboratory to support research on many aspects of technologies required for steel-making. Collaborative research programs with leading institutions in India and abroad were leveraged to augment the expertise, knowledge, and capabilities of RDCIS's scientific manpower. The text showcases RDCIS's journey in fulfilling its objectives and enhancing the steel industry's capabilities in the nation.

18.1 Genesis and Formative Phase

18.1.1 Inception History

Passion and intent were etched deep in the mind of the then Minister of Steel, Shri Mohan Kumaramangalam, who first conceived the country's first corporate R&D in the public sector steel industry as early as 1972. He was then Union Cabinet Minister of four major Ministries, that of Steel, Mining, Coal and Industry. Being ably supported by the then Secretary Steel, Shri Wadud Khan, the blue print for creation of an 'R&D Organisation' in Ranchi was approved in its 160th Board Meeting of

D. Mukherjee · S. Mishra (✉)
Formerly with Research & Development Centre for Iron and Steel, Steel Authority of India Limited, Ranchi, Jharkhand 834002, India
e-mail: mishra.sanak@gmail.com

R. Divakar et al. (eds.), *Indian Metallurgy*, Indian Institute of Metals Series,
https://doi.org/10.1007/978-981-99-5060-7_18

erstwhile Hindustan Steel Limited (HSL) in March 1972. Thus, the sapling for the premiere institute of ferrous research was planted with the avowed objectives of achieving strategic improvement in critical performance parameters of public sector steel industry besides developing niche steel products for market expansion and import substitution. The fledgling institution got its first impetus with the appointment of Dr. S Ramachandran, a respected technocrat with long-standing research and steel-making experience from USA, in January 1972 as General Manager In-charge. By the middle of 1974, he had put together a core group comprising among others Dr. V. Ramaswamy, Dr. Sanak Mishra, Dr. S. K. Gupta, Dr. R. K. Iyengar and Dr. S.M. Aeron (Sanak 2020). Government of India (GoI), by an Act of Parliament, on 24th January, 1973 created a public sector undertaking, Steel Authority of India (SAIL), to oversee and coordinate all activities in public sector steel entities. With the winding up of HSL office in Ranchi in 1976, Research and Development Centre for Iron and Steel (RDCIS) became the nodal and central R&D unit of SAIL.

18.1.2 Formulation of Plans and Programmes

The first major task carried out by RDCIS, from 1973 to 1977, was to formulate a detailed **'Industrial R&D Plan for Iron and Steel'**. The task included preparation of a document highlighting the technological gaps in the Indian Steel Industry and using that as a baseline, to formulate programmes and underlying projects for implementation. Simultaneously a task was undertaken to develop the blueprint of a world-class laboratory complex by identifying critically needed diagnostic equipment and pilot plants to be housed in it. Meanwhile, several trips were made to the steel plants in SAIL to identify areas requiring improvement and begin implementation of plant-based programmes/projects for which either the testing tools available in the Research and Control Laboratories of the plants were adequate, or, the testing could be carried out in other laboratories already existing in India. Intensive interactions were initiated with leading national laboratories/academia to utilise their in-house facilities. Efforts were also initiated to identify specifications, price and availability of sophisticated tools for procurement, once laboratory facilities were created.

The initial detailed report by RDCIS, prepared by the middle of 1977, was referred to an expert committee comprising Dr. V S Arunachalam, Director DMRL, Dr. S Ramaseshan, Dy Director, NAL and Dr. V Altekar, Director, NML. The committee strongly recommended the setting up of central R&D laboratory facility besides pursuing research to fill the existing gaps in performance between domestic steel plants and those abroad (Datta et al. 2016).

18.1.2.1 Creation of Appropriate Human Resource and World Class Laboratory

The mandate given to Dr. S Ramachandran for establishing an R&D Centre of international repute, started with careful selection and grooming of scientists and engineers with appropriate skill, talent and expertise. By 1978, a small group of dedicated research professionals was inducted primarily from three sources: steel experts from academic institutes with research experience, engineers with operating experience of steel plants and graduate engineers and research scholars with fresh PhDs and post-doctoral fellowships from foreign universities (Datta et al. 2016). The development of a blend of talent and experience became an enduring feature during the formative years and continued further in future. A simple organisational structure was created with four primary divisions, namely, Physical Metallurgy, Iron Making, Steel Making and Computer Applications. After September 1978, when Dr. S R Pramanik took charge of RDCIS, a very far-reaching and successful strategy took place of human resource optimisation—creation of 'RDCIS Plant Centres', staffed with Research Engineers, in Bhilai, Rourkela, Bokaro, Durgapur and Burnpur, to enhance the trust between the steel plants and RDCIS and to interface/undertake day-to-day co-ordination of plant-based projects.

Towards establishing world-class research support infra-structure, the requisite thrust came when on 16th March, 1978, creating a multi-dimensional R&D laboratory cutting across various technological disciplines and steel product evaluation, was approved by SAIL Board with an initial outlay of Rs 15.10 crores. Metallurgical and Engineering Consultants Limited (MECON), Ranchi transferred 20 acres of land for the purpose. The construction started for the 14,000 sq.m. laboratory complex in early 1980 and was made ready for use by the end of 1983 (Fig. 18.1). Eventually, 15 sectional laboratories along with pilot facilities (Fig. 18.2) and administrative building with auditorium, computer centre and information and documentation centre were established. Simultaneously, close by in Heavy Engineering Corporation (HEC), Ranchi, a dedicated research facility on direct reduction technology was built. In parallel, efforts were made to recruit personnel to man the equipment as well as to undertake advanced research. During 1980–82, around 250 persons including engineers/scientists and support staff were inducted. By early 1986 all workforce of RDCIS was stationed in its newly built complex. Thus, the primary requisites were fully established to initiate a vigourous and massive campaign for innovative product and process development projects in the steel plants.

18.1.3 Initial Collaborative Efforts

To train and augment expertise, knowledge and capability of its scientific manpower, RDCIS embarked, from its early days, in collaborative research programmes with leading institutions in India and abroad (Datta et al. 2016). The first collaboration agreement for 5 years was signed with TSNIICHERMET, the largest industrial

Fig. 18.1 View of the main administrative building of RDCIS, SAIL

Fig. 18.2 Pilot coke oven

research institute of USSR, in May 1978. This was aimed to update the prevailing technology in Bhilai Steel Plant (BSP). Later this collaboration was extended for another 5 years. From 1986 to 1991, an intensive collaboration with the National Science Foundation, USA was pursued with 12 research programmes. Meanwhile, from mid-eighties, collaborative efforts with assistance from United Nations Development Programme (UNDP), Council of Scientific and Industrial Research (CSIR) laboratories and Bengal Engineering College, Shibpur were pursued.

18.1.4 Project Categorisation/Verticals

From mid to late eighties, when plant-based projects got annually formulated after interaction with plant personnel, a method was evolved to categorise all projects to be pursued under five distinct verticals/groups with specific outputs and time frames (Performance Reports of RDCIS 2000–2022). These are:

(i). Plant Performance Improvement (PPI): The projects under this head, which comprised almost 70% of the total number of projects, aimed at improvement in productivity, quality, yield, reduction in material usage, energy conservation, development of refractories, etc. The duration of such projects was 1–2 years in SAIL plants/Mines. Later there was a decrease in PPI projects with a commensurate increase in Scientific Investigation and Development (SID) projects.

(ii). Product Development (PD): The sole aim is to develop new/value-added steel products followed by their commercialisation. Generally, the product development cycle was restricted to 1–2 years. Around 7–10% of projects constitute this category and are implemented in SAIL plants and later commercialised with Central Marketing Organisation (CMO).

(iii). Basic Research (BR): The scope of these projects is geared for upgradation of existing knowledge and generation of new knowledge and is pursued in RDCIS. These projects of typically 1–3 years duration sometimes led to generation of PPI/PD projects. Less than 5% of projects pursued fell under this head.

(iv). Scientific Investigation and Development (SID): Projects aimed at process investigation, mathematical modelling, instrument and equipment development, software development, etc., were under the purview of this category. These projects of a 1–2 year time span are undertaken in SAIL plants/RDCIS and can lead to generation of PPI projects and transfer of know-how. Such projects comprise around 10% of overall projects.

(v). Technical Services (TS): Feasibility and environmental studies, plant investigation, specialised testing and infrastructure development constitute projects under this vertical. They are implemented in RDCIS/ SAIL plants and are generally less than a year stretching to 2 years at the most and accounts for around 10% of the projects. These projects result in techno-economic benefits and fulfillment of customer needs.

18.2 Innovations and Applications in Steel Plants

During the sojourn of almost five decades, RDCIS has consolidated its position as the premier organisation in the country in the field of ferrous research and design by way of making significant and break-through contributions, some of them for the first time in the country, in the domain of process and product innovation. Some of the contributions under process innovation include technology demonstration initiatives. A glimpse of such salient efforts is highlighted below.

18.2.1 Process and Design Research

18.2.1.1 Lime Dust Injection

In the early eighties, for the first in India, the process was developed to inject a part of the flux as burnt lime through blast furnace (BF) tuyeres while the rest was charged from the top as free limestone and/or through sinter. This was initially tried in a low-shaft furnace in Kalinga Iron Works (KIW), Barbil. Later the technology was transferred to a BF in Durgapur Steel Plant (DSP), with a capacity to inject 125 tons per day (tpd) of flux, which resulted in productivity improvement by 8% and concurrent coke rate decreased by 7.5% (Aswath and Mukerjee 1985).

18.2.1.2 Direct Reduction (Sponge Iron) Pilot Plant

A 10 tpd rotary kiln, acquired with the barest minimum import component from M/s Lurgi, West Germany, was established in the early eighties. Various ore-coal combinations were experimented with to develop a technology for Indian raw materials. An average degree of metallisation of 88–90% was achieved consistently. RDCIS provided consultancy to various organisations for setting up non-coking coal-based sponge iron plants and tested their raw materials.

18.2.1.3 Technology of Cold Bonded Pellets

Production process for cold bonded iron ore pellets was developed in mid-eighties without resorting to high temperature induration. Commercially available cement was used as a binder. Pellet production facility was set up at Gua mines and trials at Indian Iron and Steel Company (ISP), Burnpur using the pellets, comprising 20% of the burden, indicated an increase in productivity by 18% and a decrease in coke rate and slag volume by 11.5% and 13% respectively.

18.2.1.4 Development of Combined Blowing Technology for BOF Converters

The development of inert gas purging technology with complete instrumentation and bottom blowing device was successfully commissioned, through in-house R&D efforts, in one of the Basic Oxygen Furnace (BOF) converter in Bokaro Steel Limited (BSL). Extensive modelling and simulation inputs besides immaculate design of bottom refractory porous plugs were central in developing the know-how for the process. This was implemented towards late eighties and subsequently laterally transferred to other SAIL plants as SAIL Combined Blowing (SCB) technology. The technique resulted in improvement in yield due to lower FeO content in slag, improvement in lining life, saving in ferro-alloy consumption and production of low carbon grades of steel.

18.2.1.5 Continuous Casting Technology for Cold Rolled Non-Oriented (CRNO) Steels

CRNO steels (grades M36 to M47), an important product for manufacture of magnetic cores and components of electrical motors, are very difficult to be continuously cast because of high silicon and low carbon contents. RDCIS formulated the continuous casting (CC) technology based on close control of chemistry to ensure balance between ferrite and austenite, intensification of secondary cooling, choice of suitable mould powder and warm charging of slabs. This resulted in increasing yield from 48% (ingot casting) to 78%, besides facilitating Rourkela Steel Plant (RSP) to become a plant with 100% of its products cast through CC route. This breakthrough was implemented in 2004–05.

18.2.1.6 Slime Beneficiation System

To prevent environmental degradation and recover precious iron values from slimes in Dalli mines under BSP, a very innovative process was developed consisting of fluidised bed classifier (FBC), slurry pumps, cluster of hydro-cyclones and slow speed spiral classifier (SSSC). The line diagram of the entire process is shown in Fig. 18.3. This enabled the entire quantum of slime generated to be treated through the system. Brought into operation in February 2008, the slime beneficiation system could enrich the slime from 49% Fe to 62.5% Fe at the rate of 40–45 t/hr. The enriched fines concentrate was mixed in sinter fines product.

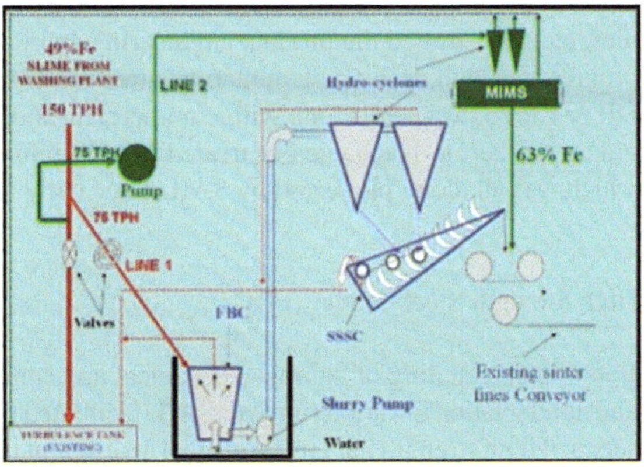

Fig. 18.3 Generation of iron concentrate > 63% Fe in slime beneficiation plant in Dalli mines

18.2.1.7 Roll Bite Lubrication (RBL) in Finishing Stand

RBL was indigenously designed and introduced in the first three finishing stands (F6, F7 & F8) of the Hot Strip Mill (HSM) at BSL in September 2008. This automated system was found to be beneficial in reducing roll force and power consumption by 5–10% and facilitated rolling of high strength LPG steel (SG295 and P310) in thinner gauge of 2.2 mm for the first time. Increase in campaign length of rolling by 10% min., decrease in grinding off-take of rolls and suppression of generation of rolled-in-scale were other accrued benefits of this innovation. Later, the technology was introduced in all the finishing stands as well as in edger stands of HSM, BSL. Horizontal transfer of the RBL technology was also made to HSM, RSP.

18.2.2 Development of Niche Steel Products and Superior Refractories

18.2.2.1 Flat and Long Categories of High Strength Steels

Scientists entrusted with development of advanced steel products in RDCIS had made remarkable contributions over the years right from inception, to conceive and commercialise state-of-art products for different market segments, often for the first time in the country. The efforts were aimed to promote self-reliance and import substitution. The thrust from inception was towards development of high strength steel products, both in flat and long category, with superior product attributes. The belief behind developing such value-added products was the adage: "Steel can best be replaced by only better steel". Innovative alloy design, optimisation of process variables, recourse to grain refining techniques, development of composite microstructures, application of dislocation and precipitation hardening measures, inclusion engineering and morphology control, selection of appropriate heat treatment schedules, etc., constituted the product engineering approach adopted by the Centre. The progress made in developing numerous grades of high strength flat products (plates, HR & CR coils/sheets) along with commercialisation of a plethora of high strength grades of thermo-mechanically treated (TMT) reinforcement bars (long product), which was all along pioneered by SAIL since early nineties, is briefly enumerated.

High Strength Steel Grades (Flat)

Better understanding of thermo-mechanical and controlled rolling process (TMCP) and its adaptation in various units of SAIL facilitated the continuous development of a series of high strength flat grades of steel at different times, from 1980 till the present, having yield strengths (YS) ranging between 300 to 800 MPa (Fig. 18.4) (Datta et al. 2016). These include SAILMA 300, SAILMA 350, SAILMA 410, SAILMA 450, SAILMA 550 and SAILMA 600 plates for earthmovers, impellers, heavy transport vehicles and construction segment. SPADE plates in 130 mm thickness through

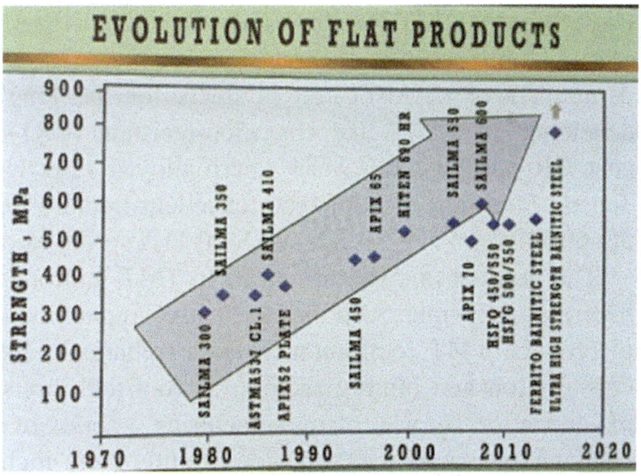

Fig. 18.4 Evolution of high strength flat products in SAIL

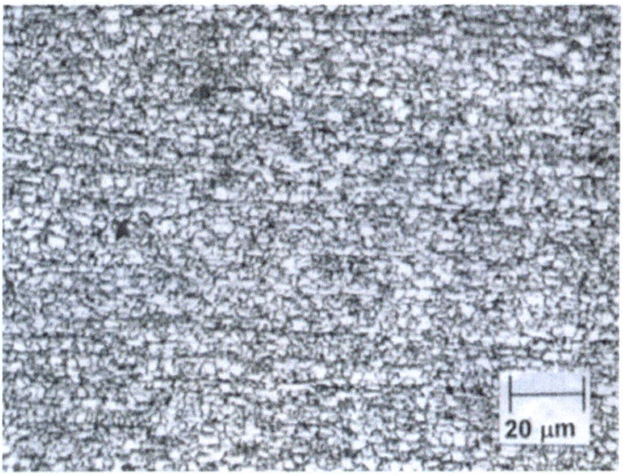

Fig. 18.5 Fine grained steel (<3 μm) in HSFQ 500 h coil

Continuous Casting (CC) route, with YS of around 900 MPa and Charpy Impact Energy (CIE) of 35 J at –40 °C and superior ballistic property, is being used for the manufacture of T-90 tanks. DMR 249 Grades A & B, in quenched and tempered (Q&T) condition, with excellent low temperature toughness (CIE of 78 J at –60 °C (Gr.A)/ 78 J at –40 °C (Gr.B)) are being used in the fabrication of warships and landing platforms for aircraft carriers respectively. API X-65 and X-70 grades for line pipes, ASTM 387 Cl. 1 & 2 for pressure vessels, HITEN 690 AR for ATM chests, high strength forming quality (HSFQ) 450 and 550 grades for auto chassis are in the product basket of SAIL. Development of high strength fine grain (HSFG) 500 and 550 grades, containing Nb and Si, with a fine grain size of <3μ across the thickness in 3.2 mm HR coil was a technical marvel (Fig. 18.5). HSFG 500/550 is suitable for weight reduction applications like auto chassis and pre-fabricated structures. Ultra-high strength bainitic steel with YS of around 800 MPa has been designed.

High Strength TMT Reinforcement Bars (Non-Flat)

A number of sophisticated products for the long product segment have been developed in SAIL, like corrosion-resistant (CR) rails (Cu–Mo/Ni–Cr–Cu low-cost CR variety), 110 UTS micro-alloyed rails with YS/UTS (UTS—Ultimate Tensile Strength) ratio of 0.60, excellent fracture toughness bainitic rails (KIC = 55–60 MPa$\sqrt{m}$ at −20 °C) and SUP 11A spring steel for automotive applications.

The product innovation related to TMT reinforcement bars will be highlighted here, as an example: SAIL has the distinct reputation for being the first in the country to produce TMT reinforcement bars (rebars) in 1991 in DSP, used primarily in construction and mining sectors (as roof/rock bolts), and thereafter led the saga of innovation by way of manufacturing a series of new grades for application in challenging environments besides getting them included under various Bureau of Indian Standards (BIS) specifications. Figure 18.6 shows the successful unraveling of the products with enhanced performance features (Datta et al. 2016). *SAIL is now an acknowledged world leader in the manufacture of the widest variety and range of TMT rebars (8–40 mm).* Starting with the installation of the Thermax line from H&K Germany in DSP, the production of SAIL TMT 415 (YS 415 MPa min.) was heralded in India. Thereafter, Thermax lines were installed in BSP and ISP. A number of grades were introduced periodically in SAIL plants: SAIL TMT 500, SAIL TMT HCR (high corrosion resistance), SAIL TMT 500, SAIL Roof Bolt FE 500, SAIL TMT EQR (earthquake resistance grade with UTS/YS of 1.18 min), SAIL TMT 500D (with superior ductility), SAIL Roof Bolt Fe 600, SAIL Roof Bolt FE 600 with better rib pattern design for enhanced bond strength and system stiffness, SAIL Fe 415 S (seismic grade with enhanced UTS/YS of 1.25 min.) and SAIL Fe 500 S. Subsequently SAIL Roof Bolt Fe 640 grade was commercialised for stringent requirements in underground coal mines. Presently, SAIL also supplies TMT rebars under the brand name of SAIL SeQR (SAIL Secure) for the retail segment with Fe 550 grade as the highest strength quality with high UTS/YS of 1.15 min. The size ranges popularly include 8 mm to 25 mm, produced from the plants of ISP, DSP and BSP.

Futuristic Products

RDCIS is engaged in development of sophisticated futuristic steel products in Indian context. While some are being attempted on its own, some are being developed in collaboration with others. Research is underway for development of cold rolled grain oriented (CRGO) steel, nano-structured bainitic steel, low density steels and corrosion resistant high strength seismic grade rebars.

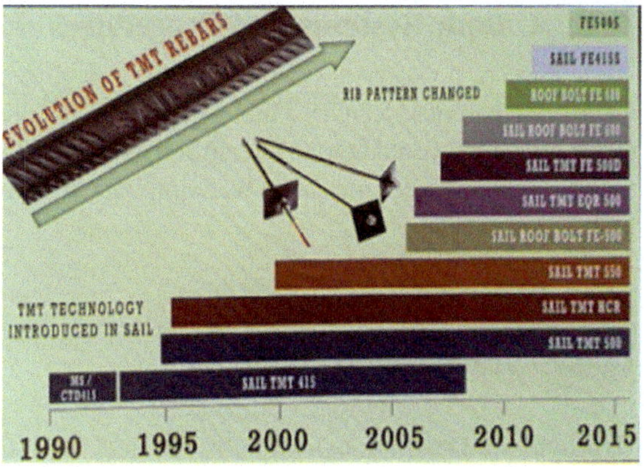

Fig. 18.6 Pioneering efforts in developing high strength reinforcement bars in SAIL

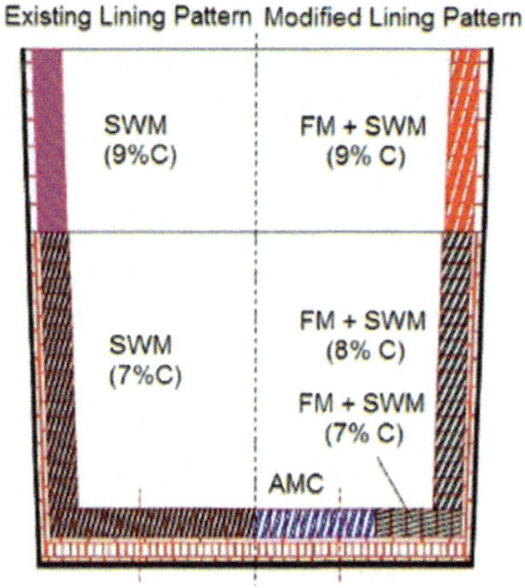

Fig. 18.7 Earlier and modified lining pattern in 150t steel ladle in SMS-II in RSP, SAIL

High End Refractory Products

Improving campaign life in converters and ladles with associate reduction in specific refractory consumption have been an area of focus for refractory technologists. Introduction of superior materials in linings of converters and ladles and judicious usage of spinel castable, anti-oxidants and additives (Al, Mg–Al, B4C) and nanomaterials in refractories have led to phenomenal increase in performance of refractory linings. A number of new refractory bricks and castables have been developed. Based on in-depth wear profile in 150t steel ladles in SMS-II, Rourkela Steel Plant (RSP), the lining design was modified. Resin-bonded Alumina-Magnesia-Carbon (AMC) bricks for 'bottom impact pad' and improved bricks with 98% MgO containing fused magnesia (FM) mixed with 97% sea water magnesia (SWM) in the ratio of 1:1 for the 'metal zone' were developed and used (Fig. 18.7). The new lining pattern resulted in achieving an increase in campaign life by 37%.

18.3 Unique Systems and Procedures

From mid-nineties, RDCIS labored to put in position a network of systems and procedures that would help to streamline its output (Datta et al. 2016). Besides being unique, some of these measures were implemented for the first time in the world for an R&D organisation.

18.3.1 ISO 9001 Accreditation

To provide and institutionalise quality services for its customers, RDCIS adopted ISO 9001 as its Quality Standard and it received its ISO 9001:1994 certificate in 1994. Later RDCIS transformed its Quality Management System as per ISO 9001:2000 Standard and 2008 Standard and currently 2015 standard. The Vision of RDCIS was '*To emerge as the best R&D organisation in the world by conceiving and developing original technological innovations and using them to: reduce cost, improve quality and manufacture value added products in order to fully satisfy its customers*'. The derived Quality Policy of RDCIS states that '*RDCIS shall provide innovative R & D solutions in the field of iron and steel making; develop market-driven and futuristic special steel products; assist SAIL in achieving technological eminence; and continually grow as a Centre of Excellence*'.

18.3.2 Certified Annual Benefit (CAB)

The detailed methodology for evaluating the monetary benefit accruing from a R&D project was conceived and implemented, **for the first time worldwide for an R&D unit**, in RDCIS in 1993–94. This procedure was phenomenal and path-breaking as it allowed the steel plants to be sensitised about the financial benefits that can be derived from an innovation. The most important part of the procedure is that a Standing Committee comprising ED(W)/head of operations in each plant/unit along with senior plant/unit officials and senior RDCIS officials jointly certify the financial benefits. Thus, the customer approves the benefits. The total CAB consists of two contributions: 'Incremental CAB' (innovation for the first 12 months after its utilisation) and 'Recurring CAB' (for each innovation after 12 months and up to 36 months from the starting date of utilisation). Very simply monetary benefit can be calculated by multiplying improvement made by an innovation (Rupees/Ton) with the volume of its utilisation in the shop (Tons). Table 18.1 provides an idea of CAB generated over the years due to innovations implemented by RDCIS in SAIL plants.

Table 18.1 CAB generated through innovations implemented by RDCIS

Year	1994–95	2007–08	2009–10	2011–12	2013–14	2015–16	2017–18	2019–20	2021–22
CAB (Rs. in Crores)	65.88	488.93	483.45	669.69	383.42	289.22	235.78	166.55	255.66

Table 18.2 Overall CSI accumulated from all project innovations year-wise

Year	2007–08	2009–10	2011–12	2013–14	2015–16	2017–18	2019–20	2021–22
CSI (In a scale of 5)	4.73	4.77	4.89	4.95	4.93	4.82	4.77	4.87

18.3.3 Customer Satisfaction Index (CSI)

This procedure, which is truly unique, forms a part of the Project Management System where customer feedback is obtained in a CSI feedback format on completion of each plant-based project. The CSI format has four factors: fulfillment of objective, adherence to schedule, impact and implementability and ease of usability of solutions. The shop heads were to grade each factor and rate them as excellent, very good, good, satisfactory and poor. CSI for each project was rated on a scale of 5. Using the average CSI value for each project, weighted average is calculated for all projects plant/unit-wise and then for all the plants/units together. The CSI data obtained for specific years is revealed in Table 18.2. The CSI figures have been very consistent over the years, reflecting that the project innovations of RDCIS were well appreciated by SAIL plants/units.

18.4 National/International Accolades and Recognitions

For its pioneering role in ferrous research in the country, RDCIS has been acknowledged for the immense contributions made, since inception, not only in India but abroad as well. The Centre is recognised today as the leader in the arena of ferrous innovations, among all contemporary laboratories/institutes in India. Some such recognitions and contributions are briefly highlighted.

18.4.1 Corporate and Individual Awards

Numerous national awards have been conferred to RDCIS as an organisation and to its individuals for the towering contributions made. The prestigious institutional awards bestowed to RDCIS includes FICCI Award (1988), DSIR National Awards (1998 and 2005), DGTD National Award (1992), Golden Peacock Awards (1996, 2007, 2008 and 2012 for R&D achievement, Product innovation, Quality and Management of innovations), SCOPE Meritorious Award (2008), Good Green Government Award (2009 and 2016), PSU Excellence Award (2010), Innovation Energy Saving Award (2011) and BT Star PSU Award (2015) (Datta et al. 2016; Performance Reports of RDCIS 2000–2022).

The most coveted individual award conferred by Ministry of Steel, Government of India is the Metallurgist of the Year Award. During the past 50 years, since RDCIS was established, 51 engineers/scientists received the recognition, which is highest among all organisations/ academic institutes/industries in India (Datta et al. 2016; Performance Reports of RDCIS 2000–2022). In addition, there were recipients who were awarded National Metallurgist Award (Industry), Young Metallurgist Award, Steel Eighties Award, O P Jindal Gold Medal Award, Essar Gold Medal Award, G D Birla Gold Medal Award, National Mineral Award, IIM-TSL New Millennium Award, etc.

18.4.2 Publications and Intellectual Property Rights (IPR)

RDCIS engineers, since the beginning, have been very alert and shared their interesting research findings by way of publications in reputed, including peer-reviewed, national/international journals. Besides, some have received best paper awards in international transactions/journals having high impact factor. In the past two decades (2000–01 to 2019–20) alone, the Centre has published 1398 papers (Performance Reports of RDCIS 2000–2022); a large majority of which were in reviewed international journals. Since inception, the number has grown to be 1984. Such a large number of publications had established the reputation of the Centre as the leading institution in ferrous research in the country.

Towards protecting its competitiveness, creativity and innovations, thrust was provided by RDCIS to file and seal patents and obtain intellectual property rights (IPR), related to process and product design. To get an idea, RDCIS in the past two decades (2000–01 to 2019–20) had filed for 610 patents (Performance Reports of RDCIS 2000–2022). In fact, from its inception, the Centre had filed 886 patents. This includes 15 patents filed abroad. RDCIS provided the lead for company-wide IPR management in SAIL.

18.4.3 Strengthening of Collaborative Efforts

As the Centre grew in strength and stature, a number of renowned organisations/ laboratories/academic institutes in India and abroad were willing to have collaborations under participative research mode. This resulted in a strong network of collaborative research and training effort in RDCIS. Among the notable collaborating partners from early 2000, fruitful interaction has taken place with CBMM, Brazil, Swerea MEFOS, Sweden, Carnegie Mellon University, USA, Deakin University,

Australia, IP Bardin, Russia, IIT's at Kharagpur, Kanpur, Bombay and Roorkee, IISc, Bangalore, MIDHANI, Hyderabad, C-DAC, Thiruvananthapuram, CSIR-NML, Jamshedpur, CSIR-IMMT, Bhubaneswar, CSIR-CMERI, Durgapur, IIT-ISM, Dhanbad, ISRO, Thiruvananthapuram, IIM Ranchi, BARC, Bombay, Usha Martin Ltd. etc. (Datta et al. 2016).

18.4.4 Priority for R&D Outlay in Steel Research in India

RDCIS and SAIL, all along, have been at the forefront, amongst all steel producers in the country, and has been consistently the highest spender in the country for R&D outlay over the years. In addition, R&D expenditure as percent of turnover of the company has also been the highest in India for SAIL (Annual Reports of Ministry of Steel, Government of India). From a meagre amount of Rs. 2.0 lakh in 1972–73 (inception of RDCIS), the expenditure rose to Rs. 10.00 crores in 1982–83, Rs. 30.00 crores in 1992–93, Rs. 50.00 crores in 2002–03, Rs. 101.86 crores in 2007–08, Rs. 147.63 crores in 2012–13, Rs. 264.20 crores in 2014–15 and Rs. 339.43 crores in 2016–17(Datta et al. 2016; Performance Reports of RDCIS). Thereafter, there was a slight tapering in outlay to Rs. 319.86 crores and Rs. 292.80 crores in 2018–19 and 2019–20 respectively (Performance Reports of RDCIS). R&D expenditure as percent of sales turnover rose from 0.22% in 2007–08 to 0.30% in 2012–13, 0.52% in 2014–15 and 0.69% in 2016–17(Annual Reports of Ministry of Steel, Government of India). In 2018–19 this figure has come down to 0.48% (Annual Reports of Ministry of Steel, Government of India). However, in 2020–21 and 2021–22, R&D expenditure as percent of sales turnover rose to 0.53% and 0.79% respectively (Performance Reports of RDCIS; Annual Reports of Ministry of Steel, Government of India). In comparison to international advanced steel manufacturers, the R&D outlay, as well as % of turnover, is quite low. Whereas, some of the reputed steelmakers like POSCO, Korea, ArcelorMittal, Nippon Steel, Japan (Nippon Steel Integrated Report 2020,) etc. have an R&D expenditure in excess of 1% of their turnover, SAIL has a much lower figure. However, within the country, SAIL has borne the mantle of being the largest spender in R&D. The other steel manufacturers mostly operate with expenditures between 0.1 and 0.30% of turnover (Annual Reports of Ministry of Steel, Government of India).

18.5 In Summary

For almost five decades RDCIS has been the flag-bearer of ferrous research and innovation outputs in India. It faced and overcame challenges in creating the requisite talent and infrastructure base within SAIL. The numerous innovations implemented in SAIL plants/units have not only augmented the techno-commercial performance of SAIL but facilitated and guided future technological investments in the company to ensure its competitiveness and viability. The Centre has the distinction of implementing unique systems and procedures, some for the first time in the world, which resulted in its phenomenal outreach and recognition as the premier R&D organisation in India. There can be no let-up in the zeal that allowed RDCIS to reach the zenith and the laurels it achieved. It has to continuously ponder and reflect on the sentiments expressed in the immortal poem of Robert Frost - *But I have promises to keep, and miles to go before I sleep.*

Acknowledgements The remarkable contributions, since inception, made by erstwhile and currently serving engineers, scientists and staff of RDCIS are sincerely acknowledged.

References

Annual Reports of Ministry of Steel, Government of India, https://steel.gov.in> annual-reports (2000–01 to 2021–22)

Aswath, H.S., Mukerjee, D.: Steel India **8**, 114 (1985)

A Journey Towards Excellence, (ed) Datta R, Agrawal B B, Bhaskar U, Chatterjee A and Prasad A, Samudran Press, Ranchi (2016)

Performance Reports of RDCIS, SAIL, (1988–89 to 2021–2022)

Nippon Steel Integrated Report 2020, https://www.nipponsteel.com> nsc_en_ir_2020_a4, pp 101–102

Mishra Sanak, An Autobiography, Notion Press, Chennai, Chapter No. 7, p 75 (2020)

Debasis Mukherjee Former Executive Director In-Charge, RDCIS, SAIL, Ranchi and Ministry of Steel Chair Professor in NIFFT, Ranchi.

Sanak Mishra Former Founder Member and Executive Director, RDCIS, SAIL, Ranchi; Managing Director, Rourkela Steel Plant, SAIL; CEO India Projects, ArcelorMittal and Secretary General and Executive Head, Indian Steel Association; Former President, Indian Institute of Metals.

Chapter 19
Applications of Plasma in Metallurgy and Vice-Versa: Indian Context

Alphonsa Joseph, Sudhir K. Nema, Amit Sircar, Paritosh Chaudhari, Upendra Prasad, Samir Khirwadkar, and Nirav Jamnapara

Abstract Plasma technologies for metallurgical applications are increasingly being adopted as they have the advantages of the unique properties of plasma. Amongst the major development priorities in many metallurgical operations and the development of new materials, plasma promises improved process control, direct utilization, improved environmental compliance and increased efficiency of energy. As a result, plasma is used for a wide spectrum of applications in materials processing like waste destruction, plasma spraying, plasma cutting, plasma welding, plasma synthesis of nanopowders, iron and steel making and extractive metallurgy for recovery of precious materials. Plasma has also recently been used as a source for the future fusion reactor, where it is confined by using the magnetic field inside the doughnut-shaped vacuum vessel. The fusion reactor which is being developed is seen as a promising, clean source of energy to solve the world's energy problem in the future as it does not use radioactive materials or pollute the air by carbon emissions. There are many challenging issues related to material constraints and plasma wall interactions which need attention as the operating environment of a fusion reactor imposes radiation damage effects. Detailed metallurgical investigation studies to select the most appropriate material are being done by many researchers. Moreover, the extremely harsh loading conditions in the fusion reactor can only be met with very diligent component design and careful selection of the best-suited material; fabricating and manufacturing techniques and efforts on developing new techniques are being carried out globally by many researchers. In this contributory article, a few examples of plasma-based technologies developed at Institute for Plasma Research (IPR) for industrial applications are described in brief. IPR has also developed, materials, technologies for blanket and fusion reactors, high temperature superconductors (HTS) for fusion magnets and they are elucidated in the below sections.

A. Joseph (✉) · S. K. Nema · A. Sircar · P. Chaudhari · U. Prasad · S. Khirwadkar · N. Jamnapara
Institute for Plasma Research, Near Bhat, Gandhinagar, Gujarat 382428, India
e-mail: alphonsa@ipr.res.in

Homi Bhabha National Institute, Training School Complex, Anushakti Nagar, Mumbai 400085, India

R. Divakar et al. (eds.), *Indian Metallurgy*, Indian Institute of Metals Series,
https://doi.org/10.1007/978-981-99-5060-7_19

19.1 Introduction

In the last two decades, plasma technologies are globally well-established in several areas of materials processing, stretching from melting and smelting in the iron and steel industry to fabrication processes such as plasma cutting, plasma welding, plasma-based deposition and plasma spraying processes. The use of plasmas for processing iron and steel has not only given the freedom in the selection of size, composition of iron ores and smelting conditions but has also eliminated the use of coke to achieve high purity levels (Ramachandran 1984). Moreover, the plasma-based fabrication processes have also provided precision and high cutting speed resulting in clean, high quality welds and coatings compared to the conventional methods (Klein et al. 2013). Further, plasma has been found to be a key enabling technology in more areas beyond these also. Recently, advanced plasma-based surface modification and waste management processes like plasma nitriding and plasma pyrolysis have been developed. The requirement of less time, low energy and an environmentally clean process have made these technologies superior and widely acceptable compared to the conventional techniques.

One of the major challenges in the development of fusion reactors is to create materials capable of withstanding the high heat loads and stresses involved in fusion devices. Materials with improved irradiation resistance and heat-removing capacity are required. Tungsten, beryllium, carbon, Reduced Activation Ferritic Martensitic (RAFM) material are being considered as the suitable material (Rowcliffe et al. 2017; Linke et al. 2019). The blanket is the key nuclear component of the fusion reactor and is responsible for the power extraction at high temperature as well as tritium breeding to ensure tritium self-sufficiency. Research on developing new techniques to fabricate nuclear components and tritium breeding material is in full swing (Abdou et al. 2015; Serra et al. 2017). It is known that high temperature superconductor (HTS) wires and tapes demonstrate remarkable current density, at very high magnetic fields, driving progress in fusion and other applications. Several techniques to manufacture $YBa_2Cu_3O_7$ (YBCO) HTS tapes have been initiated and is considered as a potential superconductor for future fusion reactor magnet applications (Molodyk et al. 2021).

In this article, a few examples of plasma technologies like plasma nitriding and plasma pyrolysis developed at Institute for Plasma Research (IPR) for industrial applications are described here. IPR has also developed HTS for fusion magnets and some recent developments along with results in material and technologies for blanket and fusion reactors are presented.

19.2 About IPR

Institute of Plasma Research largely involved in both theoretical and experimental studies in plasma science is an aided R and D institute under the Department of Atomic Energy (DAE), Government of India. This institute founded in 1986, owns two operational Tokamaks (a machine for controlling thermonuclear fusion) namely Aditya and Steady-state Superconducting Tokamak (SST). In addition, the institute also conducts 'basic experiments' on plasma and trains India's future plasma physicists and technologists. Facilitation Centre for Industrial Plasma Technologies (FCIPT) and ITER-India located in Gandhinagar and Centre for Plasma Physics (CPP) located in Guwahati are part of IPR. At FCIPT, industries can approach them for their research needs, development of specialized plasma processes, as well as equipment along with their power sources, material characterization services, etc. ITER-India has a major scientific and technical role in Indian partnership in the international fusion energy initiative ITER.

19.3 Introduction to Plasma

Plasma is an ionized form of gas and represents the fourth state of matter. When an electric arc is ignited between two electrodes in the presence of a gas, it gets partially ionized and becomes electrically conductive. In general, a plasma gas consists of electrons, ions and neutral species. As negative and positive charges compensate each other, the overall plasma becomes electrically neutral. Plasma that is produced by electrical discharge can be classified into two categories namely equilibrium/thermal plasma or non-equilibrium/cold plasma. In thermal plasma, the temperatures of the electrons and heavy particles are about the same and reach local thermodynamic equilibrium. In cold plasma, the electron temperature is much higher than the heavy particle temperature. Both these plasmas are used in metallurgy.

Thermal plasmas technology has emerged as widely used technology over the past decades in various applications like aerospace, microelectronics, automotive, material treatment and processing, melting and welding of metals, nanomaterial, plasma spraying and waste destruction. Typical atmospheric plasma devices are produced by torches for producing transferred and non-transferred arcs or by using radio frequency (RF) inductively coupled plasma discharges. The use of thermal plasma in the area of cutting, welding and spraying is well known over the four decades. However, for certain applications, thermal plasma sources are not suitable and they have been replaced by non-thermal plasmas. Non-thermal plasmas have been recently studied for a variety of industrial and medical applications such as surface modification using plasma deposition technologies, plasma sterilization, water purification, automobile exhaust emission control, volatile organic compounds removal, and surface activation to enhance surface properties such as wettability, printability and adhesion. These plasmas include low-pressure glow and RF discharges, microwave discharges and dielectric barrier discharges and can be effectively operated at low temperatures.

19.4 International Scenario

Plasma technology has been used as a novel technique globally for the manufacture of newer and better materials in recent years. The high temperature and high reactivity of the plasma make it a powerful medium to promote chemical reactions. Application of thermal plasma in coating technologies, synthesis of fine powders, waste destruction, spheroidization with densification of powders and in slag metallurgy are reported from the lab scale basis towards the industrial utilization (Samal 2017). Internationally, plasma technologies have been impactful in core metallurgy such as 'plasma cupola' for eco-friendly production of cast irons (Dighe 1996), plasma treatment for Al recovery from dross (Tzonev and Lucheva 2007), plasma tundish heating in steel making (Ye et al. 2020), plasma rotating electrode process (Cuia et al. 2020) and spark plasma sintering (Saheb et al. 2012) for powder metallurgy, etc. Plasma surface treatments are also extremely versatile and are used by a wide range of industries. It can enhance surface properties like biocompatibility, hydrophobicity and adhesion, also demonstrated as beneficial to treat pollutant gases using plasma catalysis (Hinde et al. 2020)—all from a value-driven, green alternative to harsh chemical techniques. It even extends service lifetime and reduces component downtime period. All materials including ceramics, polymers, elastomers and metals can be considered good candidates for plasma treatment.

Large efforts in plasma and fusion research in pursuit of the dream of making fusion power a reality is being done worldwide. Many research activities are being done to improve the confinement time and use of novel superconducting magnets and tungsten-based materials for handling extreme heat flux for the fusion reactor. Computer simulations and modelling of fusion plasmas have made enormous progress in the past decade and have added to the understanding of complex physical phenomena to make quantitative predictions for many of the experimental situations (Deshpande and Kaw 2013).

19.5 National Scenario

Plasma technologies involving both hot and cold plasmas have been developed nationally addressing different applications in different sectors. Thermal plasma technologies have been developed in India to address melting, smelting, and waste disposal issues (Samal 2017). Other thermal plasma technologies like nanomaterial synthesis, mineral processing, and involving direct and indirect plasma torches are in use in several industrial sectors owing to their superior properties. Plasma-based technologies like plasma-assisted physical vapour deposition and chemical vapour deposition, plasma nitriding have been in high demand in most of the heat treatment sectors as well as tooling industries as it offers an alternative solution to the conventional coating technologies in terms of an environmentally friendly process. In the sections below, various plasma technologies and metallurgical processes and new fabricating techniques for fusion reactors developed by IPR are discussed in brief.

19.6 Plasma Technologies Developed at IPR

19.6.1 Plasma Surface Modification Technologies

The plasma surface technologies developed by IPR include surface activation by plasma, plasma etching, plasma-assisted physical vapour deposition, plasma-assisted chemical vapour deposition and plasma nitriding. Surface activation by plasma enhances surface adhesion and increases the ability for paints and glue to adhere to the surface. Plasma etching has been used to improve the bonding between rubber and brass valves and is made available to industries. Plasma surface modification involving deposition of single and multiple coatings for imparting aesthetic and longer component service life has also been developed. Plasma nitriding and carburising are surface-hardening processes to make the surface hard. This process is adopted for many components like dies, automobile machinery parts like gears, camshafts and crankshafts, which desire high hardness on the outer surface with a softer core to withstand wear and corrosion resistance. Amongst these processes plasma nitriding is the most commonly used process by many industries. Plasma nitriding has several superior properties compared with conventional techniques such as gas and liquid nitriding like faster nitrogen penetration, short treatment time, low process temperature, minimal distortion, low energy use and easier control of layer formation. This process can be used for all ferrous materials including stainless steels. A glow discharge typical to a plasma nitriding process is generated around the parts as shown in Fig. 19.1. Institute for Plasma Research has developed this process and is currently demonstrating it to industries in the form of job working. IPR has also transferred this technology to industries. This technology can be used in numerous applications such as machinery parts, plastics and food processing, pumps and hydraulic machine parts, crankshafts, gears, motors, cold and hot working dies and cutting tools.

Fig. 19.1 Glow around a gear during plasma nitriding process

Fig. 19.2 Thermal plasma pyrolysis System and the plasma torch used for disposal of biomedical waste

19.6.2 Thermal Plasma Applications

Institute for Plasma Research has successfully developed and demonstrated thermal plasma processes and systems for many applications which include:

- Spheroidization of Alumina,
- Zircon sand dissociation in a single step,
- Phosphorus recovery from Phosphoric acid,
- Disposal of biomedical, organic and petroleum industry waste and
- Energy recovery from waste using plasma pyrolysis technology

Thermal plasma reduces the number of steps that are essential in conventional metallurgical processes and is mostly non-polluting. Using thermal plasma, a variety of waste can be disposed of in an environment-friendly manner and energy recovery from organic waste makes the thermal plasma technology economically attractive.

In thermal plasma, hazardous and toxic compounds are broken down into elemental constituents at high temperatures and extremely toxic compounds such as polyaromatic hydrocarbons, dioxins and furans are destroyed and organic mass is reduced by more than 99% by volume. At high temperature, using controlled oxygen or oxygen-starved environment, waste is converted to fuel gases like synthetic gas (H_2, CO) and CH_4. A plasma pyrolysis system developed by IPR and installed in a hospital in Goa is shown in Fig. 19.2. IPR has transferred Plasma Pyrolysis technology to several industries.

19.6.3 Development of Materials for Blanket Technology

Fusion reactor blanket materials research at IPR, is mainly focused on the development of structural and functional materials for Indian fusion blanket program. The ongoing R and D activities include development of technologies and characterizations of India specific Reduced Activation Ferritic Martensitic (IN-RAFM) Steel as the structural material and lithium titanate (Li_2TiO_3) ceramic pebble, lead–lithium (Pb-16Li) eutectic and coating materials (Al_2O_3, Er_2O_3) as the functional material. A brief overview of the present status and ongoing R and D activities for these material developments has been summarized below.

19.6.3.1 Indigenously Developed IN-RAFM Steel for Fusion Blanket

One of the key technological challenges for the development of a suitable structural material for fusion reactor is to have adequate mechanical properties under intense neutron irradiation (14.1 meV) and high thermo-mechanical loads during reactor operation. With this objective, India started its R and D activities to develop 9Cr-W-Ta steel for fusion application through a collaboration among IGCAR, IPR and MIDHANI. Several laboratory heats were produced of various compositions and after extensive tests for mechanical properties, finally 9 wt.% Cr, 1.4 wt.% W and 0.06 wt.% Ta composition was optimised for IN-RAFM (Indian-Reduced Activation fusion material) Steel (Laha et al. 2013).

Commercial heats of several tons of IN-RAFM steel are being made to produce plates of different thicknesses ranging from 1 to 60 mm. Extensive characterization of the microstructure, mechanical properties, thermo-mechanical and thermophysical properties of this material have been carried out as a part of materials qualification. Various welding techniques for joining IN-RAFM steel have also been developed and characterized for the fabrication of blanket module (Mistry et al. 2017). Welding Consumable development and evaluation of mechanical properties of various weld joints are currently under R and D studies.

19.6.3.2 Development of Lithium Ceramic Pebbles

In a fusion blanket, lithium ceramics are used as the tritium breeder material in the form of pebble bed. Lithium meta-titanate (Li_2TiO_3) has been chosen as the candidate tritium breeder material and its powder having purity $\geq$99.5%, is prepared in-house at IPR by solid-state reaction method (Shrivastava et al. 2014). After synthesis of the Li_2TiO_3 powder, spherical pebbles of ~1 mm diameter and with porosity of 15–20% are fabricated by both extrusion-spheroidization method and freeze granulation method. At every stage, the powder, pellet and pebbles undergo extensive characterizations to meet the desired properties. The average crushing strength of the ~1 mm pebbles is found to be $\geq$40 N. The achieved aspect ratio of the pebbles is ~0.9 as shown in Fig. 19.3.

Several experimental facilities have been set up for the measurement of various thermo-physical properties of Li_2TiO_3 pebble and pebble bed (Panchal et al. 2020). This includes effective thermal conductivity measurement of pebble bed by steady-state and transient hot-wire methods up to 800 °C under inert and air atmosphere,

Fig. 19.3 Li_2TiO_3 powder and pebbles prepared at IPR

thermo-mechanical characterization (uni-axial cyclic compaction and creep deformation) of pebble beds, etc. Packing characterization of various pebble bed assemblies (mono, binary and poly) under influence of different filling strategies (gravity, compression and vibration) are being simulated using Discrete Element Method (DEM).

19.6.3.3 Pb–Li Alloy Development

Lead–Lithium eutectic alloy (Pb–16Li) is also considered as a tritium breeder as well as neutron multiplier in the liquid breeder blanket concept. The production of this alloy is challenging due to large density difference (~20 times) in the component metals and formation of intermetallic during their exothermic chemical reactions. A very intensive mixing must be ensured to avoid stratification in the vessel and also to achieve homogeneous mixing without formation of intermetallic. Magneto Hydro Dynamic (MHD) stirring technique, using a travelling magnetic field, has been found to be an effective tool for mixing liquid metals with different densities. Pb–Li Production system, based on such technique and having a capacity of producing ~75 kg of Pb–Li alloy per batch, has been commissioned at IPR. Several batches of Pb–Li ingots have been produced and characterized for their melting temperature and elemental composition. The melting temperature of the produced material is found to be ~236 °C, which indicates the formation of Pb-16Li eutectic. The elemental composition analysis results have also confirmed the eutectic composition with Li Composition varying in the range of ~0.63 ± 0.04 (wt. %). IPR has also developed and patented a novel flow meter for liquid metals and conducting fluids along with a novel concept on cold trap to eliminate impurities in liquid metal loop.

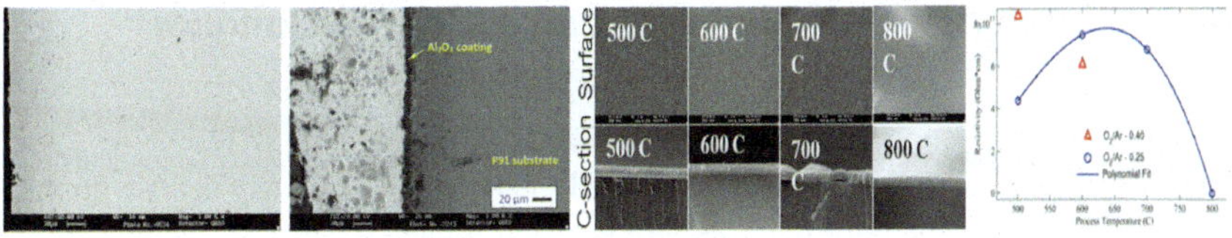

Fig. 19.4 **a** Cross-section microstructure of plasma treated aluminized 9Cr steel sample (Jamnapara et al. 2015), **b** Cross-section microstructure of plasma treated alumina coated 9Cr steel after exposure to molten Pb–Li for 1000 h under liquid metal corrosion test (Zala et al. 2019), **c** SEM Cross sections, surface morphology of erbia, **d** Resistivity of erbia deposited at different temperature

19.6.3.4 Coatings Development for Fusion Blanket Technology

Coatings like alumina (Al_2O_3) and erbia (Er_2O_3) have been reported as promising candidates as electrical insulator and tritium permeation barrier (TPB) in fusion blanket systems. However, these oxides have a large difference in the coefficient of thermal expansion (CTE) with the metallic substrate and hence such oxides tend to crack or spall during high temperature service or thermal cycling. It was therefore necessary that a bond coat exists at the interface to retain the protective nature of the oxide films. As a solution to this, hot dip aluminised coatings followed by plasma-based heat treatments (Jamnapara et al. 2015) have been developed at IPR. The novelty of Al_2O_3 coating was the plasma-assisted heat treatment, which could convert metastable θ-Al_2O_3 to stable α-Al_2O_3 phase (Zala et al. 2019). A liquid metal corrosion study was conducted on plasma-grown alumina-coated samples for 1000 h of exposure to molten Pb-16Li held at 550 °C. The results indicated excellent protection by the plasma-grown Al_2O_3 coating as visible in Fig. 19.4a, b, while the corrosion results of bare 9Cr steels were comparable to published literature (Zala et al. 2019).

Er_2O_3 coating in desired cubic crystal structure is developed and optimized through characterization for crystal structure, microstructure, the intactness of its adherence, surface morphology at nanometer scale, resistivity, etc. (Rayjada 2014). Typical characterization data are presented in Fig. 4c, d. The resistivity of such coatings is also characterized and found to be of the order of ~10^{15} and ~10^{14} Ω cm for magnetron sputtered coating and dip coating using metal–organic decomposition (MOD), respectively. These values are much more than that required for MHD control. The preliminary results on hydrogen isotope permeation studies through SS316L samples coated with Er_2O_3 using the dip coating technique have shown a permeation reduction factor of ~10^2. Dip coating technique has been utilized keeping in mind that non-flat samples will be needed for coating in future. We have also developed palladium coating on permeation-based hydrogen isotope sensors to facilitate more permeation of hydrogen isotope using magnetron sputtering. TPB coating with Er_2O_3 on the connecting tube to reduce permeation was done by hot dip using MOD.

19.7 Development of Materials and Joining Techniques for Plasma Facing Components

High Temperature Technologies Division (HTTD) of IPR primarily deals with the R and D of materials, technologies and engineering related to Plasma Facing Components (PFCs) of high temperature plasma and fusion devices. HTTD also strives to establish relevant experimental facilities that are required to perform necessary R and D work. Materials research activities of HTTD are primarily focused on development and characterization of tungsten-based materials and material joining (welding, brazing, coating) techniques for tungsten, copper and steel materials for typical thermal and structural loads on plasma-facing components. Some of the key areas include—Microscopic and Metallography Studies, Thermal Properties Testing, Non-Destructive Testing, Thermal Load Testing (for water-cooled PFCs), Thermo-Mechanical Testing, Small Specimen Testing. Specifically, efforts are being made to establish relevant test facilities for materials testing at elevated temperatures.

19.7.1 Mono-block PFC Development

High heat flux receiving PFC of a Tokamak—called as Divertor—receives quasi-steady heat flux of the order of 10 MW/m^2. Tungsten mono-block-based design of divertor is expected to withstand heat flux up to 20 MW/m^2.

Tungsten mono-block divertor target test mock-up, (shown in Fig. 19.5a) has been developed by IPR using two different techniques viz. (1) Hot radial Press (HRP) technique (in collaboration with NFTDC, Hyderabad) and (2) Vacuum Brazing Technique. These mock-ups have been successfully tested for incident heat flux up to ~20 MW/m^2 using High Heat Flux Test Facility (HHFTF) at HTTD of IPR.

Fig. 19.5 **a** (top left) Tungsten mono-block divertor target test mock-up (Area: 100×30 mm^2), **b** (top right) Tungsten macro-brush dome test mock-up (Area: 350×46mm^2), **c** (bottom) Tungsten coated test mock-up with CuCrZr substrate (Area: 440×30mm^2)

19.7.2 Dome PFC Development

Dome is a PFC of tokamak that is expected to receive quasi-steady state heat flux up to 5 MW/m^2. For such an intermediate level of heat flux, macro-brush design is preferred due to its simplicity.

Tungsten macro-brush Dome target test mock-up (shown in Fig. 19.5b) is made of multi-layered materials joints of W-Cu-CuCrZr-SS316L fabricated using Vacuum Brazing Technique. It has been successfully tested up to 1,200 thermal cycles under maximum incident heat flux of 6 MW/m^2.

19.7.3 Tungsten Coated PFC Development

First Wall of a Tokamak is a PFC that is expected to receive low heat flux of the order of 1 MW/m^2. Tungsten coatings on water-cooled metallic substrates are preferred for such applications as they can be cheaper to fabricate and easy to repair.

Thick Tungsten (W) coated test mock-up (shown in Fig. 19.5c) with tungsten thickness of ~500 μm and different metallic substrates viz. CuCrZr, SS316L, IN-RAFM Steel, has been developed using Atmospheric Plasma Spray (APS) technique in collaboration with ARCI, Hyderabad. Test mock-up with CuCrZr substrate has been successfully tested for incident heat flux up to 5.8 MW/m^2.

19.7.4 Plasma Facing Materials Development

Pure tungsten and tungsten alloys are recommended for PFM due to its excellent physical properties. IPR has developed (i) Pure W sample by powder metallurgy (ii) WCu-FGM at 1030 °C temperature and 40 MPa pressure, (iii) Fabricated Wf-W composite using 10–70% volume fraction W-fibers (W_f) at 1900 °C and 40 MPa, achieved 94% relative density with 440 HVN hardness, (iv) Fabricated W laminate (W foils of 100 μm thick) samples through diffusion bonding at 1400 °C and 60 MPa could achieve 340 HVN hardness with good interface bonding without any delamination.

19.7.5 Helium Cooled PFCs

Helium gas, supplied at high pressure and high temperature, is of future interest for active cooling of PFCs since it avoids damage to PFCs that may occur due to water leakage in a fusion device. However, due to poor heat transfer properties of helium gas, high pressure (~100 bar) and high temperature (~500 °C) of helium gas are required to generate desired heat transfer coefficient for heat extraction. This poses a challenge to maintain the thermo-mechanical stresses of helium-cooled plasma-facing components within the allowed limits of materials and joints. Development of helium-cooled diverter target test mock-up capable of extracting 5 MW/m^2 heat flux is in progress.

19.7.6 Liquid PFC R and D

Tin–Lithium (Sn–Li) alloys are of future interest for use in liquid metal-based PFCs. Their low vapour pressure appears suitable for the concept of liquid surface plasma-facing material. The Sn-Li alloy exhibits self-healing/renewability of the plasma-facing surface and also has less sensitivity to neutron damage compared to solid surface Plasma Facing Components. Activities for producing different Sn–Li alloys have been initiated. Alloys produced will be further characterized to study their relevant material properties (such as vapour pressure, thermal and electrical conductivity, viscosity etc.) for PFC application.

19.8 Fusion Magnets and Conductor Development at IPR

A tokamak is device, which initiates, confines and shape the plasma using magnetic coils in a vacuum vessel. The toroidal magnetic field coils generate steady magnetic field in toroidal direction to confine the plasma. Poloidal magnetic field coils are used to shape the plasma. The external central solenoid of this device acts as primary, in which, the change in current induces a toroidal electric field required to initiate the plasma as secondary in the vacuum vessel. The correction coils are used to correct the error fields generated due to eddy currents in cryostat, vacuum vessel and support structures of magnetic coils in the superconducting (SC) tokamak during change in currents in the primary coils. The other in-vessel coils help in controlling the plasma in steady state. Presently, IPR is working with resistive coil-based Aditya Tokamak and low temperature superconducting (LTS), NbTi-based SST-1 tokamak. R&D activity has been initiated to develop high temperature superconductor (HTS)-based high magnetic field coils for future fusion reactor, as the conventional LTS (Nb_3Sn and NbTi) cannot be operated at higher magnetic field (>13 T) due to upper critical field limitations. The development of large volume copper coils for magnetic confinement of plasma at IPR was initiated long back for basic laboratory experiments, Aditya tokamak, and NbTi superconductor-based coils for SST-1, having an operating temperature of −268.5 °C, current 10 kA and stored magnetic energy of 55×10^6 J of toroidal field (TF) coils, enclosing 35 m^3 vacuum vessel for SST-1 (Saxena 2000) first wall and plasma. TF coils of SST-1 were operated up to 2.7 T, which was 90% of the designed value at the major radius of 1.1 m of the SST-1(Prasad et al. 2019, 2013). The superconducting Cable in Conduit Conductor (CICC) for SST-1

SC coils was procured from Hitachi Cable, Japan. The in-house fabrication technologies, right from SC strands to high current NbTi and Nb_3Sn CICC have been established later. The installation and performance testing of the cryostat of volume about 64 m^3, which can house large fusion relevant SC coil of cross-sectional area about 20 m^2 for low temperature testing has been completed. Significant progress has been made in development of high temperature superconductor (HTS)-based 1 kA cable of cross-sectional area 29 mm^2, nano-ohm class HTS-HTS tape joints (Prasad et al. 2018), quench dynamics study in 2nd generation HTS tapes (Prasad et al. 2019) and a 1.1 T compact solenoid coil of inner diameter 50 mm. The in-house development of intermediate temperature mono-filament MgB_2 wire using Ex-situ technique and high temperature YBCO tapes using physical vapour deposition method with multiple magnetron sputtering targets have also been initiated. The Ex-situ Powder in Tube (PIT) technique for the fabrication of MgB_2 wires consists of direct filling of the metallic Nb tube with reacted MgB_2 powders, followed by cold drawing and heat treatment at 600–1000 °C to obtain superconductivity for an improved core density with well-connected grains. In In-situ technique, Mg and B powder is filled in metallic Nb tube in certain proportion to make MgB_2 powder. During cold working and annealing, there are chances of formation of MgB_4, MgB_7, MgB_{20} phases along with MgB_2 phases. These boron-rich phases may degrade the current density of wire. The formation of voids due to intermetallic diffusion in the place of magnesium after the formation of Mg-B phases may also degrade the quality of wire. The operating temperatures and magnetic field of HTS are higher than LTS, which makes it a potential superconductor for future fusion reactor magnet application.

19.9 Conclusion

Plasma technology has emerged as a novel technique for the production of novel and better materials in recent years. Plasma processing has provided technologies like plasma cupola, plasma treatment for Al recovery, plasma tundish heating in steel making, plasma rotating electrode process for the extractive and process metallurgy industry and spark plasma sintering for powder metallurgy, etc. It has already proven useful in the melting/remelting and refining of refractory metals. Plasma-based surface modification and waste destruction technologies have potential to offer plenty of possibilities for surface modification of almost any conceivable substrate material and the destruction of all waste with zero hazardous emission. Especially, considering the growing inclination of reduced carbon emissions, renewable energy and electric transportation, plasma technologies such as plasma catalysis, plasma-assisted CO_2 sequestration, plasma-assisted melting/smelting including hydrogen-assisted melting/smelting, plasma material synthesis of battery/energy materials, etc., are likely to find importance. The challenge today is to take full advantage of this opportunity and use science and technology to discover innovative methods and products for the future generation. Several awareness programs and collaborative efforts with research organizations and industries can strive for bringing imaginative, adaptive and cost-effective technologies to mankind at large.

In the pursuit of developing a fusion reactor in collaboration with ITER consortium, IPR has developed technologies for making superconducting coils with NbTi and Nb_3Sn multi-filamentary conductors and compact HTS coils for high magnetic field generation. New material and technologies for blanket and divertor, irradiation and other test facility development work have also been initiated in India's fusion research program. Work at several experimental test setups is in progress for validation of the material for use in different areas of nuclear fusion reactor construction.

References

Abdou, M., Morley, N.B., Smolentsev, S., Ying, A., Malang, S., Rowcliffe, A., Ulrickson, M.: Blanket/first wall challenges and required RandD on the pathway to DEMO. Fusion Eng. Des. **100**, 2–43 (2015)

Cuia, Y., Zhao, Y., Numata, H., Bian, H., Wako, K., Yamanaka, K., Aoyagia, K., Zhang, C.: Effects of plasma rotating electrode process parameters on the particle size distribution and microstructure of Ti-6Al-4 V alloy powder. Powder Technol. **376**, 363–372 (2020)

Deshpande, S. Kaw, P.: Fusion research programme in India. Sadhana **38**(5), 839–848 (2013)

Dighe S (1996) Plasma cupola iron melting, technical article in the EPRI center for materials production. Ind. Agric. Technol. Serv. CMP-1 **1**(1), 1–2 (1996)

Hinde, P., Demidyuk, V., Gkelios, A., Tipton, C., Matthey, J.: Plasma catalysis: a review of the interdisciplinary challenges faced, realising the potential of plasma catalysis on a commercial scale. Technol Rev **64**(2), 138 (2020)

Jamnapara, N.I., et al.: Phase transformation of alumina coating by plasma assisted tempering of aluminized P91 steels. J. Nucl. Mater. **464**, 73–79 (2015)

Klein, A.N., Cardoso, R.P., Pavanati, H.C., Binder, C., Maliska, A.M., Hammes, G., Fusao, D., Seeber, A., Brunatto, S.F., Muzart, J.L.R., DC Plasma technology applied to powder metallurgy: an overview. Plasma Sci. Technol. **15**(1), 70–83 (2013)

Laha, K., et al.: Development of India-specific RAFM steel through optimization of tungsten and tantalum contents for better combination of impact, tensile, low cycle fatigue and creep properties. J. Nucl. Mater. **439**(1–3), 41–50 (2013)

Linke, J., Du, J., Loewenhoff, T., Pintsuk, G., Spilker, B., Steudel, I., Wirtz, M.: Challenges for plasma-facing components in nuclear fusion. Matter Radiat. Extremes **4**(056201), 202–218 (2019)

Mistry, A.N., et al.: Status of India-specific reduced activation ferritic-martensitic steel and fabrication technologies development for LLCB TBM. Fusion Eng. Des. **125**, 263–268 (2017)

Molodyk, A., Samoilenkov, S., Markelov, A., Degtyarenko, P., Lee, S., Petrykin, V., Gaifullin, M., Mankevich, A., Vavilov, A., Sorbom, B., Cheng, J., Garberg, S., Kesler, L., Hartwig, Z., Gavrilkin, S., Tsvetkov, A., Okada, T., Awaji, S., Abraimov, D., Francis, A., Bradford, G., Larbalestier, D., Senatore, C., Bonura, M., Pantoja, A.E., Wimbush, S.C., Strickland, N.M., Vasiliev, A.: Development and large volume production of extremely high current density $YBa_2Cu_3O_7$ superconducting wires for fusion. Sci. Rep. **11**(2084), 1–11 (2021)

Panchal, M., et al.: Measurement of effective thermal conductivity of lithium metatitanate pebble bed by transient hot-wire technique. Fusion Eng. Des. **158**, 111718 (2020)

Prasad, U., Sharma, A.N., Patel, D., Doshi, K., Khristi, Y., Varmora, P., Chauhan, P., Jadeja, S.J., Gupta, P., Pradhan, S.: Fabrication of new joints for SST-1 TF coil winding packs. Fusion Eng. Des. **88**, 2945–2949 (2013)

Prasad, U., Bano, A., Srinivasan, R.: Experimental Study of Different Joint Configurations of HTS BSCCO (Bismuth Strontium Calcium Copper Oxide) Tape, IPR/RR-1005 (2018)
Prasad, U., Varmora P., Parghi B., Babaudha, M., Makwana A.R., Dhodiya, C., Raj, P., Roy, S., Kanawar, D., Panchal, A., Bhavsar, D., Kumar, N., Bano, A., Ghate, M., Singh, Y., Srinivasan, R.: High field operational performances of superconducting TF coils of SST-1. In: 27th National Symposium on Cryogenics and Superconductivity, 16–18 January 2019, Victor Menezes Convention Center, IIT Bombay (2019)
Ramachandran, S.: Iron and steel technology: options for India. Sadhana **7**(3), 181–202 (1984)
Rayjada, P.A.: Process optimization of Er_2O_3 coating by reactive magnetron sputtering for demo-relevant blanket modules. Fusion Sci. Technol. **65**, 194 (2014)
Rowcliffe, A.F., Garrison, L.M., Yamamoto, Y., Tan, L., Katoh, Y.: Materials challenges for the fusion nuclear science facility. Fusion Eng. Des. **135**(5–6) (2017)
Saheb, N., Iqbal, Z., Khalil, A., Hakeem, A.S., Aqeeli, N.A., Laoui, T., Qutub, A.A., Kirchner, R.: Spark plasma sintering of metals and metal matrix nanocomposites: a review. J. Nanomater, 13 (2012). Article ID 983470
Samal, S.: A review on thermal plasma technology: the prospective future in material processing. J. Clean. Product. **142** (2017)
Saxena, Y.C.: Present status of the SST-1 project, SST-1 team. Nucl. Fusion **40**(6) (2000)
Serra, D.D., Frano, R.L., Puccini, M., Aquaro, D., Vitolo, S., Stefanelli, E., Forgione, N., Santis, S.D., Marco, A., Ciolini, R., Seggiani, M.: Manufacturing challenges of fusion pebbles bed material. In: Proceedings of the International Conference Nuclear Energy for New Europe, Bled, Slovenia (2017)
Shrivastava, A., et al.: Preparation and characterization of the lithium metatitanate ceramics by solution-combustion method for Indian LLCB TBM. Fusion Sci. Technol. **65**, 319–324 (2014)
Tzonev, T., Lucheva, B.: Recovering aluminum from aluminum dross in a DC electric-arc rotary furnace. JOM **59**, 64–68 (2007)
Prasad, U., Bano, A., Banaudha, M.: Magnet system division and R. Srinivasan, quench dynamics study in 2nd generation HTS tapes. In: 27th National Symposium on Cryogenics and Superconductivity, 16–18 January 2019, Victor Menezes Convention Center, IIT Bombay (2019)
Ye, M., Zhao, M., Chen, S., Yang, S., Li, J.: Influence of plasma heating on the metallurgical effects of a continuous casting tundish. Metals **10**, 1438 (2020)
Zala, A.B., et al.: Delta (δ) ferrite formation in the welds of aluminized 9Cr-1Mo steels. Metallogr. Microstruct. Anal. **8**(2), 256–262 (2019)

Chapter 20
Evolution of "International Advanced Research Center for Powder Metallurgy and New Materials (ARCI)", a Unique Centre for Translating Materials Research to Technology

R. Vijay, D. Srinivasa Rao, Roy Johnson, R. Gopalan, and Tata Narasinga Rao

Abstract The International Advanced Research Centre for Powder Metallurgy and New Materials (ARCI) is a research and development institution established in 1990 as a product of India's extensive bilateral cooperation program, the Integrated Long-Term Programme on Cooperation in Science and Technology with the erstwhile Soviet Union. The centre has a mandate of developing high-performance materials and processes for niche markets, demonstrating technologies at a prototype/pilot scale, transferring technologies to the Indian industry, and training personnel in the identified fields. ARCI has emerged as a nationally important research centre, aligning its research directions with national needs and national missions. The technical progress and evolution of the institute over the last 25 years, along with new initiatives and the vision for the institute, are discussed. The centre has three main thrust areas, namely powder metallurgy and nanomaterials, surface engineering, and new materials for energy applications.

20.1 Genesis

ARCI is a unique R&D institution of the Department of Science and Technology (DST), Government of India with its motto of 'Translating Research to Technology' and its genesis as a product of India's most extensive bilateral cooperation programme, the Integrated Long-Term Programme on Cooperation in Science and Technology (ILTP), with the erstwhile Soviet Union. It started as a collaborative

R. Vijay · D. S. Rao · R. Johnson · R. Gopalan · T. N. Rao (✉)
International Advanced Research Center for Powder Metallurgy and New Materials (ARCI), Hyderabad, India
e-mail: director@arci.res.in

R. Divakar et al. (eds.), *Indian Metallurgy*, Indian Institute of Metals Series,
https://doi.org/10.1007/978-981-99-5060-7_20

project in the field of 'Powder Metallurgy' implemented by the Defence Metallurgical Research Laboratory (DMRL), India and the Belarusian Powder Metallurgy Association (BPMA), USSR and transformed into one of the best materials-based application-oriented research laboratories in the country. Due to the vision of Prof. Oleg V. Roman from the USSR and Dr. VS Arunachalam, Dr. P. Rama Rao and Shri SLN Acharyulu from India, the idea of setting up a joint research centre for powder metallurgy was first conceived in 1988, and in 1990, Indo-Soviet Advanced Research Centre (ARC) for Powder Metallurgy was registered as a society in erstwhile Andhra Pradesh.

On September 21, 1991, Prof. Rama Rao, who was Secretary, DST, took over as Chairman, Governing Council, Shri Acharyulu became Project Director (in addition to being Director, DMRL) and Shri G.S. Bhattacharjee became the Chief Project Officer (CPO). Owing to the political developments in the Soviet Union in the early 1990s, an Agreement was signed with CIS countries in 1993 and independent collaborations with republics like Russia, Ukraine, Moldova, Uzbekistan were established in various areas of materials technology leading to the signing of multilateral agreements. Consequently, on March 21, 1994, the name of ARC was changed to 'International Advanced Research Centre for Powder Metallurgy and New Materials (ARCI)' as it is known today (Fig. 20.1).

With the completion of the project phase in 1996, ARCI became operational as an autonomous institution of DST. The transition of ARCI from the project phase to a full-fledged, independent laboratory occurred through a series of events spread over a whole year, in 1996–1997. On January 1, 1997, Dr. G. Sundararajan, the first full-time independent Director of ARCI, was appointed and due to his vision and guidance, ARCI has established itself as a place for translational research resulting in many technology transfers and also made forays into new areas of research (Fig. 20.2). Dr. G. Padmanabham, after taking over as Director in 2016, provided overall leadership for the successful development of technologies in the fields of engineering materials

Fig. 20.1 (left) Building a permanent home for the Centre during 1991(Bhoomi Puja), (right) Shri G. S. Bhattacharjee and Prof. O. V. Roman—signing of multilateral agreements between ARC and CIS Countries during 1993

Fig. 20.2 (left) Dr. A. P. J. Abdul Kalam with Shri S. L. N. Acharyulu, Project Director, ARCI during INDOCIS 96 in 1996, (right) Dr. G. Sundararajan, Director, ARCI along with Joint Directors Dr. Y. R. Mahajan and Dr. R. Sundaresan

and additive manufacturing and its transfer to industry. The untimely demise of Dr. Padmanabham led to Dr. Tata Narasinga Rao taking over as Director (Additional Charge) from June 2021 and as a full time Director from November 2022. 2021–22 marks the 25th Anniversary year of ARCI and the Silver Jubilee was celebrated recently. In this journey, ARCI has emerged as a nationally important research centre aligning its research directions in line with national needs and national missions. In the present article, the technical progress and evolution of the institute over the last 25 years along with new initiatives and the vision for the institute are discussed.

20.2 The Mandate

The initial motivation was to develop and demonstrate powder metallurgy, ceramic, and surface engineering applications catering to strategic as well as industrial sectors. But, it evolved over the years to cover a wide range of materials and processes in line with the international developments and national needs with a vision to be a globally competent application-oriented R&D Centre in the field of Advanced Materials and Processing Technologies. Aiming at delivering products, processes, and equipment in a ready-for-use and commercialization, ARCI set for itself a challenging mandate of.

(a) Development of high-performance materials and processes for a niche market;
(b) Demonstration of technologies at prototype/pilot scale;
(c) Transfer of technologies to the Indian industry; and
(d) Training of S&T personnel in the identified fields and Human Resource Development in general.

The first three pillars on which the technological strengths of ARCI were built up are.

- Powder Metallurgy
- Surface Engineering
- Ceramic Processing

With a strong desire to be at the forefront of the developments in advanced materials and materials processing technologies and their effective industrial use, ARCI forayed and expanded consistently, establishing new centres of excellence in application-relevant fields such as:

- Nanomaterials and Nano-composite coatings
- Laser processing of materials
- Fuels cells
- Solar energy materials
- Automotive energy materials

20.3 Thrust Areas

20.3.1 Powder Metallurgy and Nanomaterials

Powder Metallurgy is a process for producing miniature to small and medium-sized components of complex/intricate shape using alloy or composite powders as starting raw materials, which are either not possible or extremely difficult to obtain by conventional processing. The process avoids material wastage and is economical and energy-efficient for mass-automated production.

Powder Metallurgy activity at ARCI houses expertise and infrastructure to develop innovative processes and technologies for making powders and components on prototype/pilot scale to cater to automotive, aerospace, strategic, and other industrial sectors. The centre is equipped with state-of-the-art facilities like induction melting, inert gas atomization, gaseous or solid-state reduction and chemical routes, high-energy ball mills, presses, and furnaces for the production as well as the consolidation of powders.

Technologies for the production of iron powder from blue dust, Heat pipe-based heat sinks, Tubular SS filters for the filtration of sewage water, Copper commutators, Bushes, Clutch pads, Clutch buttons, Compressor pistons of air conditioners, Tungsten heavy alloy penetrators, Lead-free copper alloys for bimetallic bearings, Magnesium based hydrogen storage materials, Titanium as well as Zirconium powders were developed. Technologies like Iron powder from blue dust, Heat pipe-based heat sinks, Tubular SS filters, Lead-free copper alloys for bimetallic bearings have been transferred to industries. Other technologies like clutch pads, clutch buttons, and zirconium powder are ready for transfer.

Over a period of time, the powder metallurgy activities are more directed towards the development of nanopowders as well as nanostructured materials. Large scale synthesis of nanopowders is key to successful application and technology development. ARCI established various techniques ranging from bottom-up chemical methods to top-down mechanical milling methods, including flame spray pyrolysis, RF Induction plasma, high energy milling, supercritical drying, arc discharge, and chemical vapour deposition methods for synthesizing nanopowders, nanotubes, and aerogels of different metals, oxides, and carbon. Nanosilver as well as nano titania suspensions for anti-bacterial and self-cleaning textile finishes, electrode materials like lithium iron phosphate and lithium titanate for Li-ion batteries, porous carbon material as an electrode for supercapacitors, oxide dispersion strengthened steels for clad tubes of fast breeder reactors and steam turbine blades, tungsten plates for jet wanes and balancing weights, silica aerogel sheets for thermal insulation, nano aluminium and nano boron powders as additives to solid propellant and jet fuel respectively and doped ZnO powders for varistor applications were developed. The technologies for nanosilver and nano titania suspensions for anti-bacterial and self-cleaning textiles respectively, and tungsten plates for jet wanes were transferred to Industries. Development of materials for energy applications is one of the major activities of the centre and in line with this; the production of both cathode (lithium iron phosphate) and anode (lithium titanate) materials for Li-ion battery was successfully demonstrated at pilot scale. The technology for the production of carbon coated Lithium Iron Phosphate (C-LFP) as a cathode material for Li-ion batteries was transferred to an Indian industry on non-exclusive basis and an agreement was signed with another industry on global exclusive basis (other than India), which is in progress. Presently, it is being incubated at Advanced Materials Technology Incubator of ARCI. Development of materials for new generation Na-ion and Li–S batteries, lightweight oxide dispersion strengthened iron aluminides for high-temperature applications, powders for additive manufacturing and coatings, biodegradable implants are some of the significant directions being pursued by the centre. ARCI has received the prestigious 'Technology Day National Award' in 2018 for successfully commercializing nanotechnologies in collaboration with its industry and academic partners.

Research on carbon materials is yet another activity carried out under Powder Metallurgy vertical. Technology for the production of exfoliated graphite, and various products like flexible graphite sheets, high-temperature seals was developed and transferred to industry. Over a period of time, these growing activities resulted in the evolution of a new centre for carbon materials. Bipolar plates for PEM fuel cells were successfully developed and field-tested. Graphene production at a scalable level for functional composites was established. Development of nanocarbon additives for lubrication and aligned carbon nanotube arrays for field emission is being pursued (Figs. 20.3, 20.4, 20.5, 20.6, 20.7, 20.8, 20.9, 20.10 and 20.11).

Fig. 20.3 (top left) Parts developed through the powder metallurgy route, (top right and bottom) Vertical shaft reduction of blue dust to obtain sponge iron briquette

Fig. 20.4 (left) Exfoliated graphite products for various applications, (centre) Oxide dispersion strengthened steels for high-temperature applications, (right) Silica aerogel based thermal insulation sheets for oil and gas industries

Fig. 20.5 Inert gas atomizer, induction melting furnace, inert gas atomizer, spark plasma sintering facility and Zoz Simoloyer high energy milling

Fig. 20.6 (left) Nanomaterials for self-cleaning textiles (Photocatalytic), (centre) Indigenously developed C-LFP (cathode) powder, (right) LTO anode powder

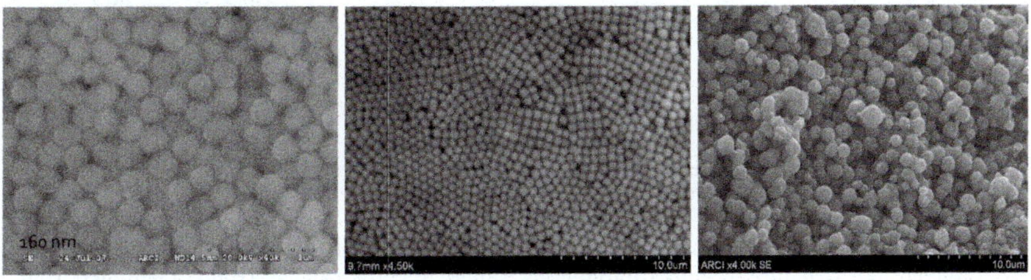

Fig. 20.7 (left and centre) Silica nanoparticles, (right) TiO_2 microspheres

Fig. 20.8 (left) Nano Al powder, (centre) ZnO nanorods, (right) Ag-TiO_2 @ Silica particles

Fig. 20.9 Technology development board national award 2016 for indigenous technology commercialization of nano-silver based antibacterial textiles

Fig. 20.10 (top) Semi-pilot plant for the fabrication of Supercapacitors

Fig. 20.11 (left) Petcoke derived porous Carbon powder and TEM image showing the presence of Graphene like structure, (centre) Indigenous Supercapacitor Device, (right) Demonstration of e-bicycle using indigenous supercapacitors

20.3.2 Surface Engineering

Surface coatings are indispensable in every industrial application with the need for corrosion protection, wear resistance, improved tribology, and a range of functionalities in different environments. The applications can range from simple corrosion-resistant electroplating to thermal barrier coatings on aero-engine components. Since its inception, surface engineering has been at the forefront of R&D activities of ARCI apart from powder metallurgy and ceramic processing. The Detonation Spray Coating (DSC)—a proprietary thermal spray coating technique, was the first equipment to be installed and commissioned in the project stage of ARCI during 1991. After the installation, optimization studies for wear-resistant coatings were aggressively carried out during 1992–93, and the technology was successfully demonstrated on LPT blades, thus sowing the seeds for a remarkable trend of working on advanced techniques and subsequent indigenization for the Indian market. Later expansion activities included multiple coating technologies such as Electro Spark Coating (ESC) and Micro Arc Oxidation (MAO). The indigenization of DSC, MAO, and ESC technologies was aggressively pursued which resulted in the first DSC system to be built by ARCI in the year 1997 and successfully transferred to Indian industries during 1999. In the case of MAO technology which is targeted for obtaining oxide coatings, predominantly on aluminum, the technology was initially demonstrated on a laboratory scale, subsequently scaled up to the industrial scale, and finally transferred to the multiple technology receivers, who now cater to the requirements of the textile, automotive and aerospace industries. In addition, an academic version of MAO with advanced capabilities for depositing coatings on new material systems has been developed and successfully transferred.

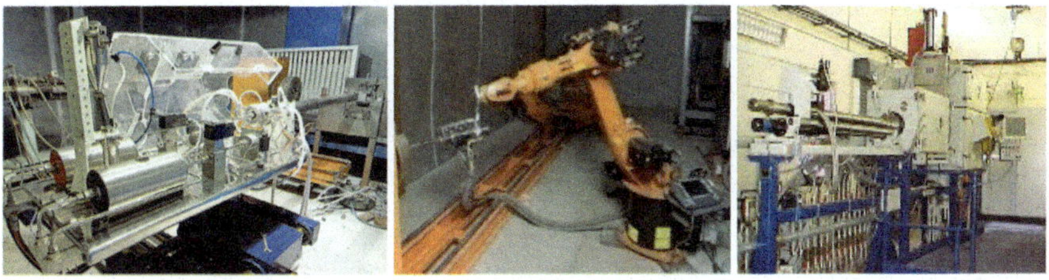

Fig. 20.12 (left) Advanced detonation spray system, (centre) cold spray coating system, (right) EBPVD facility

After the successful indigenization and transfer of the early technologies, the Centre for Engineered Coatings (CEC) further diversified to a wider range of coating technologies, including cold spray coating technology (CSC)—a low-temperature thermal spray variant for oxidation sensitive and low melting metals and alloys. This technology has now been successfully indigenized and is available for transfer with capabilities on par or better than globally available commercial systems. Concerning electrodeposition, a green technology called Pulsed Electro Deposition (PED) was successfully developed as an alternative to hard chrome coatings which involve carcinogenic hexavalent chromium. The recipe for nickel-based coating compositions has been successfully transferred to the industry. In the case of PVD-based technologies, industrial-scale Electron beam Physical Vapor Deposition (EB PVD) and Cathodic Arc Physical Vapour Deposition (CAPVD) systems were established to cater to specific applications of wear-resistant coatings on turbines and machine tools (Fig. 20.12). Numerous applications using these techniques have been developed and the coated components have been successfully put into operation in various civilian and defence applications (Fig. 20.13). Partnerships are established with the industry as well as strategic sectors in the aerospace sector. The Centre for Engineered Coatings has emerged as an international-class one-stop shop for surface engineering solutions in the country.

Under the establishment of a National Centre for Materials and Manufacturing technologies for clean coal-based energy generation, the centre houses not only the processing facilities but also the testing facilities to simulate the operating conditions of thermal power plants either on the steam side or on the fireside as well as durability assessment. State-of-the-art High-Velocity Air/Oxygen Fuel (HVAF/HVOF) coating and Axial Plasma spray coating systems have been recently commissioned to develop coatings that improve the fire-side corrosion performance of boiler tubes (Fig. 20.14).

20.3.3 Ceramic Processing

The industrial application of ceramics depends heavily on the ability to process them into different shapes and sizes. The technological challenge grows bigger with the size and geometrical complexity of the component to be fabricated. The

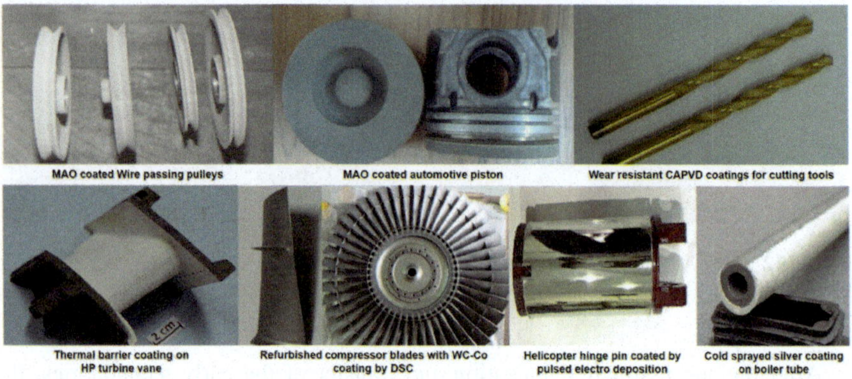

Fig. 20.13 Surface engineered components using various coating technologies

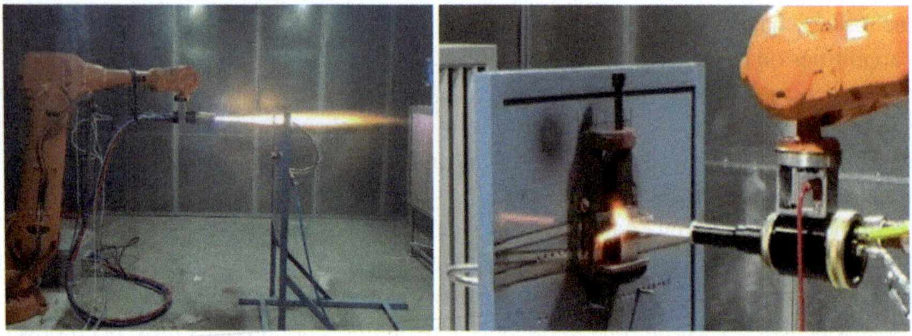

Fig. 20.14 (left) Air plasma spray system, (right) high-velocity air–fuel spray system

ceramic processing centre at ARCI has established a range of shaping techniques such as compaction, extrusion, gel casting, slip casting, pressure slip casting, chemical vapour deposition, cold and hot isostatic pressing, and 3D printing of a range of oxide and non-oxide ceramics (Fig. 20.15). Currently, application development areas span from cellular honeycomb structures to transparent and reflective optical ceramics and Solid Oxide Fuel Cells (SOFC). During the conception of the Ceramic Materials Division in 1994, cellular honeycomb ceramics was the major activity of focus. Accordingly, consistent efforts during the early years have resulted in the successful development of honeycombs and channelled ceramics with a variety of ceramic formulations. The development of ceramic honeycomb later become a platform technology that grew into developing several products such as energy-efficient air heaters, environmentally friendly 'Green Dispo' sanitary pad incinerators in partnership with Indian industry (Fig. 20.16), and anti-mine boots with the Defence Materials Stores Research Development Establishment (DMSRDE), Kanpur for the welfare of soldiers.

In the area of reflective optics, a DST-ISRO centre was established for the realization of non-oxide ceramics for application as space components. Silicon carbide (SiC)

Fig. 20.15 (left) Screw type extruder for extrusion of high symmetric ceramic parts, (centre) hot isostatic press, (right) pressure slip casting machine

Fig. 20.16 Launch of 'GreenDispo', incinerator for used sanitary napkins

is identified as the material of choice for space mirrors because of the outstanding thermo-mechanical properties for superior geometrical integrity in harsh service conditions. Later, the technology of substrate for the mirrors is developed through consolidation of SiC powder adopting powder metallurgy route. The process line established at ARCI is capable of producing high-quality sintered SiC components. Recently, the centre has also undertaken the high-temperature graphitization of carbon–carbon composites using the high-temperature processing facilities for the fabrication of parts for the Gaganyaan mission (Fig. 20.17).

Over the years, through focused efforts, the centre established specialization in the area of transparent ceramics. The candidate transparent ceramic systems developed at ARCI include polycrystalline sub-micron alumina (Al_2O_3), aluminium oxynitride (AlON), nano-yttria (Y_2O_3), spinel ($MgAl_2O_4$), and zinc sulphide (ZnS) ceramics. ARCI is working on these materials at different technology readiness levels (TRLs) for catering to diverse applications. Based on the Chemical Vapour Deposition of ZnS technology, the first commercial transparent ceramics production plant in the

Fig. 20.17 (left) Cellular honeycomb ceramics, (centre) transparent ceramic blank, (right) SiC substrate

country is established at Bharat Electronics Limited (BEL), Bengaluru catering to all the requirements of IR transparent ZnS for IR imaging applications.

As a part of enhancing the core competence in the area of optical ceramics, the latest effort is to produce near-zero expansion glass ceramics for high-end precision optical applications. A process line for producing Lithium Aluminium Silicate (LAS) transparent glass with a composition of La_2O–Al_2O_3–SiO_2 through a melt-quenching process is being established at ARCI.

A mega programme on indigenization of Solid Oxide Fuel Cells (SOFC-1 KW $\times$ 10) is also initiated in collaboration with CSIR-CGCRI (Central Glass and Ceramics Research Institute) and HPCL (Hindustan Petroleum Corporation Limited). The project envisages providing an end-to-end solution to produce ready-to-fit SOFC modules in the country and also a demonstration of Solid Oxide Electrolyser (SOEC) for efficient hydrogen generation on a prototype scale.

20.3.4 *Laser Processing of Materials*

Laser as a heat source offers several advantages in various manufacturing processes such as welding, brazing, surface cladding, surface hardening, repair by deposition, micromachining, and the recently emerging additive manufacturing. ARCI hosts the nationally unique industrial laser centre with multiple laser systems under one roof to carry out application development using laser macro as well as micro-processing. The centre has been instrumental in establishing several new laser-based applications and catalysing the adoption of laser processing by the Indian industry. In recent years a major thrust was given to establish capabilities in the disruptive technology of metal additive manufacturing by synergizing the strengths and expertise in the areas of powder metallurgy and laser processing (Figs. 20.18, 20.19 and 20.20).

Fig. 20.18 Laser-coated baffle plate and its assembly in burner-tip nozzle of 200 MW thermal power plant, (right) laser cut micro-heater

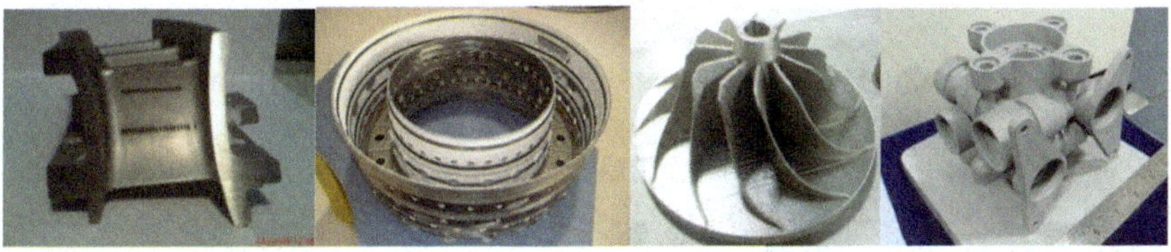

Fig. 20.19 (left to right) Laser drilled HP NGV, aero-engine Combustion liner, laser drilled HP NGV, and additively manufactured gas manifold block

Fig. 20.20 (left) Additive manufacturing facility, (centre) ultra-fast laser machining facility, (right) cardiovascular stent sample micromachined and electropolished at ARCI

20.3.5 Fuel Cell

Realizing the importance of clean energy technologies such as hydrogen as fuel, the Centre for Fuel Cell Technology was acquired from the SPIC foundation and nurtured into a full-fledged facility with the support of DST. This centre did pioneering work on Proton Exchange Membrane Fuel Cell (PEMFC) development in the country and demonstrated the capability of process know-how of its various components, testing, and viable applications. It now can make 20 kW fuel cell stacks in-house and is also well-equipped with all the required systems for running the fuel cells. Several innovations, such as exfoliated graphite-based bipolar plates, electrocatalyst, etc., are being transferred to the industry for commercialization. An automated fuel cell assembly line for the fuel cell stacks is expected to be commissioned to demonstrate the technology at a pilot scale for the industry to take it up towards

Fig. 20.21 5 kW PEM fuel cell stack

large-scale production. The activities of the centre have gained importance due to the increasing global focus on hydrogen economy and zero carbon footprint. ARCI has demonstrated electrochemical reforming of methanol for hydrogen production and pursuing to develop electrolyzers for the production of green hydrogen through water electrolysis (Fig. 20.21).

20.3.6 Nanocomposite Coatings

Nanostructures can offer several functionalities, including on surfaces. Sol–Gel coating technique is a wet chemical route by which inorganic–organic nanocomposite coatings can be realized on a wide variety of substrates. The composite nature of the coating allows advantages in terms of thickness from hundreds of nanometres to micrometres with good mechanical properties combined with in situ generations of nanostructures that offer surface functionalities such as hydrophobicity, anti-bacterial, anti-reflective, scratch-resistant, and self-healing corrosion-resistant coatings. ARCI's Centre for Sol–Gel Coatings (CSOL) has a comprehensive pilot-scale facility starting from reactors for sol preparation to substrate surface preparation, coating, and post-heat treatment, supported by an array of characterization techniques. Self-healing corrosion-resistant coatings on aluminium and magnesium alloys for aerospace applications and corrosion-resistant coatings on automotive steels is a major direction pursued, apart from anti-bacterial, (super) hydrophobic coatings for biomedical applications and scratch-resistant coatings on transparent plastics (Fig. 20.22).

20.3.7 Solar Energy Materials

The launch of the national solar mission triggered the establishment of an exclusive Centre for Solar Energy Materials (CSEM) at ARCI. CSEM is actively pursuing

Fig. 20.22 (left) Coloured coatings on designer glass for interior applications, (centre) plasma treatment, (right) hard coatings on plastics

research and technology development activities on the entire value chain of solar thermal and photovoltaic systems, emphasizing indigenous design, development and manufacturing of solar energy systems and components. Solar selective absorber coating for low and medium-temperature solar thermal applications and Easy to Clean Coating technology for easy maintenance of silicon solar PV panels were developed and tested in field conditions. The development of next-generation solar cells is intensively pursued using a full-fledged lab-scale line to fabricate and demonstrate the practical applications of low-cost, high-efficiency perovskite solar cells (Fig. 20.23).

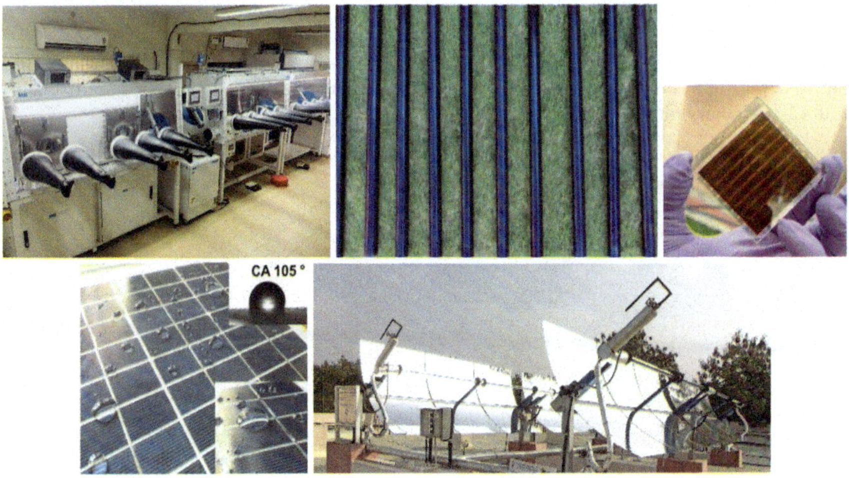

Fig. 20.23 (left to right, top to bottom) Multi port glove box integrated with thermal evaporator, solar selective absorber coatings on 1-m prototype receiver tube, prototype perovskite module, superhydrophobic effect on solar PV panel, and parabolic trough test rig facility to validate the receiver tube and thermic Fluid

20.3.8 Automotive Energy Materials

Electric mobility is the need of the hour to overcome the problems due to vehicular pollution, and Lithium-ion batteries play a key role in EV technology. A Centre for Automotive Energy Materials (CAEM) is established at IIT Madras Research Park, Chennai. CAEM has set up a Li-ion battery fabrication line and demonstrated the fabrication of Li-ion cells with different dimensions and configurations at practical sizes (Figs. 20.24 and 20.25). Over the years, the centre gained expertise in fine-tuning various fabrication steps for better throughput of the cells with high performance. CAEM also focuses on the development of hard and soft magnets for EV motor applications (Fig. 20.26) in addition to materials for sodium-ion batteries, Zn-Air batteries, thermoelectric heat recovery devices, etc. With the idea of developing safer energy storage devices, activities have been initiated to develop solid-state batteries as well.

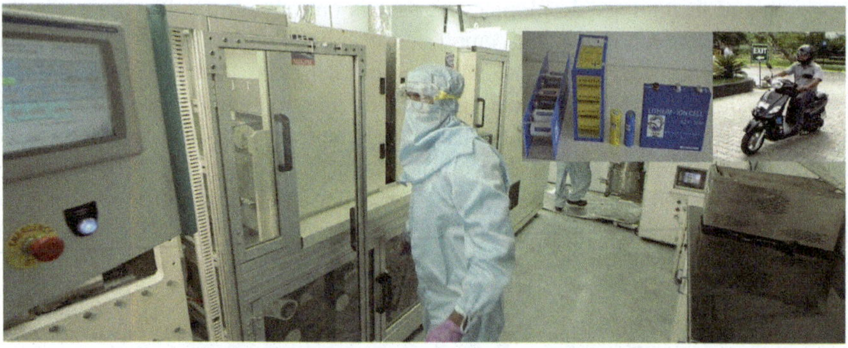

Fig. 20.24 Li-ion battery fabrication facility. Insets show batteries fabricated at ARCI and e-scooter with Li-ion battery module developed at ARCI

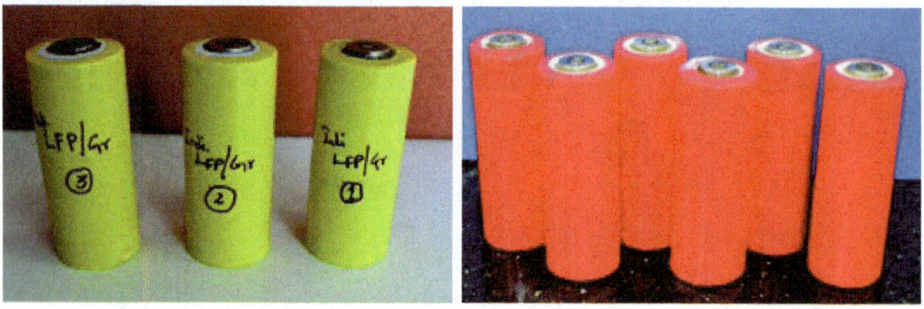

Fig. 20.25 LFP/graphite cells and NMC/LTO cells with indigenously developed LFP and LTO powders

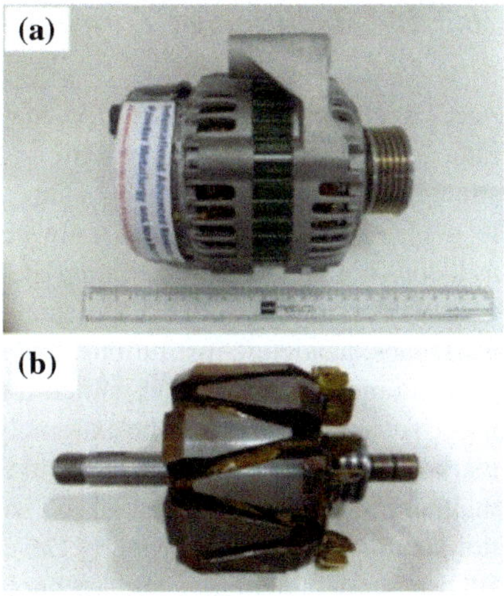

Fig. 20.26 Prototype claw pole alternator and claw pole rotor developed using soft magnetic material

20.3.9 Materials Characterization and Testing

The ability to understand the materials at all scales from atom to bulk is key to the successful application for a given purpose. ARCI has established state-of-the-art facilities and developed comprehensive capabilities to obtain a complete picture of material by performing microstructural, structural, chemical (elemental), surface and mechanical characterization. The advanced material characterization and testing facilities at ARCI are not only used internally but also extended to external users and research students at highly subsidized charges (Fig. 20.27).

Fig. 20.27 Small-Angle X-ray Scattering (SAXS) system

20.3.10 Technology Transfer Models

ARCI's commitment to provide materials-based technologies and solutions to industry and supporting the creation of technology-led ventures is reflected by the strategies that are continuously being evolved. In addition, ARCI is open to achieving its goals in collaboration with renowned national and foreign institutes/laboratories as it does not believe in reinventing the wheel. In the process, ARCI has been continuously evolving the formats by which it may engage with stakeholders from other R&D labs, academic institutions, and industrial organizations to develop technologies and innovative solutions suited for the changing world. An optional agreement is one such model found to be extremely useful for small companies to establish their market feasibility before investment. Similarly, to further streamline the decision-making process, the progress of the work from proof of concept to commercialization, unique Intellectual Property Development Indices (IPDI) from level 1–10, and IPDI-based Collaborative and Technology Transfer Models have been developed.

20.3.11 Setting-Up of National Centres

Core technological strengths and capacities developed over the years are effectively combined to address requirements in broader application domains not only at the level of the material but at the systems level. Major Programmes at ARCI have been aligned to meet National Missions like Make in India, which started as early as 1994, Atmanirbhar Bharat, Solar Mission, Nano Mission, Swachh Bharat, Electric Mobility Mission, Hydrogen Mission, etc. Two such major projects implemented by ARCI are the *Technical Research Centre for Alternative Energy Materials and Systems*, and the *National Centre for Materials and Manufacturing technologies for clean coal-based energy generation.*

20.3.11.1 Technical Research Centre for Alternative Energy Materials and Systems

This centre has been set up with a mission to develop unique, novel, and techno-commercially viable technologies in the area of 'Alternative Energy Materials and Systems' and subsequently transfer them to Indian industries. The research programmes are chosen in such a way that they would contribute to the automotive industry as well as to other sectors, including solar energy (CSP, PV), energy storage (batteries and supercapacitors), energy efficiency (magnets for motors and alternators, waste heat recovery, turbines, energy conversion (fuel cells) and electric/hybrid vehicle systems (involving Li-ion batteries with battery management systems, electric motors, etc.). Some of the major outcomes of this important initiative are:

- Solar selective absorber coatings for medium temperature (250 °C) solar thermal applications with excellent solar absorptance, low thermal emissivity, excellent corrosion resistance have been developed and the technology has been successfully transferred to the industry for process heating applications
- ARCI established a unique parabolic test rig for the evaluation of various components in a solar thermal system such as absorber tubes, thermic fluids, reflectors, etc.
- Easy-to-Clean and super hydrophobic coatings for solar PV panels have been developed to mitigate soiling loss and reduce the human efforts for cleaning of solar panels. Technology has been transferred to many industries.
- PEM Fuel cell stack up to 10 kW has been demonstrated in several field conditions, including disaster management
- 2.5 Nm^3/hr of hydrogen electrochemical methanol-reformer (ECMR) stack has been demonstrated
- Technologies for the production of both cathode (LFP) and anode (LTO) materials for Li-ion batteries were developed, and LFP technology licensing agreement has been signed with an industry (Fig. 20.28)
- Porous carbon as an electrode material for supercapacitors as well as supercapacitor devices were indigenously developed and successfully demonstrated
- Know-how transfer agreement with an Indian battery industry for Li-ion battery plant establishment and training the manpower of the company has been signed

20.3.11.2 National Centre for Materials and Manufacturing Technologies for Clean Coal-Based Energy Generation

India is likely to depend on coal for energy generation for a few more decades, and the idea is to improve the efficiency and performance of the conventional thermal power plants as well as build power plants operating at higher temperatures such as the Advanced Ultra Supercritical power plants. High-performance materials, coatings, and advanced manufacturing technologies are key to achieving this goal. ARCI, with its strengths in coatings, industrial laser-based manufacturing, and high-temperature materials, established this national centre in a consortium mode involving several IITs, NITs, National Labs, and industries. Already an improved baffle plate has been developed for burner tips of boilers and is currently being tested at NTPC Farakka power plant. The feasibility of laser hybrid welding of several important components, development of oxide dispersion strengthened iron aluminides for high-temperature applications, and advanced machining technologies are being addressed.

Fig. 20.28 (top) Know-how transfer agreement for the production of battery grade LFP cathode materials for Li-ion Batteries signed with an Indian Industry, (bottom) ARCI and an Indian Industry signed an agreement for technical know-how transfer and training of personnel to set up the Li-ion battery fabrication lab

20.4 Conclusion

Throughout its journey of over two and a half decades, ARCI has taken significant steps forward and evolved into a modern-day technology development centre providing materials-based technologies and solutions to a range of application sectors such as Aerospace, Power, Alternative Energy, Defence, Nuclear, Automotive, Manufacturing, Oil and Gas, Biomedical, Electronics, Textiles, Architectural, Water and several others. ARCI with its unique mandate of 'Translating Research to Technology' has successfully transferred technologies to 40 industrial organizations, including SMEs and start-ups and provided over 250 technological solutions for the industrial and strategic needs of the country. ARCI has 150 patents in its portfolio, and provided training to 10,000 scientific and technical personnel.

In line with *Atmanirbhar Bharat* and Make in India, ARCI has concrete plans to establish national centres for additive manufacturing, electrode material production, fabrication of cells as well as testing of batteries including the development of indigenous equipment. As a way forward, development of technologies for solid oxide fuel cells, extraction of rare-earth metals, rare earth magnets for motors of electric vehicles, high grade carbon fibre, hydrogen production, storage, and transport are some of the activities that ARCI will be pursuing.

Acknowledgements The authors thank all scientific, supporting staff and research scholars for their unstinting efforts throughout the journey of ARCI. The contribution of Mrs. N. Aparna Rao in formatting this article is acknowledged.

Part III
Institutes

Chapter 21
Historical Development and Current Status of Metallurgy Education in Banaras Hindu University

N. K. Mukhopadhyay, B. Nageswara Sarma, and Sunil Mohan

Abstract A sequential growth of Metallurgical Engineering education at Banaras Hindu University is discussed. The far-sighted vision of the founder of this great centre of learning is ascribed to the creation of the first-ever Department of Geology, Mining and Metallurgy as early as 1919. The beginning of a full-fledged undergraduate curriculum in metallurgy dates back to 1923. Three distinct phases of materials education in this centre of learning are deliberated. The erstwhile Institute is recently upgraded to the Indian Institute of Technology–a premier group of institutions that still maintains its heritage of materials education. The setting of the Department in the centre of the University is shown to have facilitated an effective interface with the physical, medical and life sciences Departments. This feature of the campus strategy in providing a great uniqueness in the trans-disciplinary teaching and research programs in materials relevant to society is clearly brought out.

Keywords Metallurgical Engineering · Metallurgical Education & Research · IIT (BHU); Banaras Hindu University · Prof. N.P. Gandhi · Prof. Daya Swarup · Prof. T.R. Anantharaman

21.1 Introduction

The Banaras Hindu University (BHU), a National University established in 1915 came into existence due to the far-sighted vision of the great patriot and educationist, **Mahamana Pandit Madan Mohan Malaviyaji**, who invited, on the recommendation of Mahatma Gandhi, **Professor Nagardas Purushottam Gandhi** (a product of the Royal School of Mines, London) to establish the Department of Geology, Mining and Metallurgy (GMM) in 1919. Under the benevolent and thorough guidance of Prof. N.P Gandhi, the Department made significant progress and eventually

N. K. Mukhopadhyay (✉) · B. N. Sarma · S. Mohan
Department of Metallurgical Engineering, Indian Institute of Technology (Banaras Hindu University), Varanasi 221005, India
e-mail: mukho.met@iitbhu.ac.in

R. Divakar et al. (eds.), *Indian Metallurgy*, Indian Institute of Metals Series,
https://doi.org/10.1007/978-981-99-5060-7_21

grew to be two separate Departments of Mining and Metallurgy respectively under the College of Mining & Metallurgy (MINMET) in 1942. The well-known Banaras Engineering College (BENCO) and College of Technology (TECHNO) and College of Mining & Metallurgy (MINMET) in BHU were amalgamated to form the Institute of Technology in 1971, which came to be known as IT-BHU, and was converted to Indian Institute of Technology (BHU), Varanasi by an Act of Parliament with effect from June 29, 2012.

Metallurgy education in India originated on this campus in 1923 under the visionary and methodical leadership of **Professor N.P. Gandhi** (1886–1960), laid the foundations for the first-ever B.Sc. (Metallurgy) program in this country nearly a hundred years ago. He was also a Founder Member of the Indian Institute of Metals (IIM). In grateful appreciation of his services, the Department instituted in 1974 the Gandhi Jubilee Medal to be awarded annually to the best outgoing undergraduate student of the Department. Subsequently, IIM instituted the Professor N.P. Gandhi Memorial Lecture, supported by the N.P. Gandhi Memorial Metallurgy (NPGMM) Trust, Department of Metallurgical Engineering, IIT (BHU).

Professor Daya Swarup (1904–1983), an alumnus of this Department (1928 batch, a student of Professor Gandhi himself and a Ph.D. from Sheffield University) succeeded him in 1937 as Head of Department and Principal of the College of Mining and Metallurgy (1944–1962). Under his dynamic guidance, the BHU-Metallurgy (BHUMET) Department initiated the postgraduate programme in 1957. The very first Ph.D. degree in Metallurgy in India was also awarded here sixty-six years ago, in 1955. In recognition of his inspiring contributions to metallurgical education, the Department instituted in 1974 the Swarup Jubilee Medal to be awarded annually to the best outgoing postgraduate student of the Department. The IIM instituted the Professor Daya Swarup Memorial Lecture.

The enlightened and inspiring leadership of **Professor Tanjore Ramachandra Anantharaman** (1927–2009), D. Phil, D. Sc. (Oxon) for over two decades witnessed the emergence of the Department to international stature. The illustrious traditions of excellence set by Professors N.P. Gandhi, D. Swarup and T. R. Anantharaman continued to be maintained by successive Heads of the Department (Table 21.1).

The dedicated and high-quality work of the faculty, researchers and students resulted in the recognition of our Department under the Special Assistance Program (SAP) by the University Grants Commission (UGC) in 1972. Our Department has the unique distinction of becoming the first-ever engineering Department in the country to be recognized as Centre of Advanced Study in Metallurgy (CAS) by UGC in 1980. The Department has a unique distinction of receiving assistance under CAS for four consecutive phases. After peer reviews, the UGC has sanctioned the second and third phases of CAS and Committee for Strengthening of Infrastructure in Science and Technology (COSIST) phases I & II. It also supported by the Department of Science and Technology (DST) under the "Fund for Improvement of S&T Infrastructure (FIST)" scheme and completed the FIST levels I and II. From 1981, it was also recognized as a Centre for Quality Improvement Program (QIP) by the Ministry of

Table 21.1 Heads of the department of metallurgical engineering, IIT (BHU)

S. No	Name	Period
1	Prof. N.P. Gandhi	1923–1937
2	Prof. D. Swarup	1937–1962
3	Prof. T.R. Anantharaman	1962–1977
4	Prof. S.L. Malhotra	1977–1980
5	Prof. S. Ranganathan	1980–1981
6	Prof. S. Bhan	1981–1983
7	Prof. B. Prakash	1983–1985
8	Prof. V. B. Tare	1985–1987
9	Prof. P. M. Prasad	1987–1989
10	Prof. S. N. Tiwari	1989–1991
11	Prof. D. S. Sarma	1991–1993
12	Prof. V. V. Kutumbarao	1993–1995
13	Prof. S. Lele	1995–1998
14	Prof. Vakil Singh	1998–2001
15	Prof. S. N. Ojha	2001–2004
16	Prof. R. C. Gupta	2004–2007
17	Prof. Shamsuddin	2007–2010
18	Prof. A. K. Ghose	2010–2011
19	Prof. G. V. S. Sastry	2011–2014
20	Prof. R. K. Mandal	2014–2017
21	Prof. N. K. Mukhopadhyay	2017–2020
22	Prof. Sunil Mohan	2021-Continuing

Human Resource Development (MHRD) and All India Council for Technical Education (AICTE). Another exemplary achievement of this Department was the establishment of the National Electron Microscope Facility (NELMIF) in 1982 funded by DST which had sophisticated facilities like transmission and scanning electron microscopes along with scanning auger spectroscope. These facilities catered to the needs of many institutions of the country besides its own faculty. The Department has continued to maintain microscopic facilities by attracting grants from other funding agencies. As a result of sustained efforts by the interdisciplinary materials group of the Department, the School of Materials Science and Technology was established at IT-BHU in 1978.

21.2 Progressive Growth of the Department

In the initial years of the Department (during 1930s), a small number of graduates proceeded for higher education preferably in British Universities. Some of them returned on the persuasion of Mahamana Malaviyaji to join as faculty members. The postgraduate program started after the country became independent, with a small grant from the Government. In the initial years, the teaching and research programs were centred on ore and minerals processing to metals extraction and processing of common metals and alloys. In the later stage (during 1970s), establishment of a school of materials science for the post graduate education and Ph.D. programs further added impetus to materials research and teaching programs that incorporated even ceramic and non-metallic materials. A number of research collaborations and visiting exchange programs of the faculty with several institutions world over created another landmark in materials education. The subsequent years witnessed the emergence of the Department to international status. The erstwhile Institute of Technology, BHU is recently upgraded to the Indian Institute of Technology (BHU), which still maintains its heritage of materials education. The setting of the Department in the centre of the university has facilitated an effective interface with various other Departments of BHU. This feature of the campus provides a great opportunity in the trans-disciplinary teaching and research programs in all aspects of materials science and engineering.

The Department is unique in its size and composition. Presently, there are six Professors, nine Associate Professors & ten Assistant Professors in the faculty, encompassing all important areas of the discipline. There are many well-equipped laboratories that cover the whole breadth of metallurgy, including heat treatment, metallography, electron microscopy, XRD, mechanical metallurgy, thermodynamic investigations, kinetics, transport phenomena, corrosion, and other areas covering ferrous and non-ferrous metallurgy, foundry, welding, tribology, rapid solidification and waste utilization apart from central facilities such as computer laboratory, library, workshop, etc. The wide range of laboratories forms the backbone of quality teaching and research in Metallurgical and Materials Engineering disciplines. In recent years, the Ministry of Steel, Government of India has sanctioned substantial funding for the creation of an Advanced Research Centre for Iron and Steel (ARCIS). The Centre will house state-of-the-art equipment for research in the area of Iron and Steel and will work as a consortium of industries, R&D laboratories and international research organizations to make the Centre self-sustaining in future. A similar noteworthy addition is the creation of Malaviya Chair for Railway Technology in our Institute, which is being coordinated by our Department. A number of collaborative programs have already been initiated between the faculty members of the Institute and Indian Railways.

21.3 Undergraduate and Postgraduate Programs

The undergraduate (UG) program is well designed to cater to the needs of Indian Industries and R&D organizations and the curriculum is periodically revised to keep up with the emergence of new disciplines of science and technologies. Since 1971, admission to our four-year B.Tech. Program is being made through the All India Joint Entrance Examination (JEE). The eight-semester UG program with an annual intake of more than hundred consists of a broad-based course structure and covers core subjects in physical, chemical, extractive, foundry, mechanical, structural, industrial metallurgy, non-metallic materials, mathematical & computer modelling, materials development, metal joining, energy and environment, etc. Most of these courses are associated with laboratory-oriented practical classes along with Elective Project work. Educational tours, six weeks of industrial training, Seminar and Group Discussion also form part of this rigorous UG training program.

Since 2005, Integrated Dual Degree (IDD) program of 5 years duration has been started which awards both B. Tech. and M. Tech. degrees to students at the end of the course. Admission to this program with an annual intake of twenty is also made through JEE. Admission to our two-year M.Tech. program with an annual intake of forty-seven is made through the All India Graduate Aptitude Test in Engineering (GATE). At present, two specializations are being offered: (i) Extractive Metallurgy (ii) Alloy Technology. The four-semester course structure comprises of a few core subjects, several electives, practicals, seminar and with a major focus on dissertation work.

The Department also pioneered doctorate degree (Ph.D.) program in the country, for which, admissions are made twice a year. This program is open to Metallurgical as well as other Engineering graduates and also postgraduates in allied science subjects. The Department of Metallurgical Engineering has so far produced a record number of 2930 B.Tech. graduates, 707 post-graduates including M.Tech. dual degree and 205 Ph.D. graduates.

21.4 Research Highlights

The members of staff have been actively engaged in research and have undertaken many research/consultancy projects. The current research activities span over wide-ranging fields of synthesis and characterization of nanomaterials, quasicrystals, spray forming, powder metallurgy, modelling of phase diagrams by cluster variation method, ultra fine-grained materials, phase transformations in steels and non-ferrous alloys and structure–property correlations, creep and fatigue behaviour of materials, bio-implant materials, thermodynamics of semiconducting intermetallics and ternary alloys, hydrometallurgy, bio-leaching, kinetics of reduction of oxides in the presence of catalysts, processing of ferrous and non-ferrous alloys, foundry and welding, wear studies of composites, waste utilization, processing of electronic waste and energy management, etc.

The UGC identified the Department as Centre of Advanced Study for four consecutive phases wherein research goals in the six identified thrust areas were vigorously pursued. The research work of the faculty and students brought several laurels and great recognition for the Department. The salient features of the research achievements are presented in the following sections:

In the area of rapid solidification, modelling and experimentations of the spray forming process have been a significant contribution in the development of industrially important alloys. Their improved tribological properties are being studied for automotive applications. In the field of quasicrystals, a quaternary series of alloys with systematically varying electron/atom (e/a) ratio between the two ternary end-members Al-Cu-Co and Al-Co–Ni decagonal quasicrystals have been designed and unequivocally established the role of e/a ratio, a concept due to Hume-Rothery in crystalline compounds, in stabilizing decagonal phases of different symmetry group at different e/a values. Studies were also undertaken to establish the correlation between the existence of a B2 (bcc) phase in the equilibrium diagram and the solidification. The models of indexing electron diffraction patterns from quasicrystalline phases developed earlier by the group have been perfected. Nano-indentation behaviour of quasicrystalline phases has also been studied. In the thrust area of Deformation and Fracture, salient contributions related to the Low Cycle Fatigue (LCF) behaviour of several Ti alloys and Ni-based superalloys have been made. Cooling rate subsequent to solution treatment of alloy IMI 834 was found to have great influence on the microstructural control. Faster cooling results in reduction in volume fraction of α-phase and suppression of Ti_3Al and silicide precipitation. The LCF behaviour of the alloy exhibited a dual slope in the Coffin-Manson relationship at room temperature that could be attributed to the planarity of slip. The hold time in tension/compression—more pronounced in compression, was observed to be detrimental to the LCF life. Further, prestressing, by cold rolling, also was found to enhance the fatigue life of IMI 834. A detailed study was undertaken to explore the role of pre-existing twin boundaries on the dual-slope behaviour of Ni-base superalloys on the basis of earlier studies by the group. Comparative studies between solution-treated and over-aged samples showed a lack of any role of pre-existing twins. Tensile and LCF properties and dynamic fracture toughness of 9Cr-1Mo steel were studied in detail.

In the field of thermodynamics of materials, significant contributions to the modelling of solutions and computation of phase diagrams have been made using the Cluster Variation Method (CVM). CVM is shown to be capable of reliably estimating thermodynamic, phase equilibria and structural data such as short-range order parameters, etc. Several classical thermodynamic results pertaining to dilute solutions have been obtained using CVM. Using these results, polynomial and rational functions have been obtained to approximate the equilibrium values of the microscopic variables/cluster probabilities/short range order parameters without having to numerically solve the system of non-linear/transcendental system of equilibrium equations. This, together with many other developments made in the algorithms leads to a very significant reduction of the computational burden of the method. Determination of thermodynamic data for group II-IV semiconducting chalcogenides is another significant activity under this thrust area.

In the area of heterogeneous kinetics, significant contributions have been made to the reduction of metal oxides and leaching of metals from concentrate/sea-nodules. Whereas alkali carbonates were found to significantly enhance the rate of carbothermic reduction of cupric oxide, the fluorides were found to have negative effect. For carbothermic reduction of NiO, the autocatalysis of the product metal was found to govern the rate of reduction and the presence of alkali carbonate was found to influence the rate but not the mechanism of reduction. For the bio-dissolution of metals like Cu, Ni & Co from sea-nodules using *thiobacillus ferroxidans* and *thiobacillus thioxidans* in presence of pyrite and sulphur, the leaching was found to follow shrinking core kinetic model and is controlled by diffusion of lixiviant through the product metal. A similar mechanism was found for the leaching of zinc from Sphalerite concentrate in sulphuric acid with sodium per-sulphate as an oxidant.

In the area of process metallurgy, extensive studies have been conducted on the recovery of metals from waste, by-products and lean ores. Investigations on bioleaching of Cu, Ni and Co from copper converter slag, polymetallic nodules and copper concentrate were conducted using *thiobacillus ferroxidans* and *thiobacillus thioxidans* in presence of pyrite and sulphur. In all three cases; the conditions for recovery of metals were optimized and mechanism of leaching was elucidated. Studies were also undertaken to recover fluorides by selective precipitation from the leach liquor obtained during the processing of a low-grade molybdenite concentrate. From the fluoride-depleted solution, attempts were made to recover copper and nickel by solvent extraction process.

Ordered Ni_3Al precipitates in platelet form were identified to be associated with improved strength of the precipitation hardening type austenitic stainless steels. The low cycle fatigue behaviour of VT9 Ti-alloy in several heat-treated conditions has been attributed to the combination of the different types of dislocations (a, c, a + c) existing in the heat-treated samples. Optimum conditions of heat treatment have been arrived at for improving the formability of Ti-48Al-16Nb alloy. In the area of high nitrogen steels, the role and kinetics of discontinuous precipitation of Cr_2N were explored. Modelling of microstructural evolution during controlled crystallization of glasses using static Monte Carlo simulation was another important contribution in this area.

In the area of development of materials, major contributions are related to a wide spectrum of materials and applications. Several types of nanostructured materials and their methods of production are being developed which are summarized in a separate section as an emerging thrust area. While the emphasis is laid on understanding the basic mechanisms and role of process parameters, efforts are also focused on the development of materials, which are application-oriented. These include bearing materials through spray forming, metal-quasicrystal composites, metal-matrix composites, dental and surgical implants from indigenous titanium and titanium alloys, creep-resistant steels, low cycle fatigue-resistant alloys, damping alloys, pre-reduced pellets and directly reduced iron, solid lubricants such as MoS_2 and biomass.

21.5 Linkages with Industry and R&D Organizations

One of the objectives of the creation of this educational edifice is to establish linkages with industries and help them in solving their problems as well as develop an aptitude to learn new technologies. The Department has conducted several programs to impart theoretical as well as practical expertise to a large number of executives from several industries as well as faculty and researchers from educational and research organizations.

The Department had been receiving continuous inflow of prestigious R & D projects funded by several Governments institutions, e.g., Department of Science and Technology, Council of Scientific and Industrial Research, Board of Research In Nuclear Sciences, Aeronautics Research & Development Board, Defence Research and Development Organisation, UGC as well as from industries on diverse topics. The annual rate of publications has been considerably high that is at par with the premier institutions in the country and abroad. Every year several faculty members deliver large number of invited lectures in India and abroad. The Department has organized innumerable number of National and International Conferences/Seminars, Workshops and Industrial Clinics, generally two to three every year. The Department is proud that several authoritative reviews, books and conference proceedings have been published by members of the faculty.

It is worth mentioning that many of our faculty members are reviewers and members of Editorial Boards of several National and International journals of repute. The Department also has large number of its faculty as Fellows of many professional as well as prestigious Societies. Several of our Professors are also Members of reputed National & International Commissions and Boards of many Government, Academic Institutions, R & D Laboratories as well as Industrial houses. The Department has been a pioneer in setting up an Industrial Consultancy and Liaison Cell nearly a quarter century ago. This is now part of the Industrial Consultancy and Testing Services Centre of IIT (BHU). It provides metallurgical testing and consultancy services, undertakes specific failure analysis and know-how development work. Recent beneficiaries of such services include M/s TATA Steels, (Jamshedpur), Hindalco Industries Ltd. (Renukoot) and Renusagar Power Plants, Naval Materials Research Laboratory (Mumbai), Mishra Dhatu Nigam Ltd. (Hyderabad), Steel Authority of India Ltd., Maruti Udyog Ltd., Singrauli Coalfields, R & D Centre for Iron and Steel (Ranchi), Usha Martin (Jamshedpur), Obra Thermal Power Plant, etc., and innumerable number of small industries of Varanasi, Mirzapur and other eastern regions of the state.

The Department established a number of collaborative programs, namely, Indo-US, Indo-British, Indo-German and Indo-French exchange programs on academics and developing research areas. In recent years, there have been some 100 collaborations, covering the fields of quasicrystals, bio-implant materials, trace elements in humans, rasa-shastra, environment-assisted failures, waste utilization and other various industrial-oriented problems. It has also contributed significantly towards "Disposal and utilization of Indian Red Muds" for HINDALCO, JNRDC. Significant work has also been done in bio-leaching, waste utilization and fundamental studies on ferrous alloys with CSIR-National Metallurgical Laboratory (Jamshedpur) which are noteworthy.

21.6 Awards and Distinctions

Faculty members and students have won an unusually large number of awards and recognitions for their meritorious contributions in diverse fields of metallurgy and materials. So far, four teachers of the Department have won the Shanti Swarup Bhatnagar Awards from both CSIR and INSA, two G. D. Birla Awards, fifteen National Metallurgist's Day Awards, six MRSI Medals, thirteen IIM Platinum and Gold Medals, five Young Scientists' Medals of INSA, two Binani Medals (IIF) and one each of Henry C. Sorby Medal (International Metallographic Society), VASVIK Award, H. J. Bhabha Award (Hari-OM Ashram), Honorary Membership of the IIM, Howe Medal, Al Kharazmi Award (Iranian Government), MOSCOT and Ramachar prize (ECSI), Alexander von Humboldt Fellowships, Young Metallurgists' Awards, INSA Medals for Young Scientists, ISCA Young Scientist Awards, Dr. R. H. Kulkarni Memorial Fellowships, Prof. C.N.R. Rao Award besides several best paper Awards. The faculty members have distinctions of receiving Fellowships (FNA, FASc, FNASc, FNAE, FIIM, FIE, FEMSI) of various professional societies such as INSA, IASc, NASc, INAE, IIM, IE (I) and EMSI as well as INAE Chair Professorship and ASM-IIM North America Lectureship. The Department feels genuinely proud whenever any of its alumni achieves distinction in the professional field. Students of the Department have not lagged behind. In recent years, our undergraduates have won twelve coveted Vidya Bharati Prizes & best paper awards from IIM and other Institutions, besides two A. K. Bose Gold Medals for outstanding Masters' dissertations (IIM) and two TMS-AIME Best Student Paper Awards.

21.7 Co-professional Activities

Departmental members participate very actively in several administrative, co-curricular, cultural, sports, and other activities of Banaras Hindu University (BHU), South (Barkachha) campus of BHU, our Institute as well as of the Department. Every year, a number of invited lectures, seminars, get-togethers, sports competitions, quiz/debating contests, etc., are organized by professional bodies such as the Metallurgy Society of BHU and the Student Affiliate Chapter & Varanasi Chapters of the Indian

Institute of Metals as well as Materials Research Society of India. The Prof. N. P. Gandhi Memorial Metallurgy Trust was established in 2001 to promote professional and many other activities of faculty and students. It also supports inviting distinguished persons in the field of metallurgy and materials to spend some time in the Department for interaction with faculty and students and for carrying out collaborative research work. Alumni are honoured regularly by the conferment of Distinguished Alumnus Awards as well as Industry-Institute-Interaction Awards. The Department also encourages its UG, PG and Non-teaching staff by awarding them with Gandhi Jubilee Medal, Swarup Jubilee Medal & Bhat prize respectively, for their best overall performance. The Department also organizes an Annual Technical Fest—"ANVESHAN" which is essentially managed by the students.

21.8 Concluding Remarks

We feel happy that the glorious and pioneering contributions of the three illustrious peers of our Department have been recognized throughout the metallurgical arena. The Indian Institute of Metals has instituted the two very prestigious Memorial Lecture series for Professor N. P. Gandhi and Professor Daya Swarup. The Lecture series are regular features of NMD celebrations. Professor T. R. Anantharaman has served as President of the IIM and is also the recipient of a very large number of national & international awards and distinctions. The IIT (BHU) has recently instituted Prof. T. R. Anantharaman Memorial Lecture series as an annual feature. The Department, with the support of the NPGMM Trust, plays an important role in organising "Professor P. Ramachandra Rao Memorial Lecture" in three different institutions by rotation, namely (i) Department of Metallurgical Engineering, IIT (BHU), Varanasi, (ii) CSIR-National Metallurgical Laboratory, Jamshedpur and (iii) Department of Materials Engineering, IISc, Bangalore.

A Department is naturally judged and becomes known by the performance of its alumni and the present as well as former faculty members. Our graduates are readily accepted for diverse jobs not only in India but also for varied employment as well as higher studies in several advanced countries. An exceptionally large number of our alumni and past faculty members had been and still continue to occupy the very top positions in all spheres of our profession. While these can be a matter of great satisfaction, one cannot rest on the past laurels. The Department has been continuously striving to maintain excellence and compete internationally, by producing high quality graduates. It is also involved in conducting focused R&D work in frontier areas at its premises, inter-departmental as well as with several national and international institutions.

The Department had been fortunate to have the inspiring guidance and blessings of Mahamana in serving the country from pre-independence era. Being the pioneering Department in Metallurgical Engineering in the country, the Department has turned out outstanding academics, entrepreneurs, administrators, thought leaders and professionals—many of whom not only led universities, research labs and organizations, industries and Government Departments but also played pivotal roles in creating new institutions and guiding scientific, academic and research policies of our country as well as other countries. As the Department completes a century of dedicated and exemplary services to the society, it is poised to achieve greater levels of excellence, by combining the wisdom of the senior faculty and the brilliant exuberance of the well-trained young faculty members who joined us in the recent past.

Chapter 22
Platinum Years of Metallurgical and Materials Education and Research—The IISc Saga

S. Subramanian, S. Suwas, R. Ranjan, and P. Kumar

Abstract The history of the Department of Materials Engineering, Indian Institute of Science (IISc), Bengaluru has been enumerated from its conception to mid 2020. The evolution of the undergraduate, post-graduate and Ph.D. programmes has been detailed with respect to the curricula and research investigations carried out. The profiles of the faculty and scientific staff, who have served the Department and those currently in service, have been briefly outlined. The landmark achievements and faculty accomplishments in the annals of the Department have been highlighted. The close linkage between IISc and the Indian Institute of Metals (IIM) has been traced.

Keywords History · Education · Research

22.1 Conception and the Formative Years

In 1910, Prof. Morris Travers, the first Director of the Indian Institute of Science (IISc), Bangalore, prepared a list of what he called more departments and sub-departments. Metallurgy found a place in this list. In 1915, Burjorji Padshah, a member of the first Council, suggested that a Department of Metallurgy, which could work in cooperation with the Iron and Steel Works at Sakchi, would be a useful addition to the Institute. The Council accepted this proposal in principle and sought an endowment from the Iron and Steel Works. On August 1, 1939, Sir Jnan Chandra Ghosh took over as the Director of IISc. He sought to give an industrial orientation to the Institute and was much influenced by the discussions at the then-newly created Board of Scientific and Industrial Research, the predecessor to the Council of Scientific and Industrial Research. In March 1941, the IISc Court, under the Presidency of Sir M. Visvesvaraya, resolved to establish the Department of Metallurgy. A few years were to pass before the Department was established in 1945. The Government of India granted Rs. 1 Lakh for building/equipment and Rs. 30,000 for staff. The

S. Subramanian · S. Suwas (✉) · R. Ranjan · P. Kumar
Department of Materials Engineering, Indian Institute of Science, Bengaluru 560012, India
e-mail: satyamsuwas@iisc.ac.in

R. Divakar et al. (eds.), *Indian Metallurgy*, Indian Institute of Metals Series,
https://doi.org/10.1007/978-981-99-5060-7_22

Mysore Government provided Rs. 50,000 as a non-recurring grant and an annual recurring grant of Rs. 15,000.

Dr. Frank Adcock from the National Physical Laboratory, Teddington, UK was appointed by invitation as the first Professor and Head of the Department (HOD) for a period of three years and he joined duty on September 24, 1945. It is worthy of mention that long before he came to India, Dr. Adcock had already distinguished himself in England, first as a physicist through the development of the 'Adcock Antenna' and later as a metallurgist through the establishment of the Iron-Chromium phase diagram. As the founding HoD, Professor Adcock drew up meticulous plans for the construction of the building and laboratories as well as for the recruitment of the technical staff and faculty. The first batch of students was admitted in August 1947 to the Diploma of the Indian Institute of Science, specializing in metallurgy. The first set of faculty recruitment as lecturers was made in 1947 in the areas of fuels (S. S. Ghosh), electrometallurgy (J. Balachandra), physical metallurgy (E. G. Ramachandran), mineralogical chemistry and ore dressing (K. K. Majumdar). Additionally, the students had to take courses in electrical engineering, mechanical engineering, engineering drawing, workshop training, chemical kinetics, thermodynamics, electrochemistry, chemical technology, etc., offered by other departments of IISc. Practical training in summer during April to June, both in ferrous and non-ferrous process metallurgical industries, such as Tata Iron & Steel Works, Jamshedpur, IISCo, Burnpur, Mysore Iron & Steel Works, Bhadravati, Indian Aluminium Co., Alwaye, Indian Copper Corporation Limited, Ghatsila, Kolar Gold Fields, Ordnance Factory, Ambarnath, Railway Workshop, Bombay, to name a few, was compulsory. It is revealing to note that one of the students of the first batch (T. R. Anantharaman) was selected for one of ten Nuffield Foundation Research Scholarships in Extractive Metallurgy, to undertake practical training during the summer of 1949 at the Broken Hill Proprietary (BHP) Steelworks, Newcastle, Australia. Prof. Adcock played an important role in the founding of the Indian Institute of Metals (IIM) and the National Metallurgical Laboratory (NML). He was a Founder Member of IIM and a Member of the Planning Committee of NML. Prof. Adcock retired in 1948 on the completion of the 3 years.

During the period from June 1948 to September 1949, Mr. E. V. Ganapathi Iyer, the acting Director, IISc was also the acting HOD of the Department. Prof. M. S. Thacker took charge as the Director of IISc in September 1949 and was also acting HOD of the Department. To complement the teaching programme, simultaneously research projects were initiated dealing with carburization of steels, etch figures of various pure metals, cemented carbide tools, electrodeposition from fluoborate baths, stresses on electrodeposition, to cite a few examples. Additionally, projects of national importance to utilize the mineral resources of India were taken up dealing with investigations on low temperature carbonization of low grade Indian coals and various properties of the coke derived therefrom, beneficiation of ilmenite and chromite ores, as well as studies on Be and Be–Cu alloys from beryl and rare metals. The vacancy created by the resignation of Dr. Majumdar, who moved to the then Atomic Energy Establishment and later went on to head the Ore Dressing section there, was filled up by the appointment of Dr. N. R. Srinivasan. In August 1950, K. V. Aiyer was appointed as a Lecturer in Physical Metallurgy, to fill the vacancy created by the promotion of Dr. E. G. Ramachandran as Assistant Professor. In November 1949, the Fuels section of the Department was transferred to the Chemical Engineering Department at IISc.

22.2 Consolidation of the Undergraduate Programme and Nascent Research Projects of Strategic Importance

The dawn of 1951 witnessed the appointment of Dr. Brahm Prakash as the first Indian Professor and HOD of the Department. He joined the Department after obtaining his Sc.D. from the famous Massachusetts Institute of Technology, USA, working under the supervision of Reinhardt Schumann (Jr.), specializing in the disciplines of Mineral Engineering and Metallurgical Thermodynamics. As he also had an offer from the then-nascent Atomic Energy programme, it was proposed by Dr. Homi J. Bhabha, who was a member of the Court of IISc, that Dr. Brahm Prakash will take up the headship in the Department of Metallurgy at IISc, Bangalore, on the specific understanding that he will return to Bombay, as soon as the programme of work picked up momentum there. During the leadership of Prof. Brahm Prakash from January 1951 to March 1957, a comprehensive course curriculum in Metallurgy was clearly formulated and the semester system was introduced, matching the graduate school system in front-line institutions in the USA. There were additions to the faculty, Mr. D. L. Bhattacharyya (Electron-Optics) and Dr. R. C. Deshpande (Mechanical Metallurgy). The courses taught in the first two years included metallurgical thermodynamics, general extractive metallurgy, mineralogy, principles of ore dressing, refractories, physical metallurgy, X-ray metallurgy, experimental methods in metallurgy, metallurgical analysis, electrometallurgy, furnace technology, advanced non-ferrous metallurgy, production metallurgy of iron and steel, electron microscopy, mechanical metallurgy, light alloys, fluid mechanics, heat transfer and diffusion in metals. Several practical courses were also introduced. Prof. Prakash induced and encouraged his colleagues to present courses in new areas, such as reactor physics and metallurgy, brass mill practice, recrystallization phenomena in metals and alloys, christened as special topics in metallurgy. The students also took courses in industrial psychology, industrial economics and foreign languages such as German and French. The last year of the 3-year metallurgy course was devoted to a research project. During his tenure, there was a steady expansion in the volume and variety of research programmes in the Department. In a landmark development, the first-ever research conferment in metallurgy, the Associateship of the Indian Institute of Science, AIISc, was awarded to Mr. V. Aravamuthan in 1951. There were externally sponsored projects led by Prof. Brahm Prakash and Dr. E. G. Ramachandran, through the Council of Scientific and Industrial Research and the Department of Atomic Energy, namely separation of zirconium from hafnium, metallurgy of nuclear grade zirconium, Al coating of steel, development of Al alloys for substituting Cu in electrical machinery, preparation of Al by the electrolysis of aluminium chloride, measurement of elastic modulus of metals, to cite a few examples. As part of the Sisterhood Relationship Programme with the University of Wisconsin, USA under the technical cooperation mission of the USA, Dr. H. L. Walker joined the Department as a Visiting Professor in July, 1954. A special feature of industrial cooperation that year was the organization of a Refresher Course for the engineers of HAL, Bangalore, in which Prof. Walker gave a series of 10 lectures on the heat treatment of industrially important ferrous and non-ferrous metals and alloys. A sample of the early research publications include studies on niobium and tantalum, studies on lower grade Indian coals, determination of the elastic constants of metals by an ultrasonic method (Ramachandran and Sreenivasan 1953), an electron microscope study on the effect of cationic wetting agents on aqueous stearic acid sol, beneficiation of vermiculite, pressure welding of metals and alloys.

Taking into account India's stake in nuclear power development, and the zircon resources in the Indian beach sands, a project was referred to Prof. Brahm Prakash to develop a process for producing hafnium—free zirconium for nuclear applications. Many aspects of zirconium chemistry and metallurgy were investigated, in which the vapour-phase de-chlorination of $Zr(Hf)Cl_4$, to produce low-hafnium zirconium oxide, followed by selective reduction of $Zr(Hf)Cl_4$ with aluminium for removal of hafnium were key steps. It is noteworthy that the results on the vapour-phase de-chlorination (Prakash and Sundaram 1955) were presented by Prof. Brahm Prakash at the First UN Conference on the peaceful uses of Atomic Energy, in Geneva (1955) and those on the selective reduction of zirconium chloride were presented in the next Geneva Conference in 1958, and were highly acclaimed for their originality (Fig. 22.1). The achievements in the zirconium project in IISc truly laid the foundation for the larger programme subsequently undertaken in Bombay, which eventually led to the setting up of the Nuclear Fuel Complex in Hyderabad in 1971, for the large scale production of zirconium sponge, zirconium products and zircalloy clad ceramic uranium oxide fuel elements.

Between September 1956 and March 1957, the new faculty members recruited included Dr. T. R. Anantharaman, who returned to India after obtaining his D.Phil. from Oxford University, as an Assistant Professor of Physical Metallurgy, as well as Mr. R. Mallikarjunan (Mineralogical Chemistry and Ore Dressing), and Mr. B. S. Subramanya (Mechanical Metallurgy) as Lecturers. Further, Dr. E. G. Ramachandran and Dr. N. R. Srinivasan resigned during the same period. By early 1957, Prof. Brahm Prakash was recalled to the Department of Atomic Energy, Bombay and he relinquished his charge of office at IISc. In March 1957, Prof. S. Bhagavantham was appointed as the Director of IISc, and in June 1957, Prof. A. A. Krishnan (Ph.D. in Metallurgy from Ohio State University) succeeded Prof. Brahm Prakash as the Professor and HoD of the Metallurgy Department. The D.IISc programme was further consolidated by the introduction of new subjects dealing with corrosion and protection, powder metallurgy, instrumentation and process control, pyrometry, metallography and heat treatment, foundry technology, principles of mechanical testing, shaping and fabrication, as well as industrial metallurgy. Additionally, factory visits

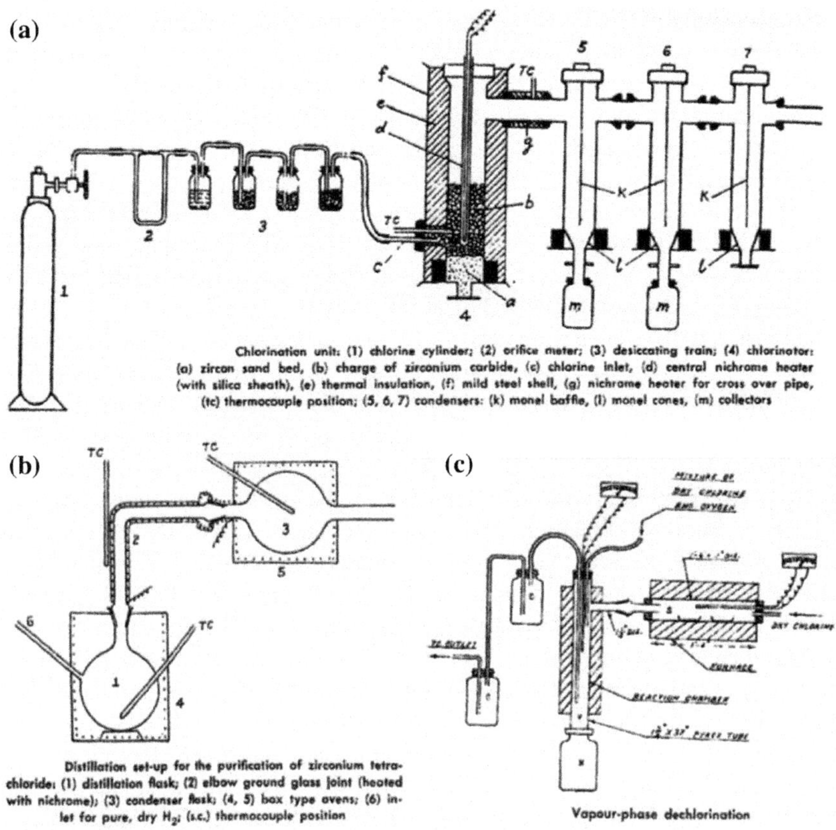

Fig. 22.1 Dr. Brahm Prakash's indigenous apparatus for separation of hafnium from zirconium (Prakash and Sundaram 1955)

and short study tours to Indian Telephone Industries, Government Electric Factory, Government Porcelain Factory, Kirloskar Electric Co., Mysore Lamp Works, Mysore Glass and Enamel Works, etc. were part of the B.E. training programme. In January 1958, a special one-year intensive course in iron and steel making sponsored by Hindustan Steel was offered, and the first batch of students completed the course in December 1958. Newer research projects dealt with the study of stacking faults in close-packed metallic systems, X-ray line broadening of plastically deformed metallic lattices, formation of alpha-massive phase from beta-brass in various brasses, development of age-hardenable alloys from Indian commercial aluminium, interface energy studies in non-ferrous alloys containing liquid phases (Ramachandran and Krishnan 1960), the reactivity of double bond of unsaturated fatty acids in mineral flotation, direct reduction of sulphides, electrowinning of metals from sulphides, electrodeposition of Cu–Sn alloys from a fluoborate bath, sintering vis-à-vis intermetallic diffusion in alloys, alloy powder production by direct reduction of mixed oxides or chlorides, spreading characteristics of soft solders, to cite a few examples.

The Government of India notified on 12th May 1958 on the advice of the University Grants Commission, that the Indian Institute of Science, which is an institution of higher education, shall be deemed to be a university for the purposes of the University Grants Commission Act. Thus the nomenclatures of the degrees awarded were changed to B.E. (Bachelor of Engineering), M.E. (Master of Engineering) and for research, Master of Science (M.Sc), Doctor of Philosophy (Ph.D.) and Doctor of Science (D.Sc). As part of the technical programme of the Golden Jubilee of the Indian Institute of Science which was celebrated in February 1959, the Department of Metallurgy arranged a Symposium on 'Structural Changes in Metals and Alloys' in conjunction with the 13th Annual Technical Meeting of the Indian Institute of Metals during the first week of December 1959, at the Indian Institute of Science, Bangalore.

The period 1959–'60, witnessed changes in the faculty in terms of the recruitment of Dr. K. P. Abraham (Ph.D., Imperial College, London) as Assistant Professor and Dr. M. Ramakrishna Rao, Mr. V. Ramachandran and Mr. A. V. Ramana Rao as Lecturers and the resignations of Mr. D. L. Bhattacharyya, Dr. B. S. Subramanya and Mr. J. Balachandra. Dr. Abraham's early initiatives laid the foundation for the establishment of a strong research school in chemical thermodynamics and process metallurgy. The induction of new faculty-led to additional courses being taught in industrial electronics, electron optics, plastic deformation of metals, plant design and layout, phase diagrams and phase transformations, experimental methods in physical and process metallurgy, to name a few. During 1962–63, Dr. Anantharaman resigned from IISc to join the Banaras Hindu University as the Head of the Department of Metallurgy. In the same period, two appointments as Lecturers were made (Mr. R. K. Ramamurthy and Mr. K. S. Raman).

22.3 Introduction of M.E. and Ph.D. Degree Programmes

As the year 1963 was ushered in, Prof. Satish Dhawan of the Aeronautical Engineering Department, was appointed Director of IISc. With the B.E. degree programme having been firmly established, it thus became logical to introduce the M.E. degree programme. After gaining the necessary approvals, a structured M.E. course with specialization in Physical and Chemical Metallurgy of the duration of two years was started in August 1964, during the headship of Prof. Krishnan. With an emphasis on the scientific content of the technological subjects, mathematics, instrumentation and electronics and materials science and technology formed an essential part of the new courses. In addition, each Master's degree student undertook an individual project and submitted a written dissertation on his/her work. Realizing the emergence of materials science as an important area of research, pervading across disciplines, a young faculty member, Mr. K. S. Raman, was deputed to various universities in the USA in 1964.

During the period 1965–'68, several faculty members were recruited, who formed the nucleus for the growth and development of the Department. Dr. K. I. Vasu (Ph.D., Imperial College, London), Dr. M Mohan Rao (Ph.D., Purdue, USA) and Dr. R. M. Mallya (Ph.D., Bombay), having carried out his research at the Inorganic and Physical Chemistry Department, IISc, joined as Assistant Professors, while Mr. E. S. Dwarakadasa, Mr. D. H. Sastry, Mr. C. R. Krishnamurthy, Mr. R. G. Satyanarayana, Mr. Y. V. R. K. Prasad, Mr. K. Narasimha Murthy, Mr. M. V. Bhat, Mr. S. C. Gupta and Mr. Kishore were recruited as Lecturers. The doctoral research projects and dissertation work of the M.E. students resulted in the development of new apparatus and provided the basis for publications in scientific literature. It is worthy of mention that many of the newly recruited faculty members obtained their Ph.D. degrees as staff registrants, under the supervision of Dr. Vasu, and the Materials Research Group was formed. About the same time, the Ph.D. programme was also commenced at the Department, with the enrollment of a limited number of students. Recognizing the importance of scientific report writing and lecture presentation, the Foreign Language Section developed a special programme of assistance in English for students at the Institute, with the assistance of Prof. L. I. Lewis, Visiting Professor of English and a handbook 'English for Scientists' was prepared. A significant decision of the Institute was the introduction of the Unit or Credit System in August 1968, whereby all the courses offered were classified at 3 levels, namely the 100 series at the B.E. level, the 200 series at the Master's level and the 300 series at the research or specialized level. Each course carried a certain number of credits at the appropriate level. This facilitated the students to take advantage of the various courses offered by the different Departments of the Institute.

The research activities gained momentum and encompassed the broad areas of mineral beneficiation, thermodynamics and kinetics, process metallurgy, physical and mechanical metallurgy. Several equipment was designed and constructed namely, 25 KVA electric furnace for matte smelting of copper concentrate, oxygen meter for the measurement of oxygen pressure in industrial furnaces at high temperature, which works on the principle of the galvanic cell, an improved solution calorimeter for measuring the heats of formation of metal oxides, thermogravimetric balance, 50 KVA AC electroslag refining unit, Wohler type fatigue testing machine, stress corrosion testing unit, a constant stress creep unit and a variable speed, constant strain-rate tensile testing unit, low frequency internal friction wear testing unit and an apparatus for measuring the surface emissivity of metallic discs. New research projects sponsored by CSIR and DRDO dealt with the development of magnetostrictive materials and high manganese, copper alloys for marine applications of the Indian Navy. Another project with funding from the Central Machine Tools Institute pertained to the development of Al-base-bearing alloys, leading to import substitution. The research studies included the development of new flotation reagents, beneficiation of low grade fluorspar ore, vapour phase copper deposition on metallic and non-metallic substrates, chlorination of low grade Mn ores, smelting of non-ferrous ores, thermodynamic properties of slags and oxide systems, activity measurements in metallic and metal oxide systems (Seetharaman and Abraham 1969), electrodeposition from the fluoborate bath, solute vacancy interaction (Dwarakadasa et al. 1968; Kishore and Vasu 1971), clustering phenomena, thermally activated deformation of metals and alloys (Prasad et al. 1969), precipitation hardening in Al alloys and hexagonal metals, solid-solution strengthening, stress corrosion properties of Cu and Al alloys and oxidation kinetics, among other topics.

The year 1971 was a watershed in the annals of the Department, with the first batch of Ph.D. conferment being made during the year, namely K. S. Raman, E. S. Dwarakadasa, D. H. Sastry, Y. V. R. K. Prasad and S. Seetharaman. During 1971–72, the new faculty appointments included that of Dr. G. N. K. Iyengar (Ph.D., Imperial College, London) and Dr. U. B. Nayak, (Ph.D., University of Bombay), with his research work carried out at the Department of Chemical Engineering, IISc, as Assistant Professors and Dr. J. P. Hajra (Ph.D., University of Strathclyde, Scotland) and Dr. A. K. Lahiri (Ph.D., IIT, Kanpur) as Lecturers, which strengthened the process metallurgy research activities. To cater to the enhanced faculty strength and research activities, an extension building of the Department was constructed. The External Registration Programme for research conferment was approved by the Institute and the Quality Improvement Programme was introduced to facilitate teachers working at other institutions to pursue their Ph.D. at IISc.

22.4 Silver Jubilee of the Department and Strengthening of Research

The Silver Jubilee of the Department was celebrated in a befitting manner during 1971–72 by the holding of Seminars, technical exhibitions and culminating in the organization of twin International Symposia on 'Defect Interactions in Solids' and 'Industrial Metallurgy', during 9–13 May, 1972, under the Convenership of Prof. Vasu. The themes of the Symposia were a critical appraisal, analysis and evaluation of the theoretical and experimental investigations involving interactions of point defects, dislocations and other imperfections and the fundamental as well as applied aspects of industrial metallurgy, particularly with respect to casting, joining and forming techniques, treatments, testing and analysis and development of special alloys. The overwhelming response to the Symposia with the participation of leading experts from both within and outside the country was a testimony to the high quality of research pursued at the Department and its strong involvement in the R&D problems of the metallurgical industries in India.

The Department and the Institute suffered a loss by the passing away of Prof. A. A. Krishnan and Prof. Abraham took charge as the Head of the Department in 1972. During the period 1973–75, major projects of industrial relevance undertaken by the Department included the development of oxygen potential probe for Tata Iron & Steel Co. Ltd., hydrometallurgical recovery of copper and sulphur from chalcopyrite concentrate at the pilot plant level for the Chitradurga Copper Co. Ltd., development of quality steels by electroslag refining for Mysore Iron & Steel Ltd., reclamation of high temperature Ni-base turbine alloys from scrap for the Indian Railways, development of heat treatment salts, designing of oil filter for Tata Mercedes Engine, ceramic moulding process, setting up of SG iron foundry, quality improvement in continuous casting, to cite a few examples. The changes in the faculty pertained to the resignation of Dr. K. Narasimha Murthy and the appointment of Prof. P. K. Rohatgi (Sc.D., MIT, USA) as a Professor in both the Departments of Metallurgy and Mechanical Engineering, during this period. From the start of the 1974 academic year, the Institute decided to make all admissions to the various programmes, based on an All-India competitive Entrance Examination. An interlocking programme of special lectures, seminars, research discussions and visits was built up at the Institute, which facilitated exchange visits of academic personnel between research institutions and universities, both within and outside the country. Under this programme, distinguished scientists and Nobel laureates from the USSR, UK and USA visited and lectured at the Institute. In 1978, Prof. Abraham relinquished the Headship and Prof. Mallya was appointed as the Chairman. During 1977–78, Mr. B. V. Narayana was appointed as a Scientific Assistant. A Curriculum Development Centre was established at the Institute with the approval of the Ministry of Education, Government of India, with a view to developing curriculum in the areas of instrumentation and metallurgy in the initial stages. An emphasis was given to book writing and a first of its kind textbook on 'Extraction of Nonferrous Metals' was co-authored by K. P. Abraham with H. S. Ray and R. Sridhar, which became a widely used undergraduate text in many Institutions, fulfilling a long-felt need in the Indian context.

In the seventies, the tempo of the research and teaching activities was increased, and new research projects were initiated by Prof. Rohatgi dealing with metal-matrix composites and metal casting. He also started a programme on Technological Forecasting or 'Futurology' and R&D planning in the areas of food, steels, population control, aluminium industries, energy, materials, transportation and health services. In pioneering studies, an impression creep testing technique was developed using an indigenously built apparatus, an ion nitriding or glow discharge unit was designed and developed for nitriding of ferrous systems, and a solar furnace was fabricated to melt zinc and aluminium and the first casting in our country using solar melted metal was produced. A fruitful development was the productive collaboration between the extractive and mechanical metallurgy faculty members in the field of electroslag remelting. In the area of process metallurgy, the research was focused on the kinetic aspects of gas–solid reactions of metallurgical interest (Deb Roy and Abraham 1974; Sivan and Iyengar 1976), studies on electroslag melting of quality steels and Ni-based alloys, hot working, fatigue, fracture and deformation mechanisms in electroslag refined materials (Hebsur et al. 1980), analysis of blast furnace process (Lahiri 1977), studies on reduction of iron oxides by carbon, kinetics of chemical vapour deposition and a new formalism for representation of binary thermodynamic data (Hajra 1980). The research work in physical and mechanical metallurgy dealt with resistometric studies of precipitation kinetics in Al–Ag alloys, clustering and strengthening in a quenched and aged Al–2%Ge alloy, kinetic studies in oxide-dispersed Cu–Ni composites, low temperature deformation in hard HCP metals, diffusion in transition metal alloys, deformation characteristics of BCC metals, phase equilibria and transformations in Ti–Al–Mo alloys, effect of crystallographic texture and hardness on the mechanical properties of Cd alloys, dilatometric studies of austenite reversion in Fe–12Ni–4Mn maraging steel, wear properties of cast aluminium alloy-mica particulate composite and heat transfer of liquid metals. In the field of corrosion, the studies pertained to stress corrosion of Al–Mg alloy, corrosion and electrochemical studies on zirconium, superplasticity and stress corrosion cracking behavior of Zn–Al, Cu–Al alloys (Bhat and Vasu 1983), Zr and zircaloy, stress corrosion of commercial titanium and Ti alloy in methanol-bromine solution, stress corrosion cracking of nickel silver in seawater, electrometallurgy of chalcopyrite leaching, as some examples.

22.5 Special Assistance Programme of the University Grants Commission and Transformation of the 3 Year B.E. Degree to a 4 Year Integrated M.E. Course

The eighties witnessed a paradigm shift in the research infrastructure and activities of the Department. During 1981–82, Prof. R. V. Tamhankar (Ph.D., Paris) joined the department as a Visiting Professor and Mr. S. Sasidhara was promoted as a Scientific Assistant. In August 1981, Prof. S. Ramaseshan was appointed as the Director of IISc, following the retirement of Prof. Dhawan. The retirement of Prof. Abraham was marked by the organization of the International Conference on Metal Sciences (ICMS-81), in July, 1981, on the theme 'Thermodynamics and Kinetics of Metallurgical Processes'. This Conference was the third in the series, and was jointly organized by the Department and the Metal Sciences Division of the Indian Institute of Metals, which was a grand success, with the participation of several leading experts in the field from India and abroad to honour the contributions of Prof. Abraham.

Prof. S. Ranganathan (Ph.D., Cantab), of the Banaras Hindu University, was appointed as Professor and took charge as the Chairman of the Department in 1982. The Department received generous support from the University Grants Commission under its Special Assistance Programme in 1983, initially for a period of five years, and this facilitated major equipment to be procured and faculty and scientific manpower to be inducted. The two-year M.E. degree was shortened to a 3 Semester programme in the 1983–84 session. During the period 1982–85, the appointments included those of Dr. K. T. Jacob (Ph.D., Imperial College, London), Dr. K. A. Natarajan (Ph.D., Minnesota, USA) as Professors, Dr. K. Chattopadhyay (Ph.D., BHU) and Dr. A. M. Gokhale (Ph.D., Florida) as Assistant Professors, two Scientific Officers (Mr. S. Subramanian and Ms. B. N. Pramila Bai), and three Scientific Assistants (Mr. Shashidhara Pandit, Mr. P. Suresh and Mr. R. J. Deshpande). Dr. S. Ramachandran (Sc.D., MIT, USA) joined as an Honorary Visiting Scientist. The year 1984 marked the Platinum Jubilee of IISc and in August, Professor C. N. R.

Rao of the Solid State and Structural Chemistry Unit of the Institute took charge as the Director of IISc. The Department celebrated the occasion by the organization of the Symposium on 'Advances in Metal Processing'. To cater to the requirements of additional laboratory and office space, a first floor was added to the middle wing of the department. Prof. Brahm Prakash, the first Indian Head of the Department, passed away in January 1984, and as a tribute to him, the Department instituted a Visiting Chair Professorship in 1989. A new initiative, a first of its kind at the Institute, was the commencement of the Annual Symposium on 'Metallurgical and Materials Research' in 1986, organized entirely by the students of the Department to showcase their research work, with a faculty coordinator. In 1986, the Department received support under the COSIST programme of the University Grants Commission. In the same year, Prof. K. T. Jacob was awarded the Doctor of Science (Engineering) by the University of London.

The research activities flourished, and strong internationally recognized groups emerged in the areas of biohydrometallurgy, thermodynamics of materials, solid state sensors, solidification processing, quasicrystals, deformation processing, as well as modelling and simulation. In the area of mineral processing, the studies were concerned with the beneficiation of low-grade phosphate, molybdenite and wolframite ores (Deshpande and Nayak 1989), selective flotation studies on scheelite and calcite, serpentine and magnesite, grinding media wear and flotation studies on base metal sulphides (Yelloji Rao and Natarajan 1989) and iron ores and bioleaching of mixed sulphides. In the field of extractive metallurgy, an extension of Darken's quadratic formalism to dilute multicomponent solutions was developed (Srikanth and Jacob 1989), thermodynamic investigations were carried out on binary oxide solutions using solid electrolyte technique for Ni–Mn and Ni–Mn–Co alloys, Cr–Mn, Cr–Mn–O, Co–Mn–Cr (Ranganathan and Hajra 1988) systems, to cite a few. A new concept of 'thermodynamic capacity' was formulated and a central atoms model was developed for silicate and aluminate melts. Thermodynamic studies on cobalt (II, III) oxide provided evidence for a new and interesting phase transition in this compound, coupling spin unpairing with cation rearrangement between tetrahedral and octahedral sites (Kale et al. 1988). The use of metastable equilibria for thermodynamic measurements was explored. An SO_x ($x = 2, 3$) sensor using β-alumina/Na_2SO_4 couple (Akila and Jacob 1989) and a novel ZnO-based varistor for surge arrestor applications were developed. Other investigations dealt with ion nitriding of En 40B steel (Kurny et al. 1986), X-ray photoelectron spectroscopic studies of surface segregation and oxidation of Cu–Ni–Sn and Ag–Cu–Ge ternary alloys (Sampath Kumar et al. 1988), mechanical alloying of Ni–20Cr–2ThO_2 system and combustion and thermal decomposition of ammonium perchlorate.

A significant international acclaim for research carried out in the Department is the pioneering work carried out in the field of quasicrystals. The exciting discovery in 1984 in the USA of icosahedral quasicrystal exhibiting the cystallographically forbidden five-fold symmetry was reported by Prof. D. Shechtman and co-workers, for which he received the Nobel Prize in 2011. Barely six months after the publication of this paper, a second class of two-dimensional quasicrystals, named as 'decagonal quasicrystals' was discovered in the Department through systematic high resolution

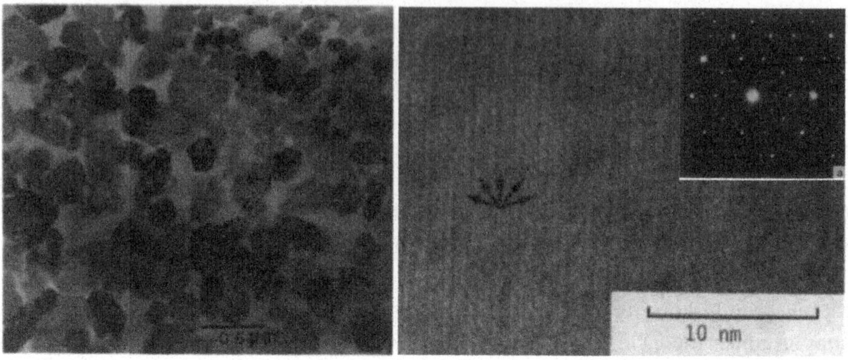

Fig. 22.2 Bright field micrograph showing the morphology of a Al–14%Mn quasicrystal (left), high resolution image of a quasicrystal obtained using the diffraction pattern shown in the inset (Chattopadhyay et al. 1985)

electron microscopy and electron diffraction in Al–Mn and Al–Pd alloys (Chattopadhyay et al. 1985). The flurry of activity following this led to a series of important discoveries that helped in a major way to establish the field. Some of the fundamental discoveries that set the future trend in quasicrystalline research and helped to establish the field include the discovery of one-dimensional quasicrystal, ordering transition in quasicrystals and experimental validation of the relation between crystal and quasicrystal through higher dimension projection formalism and establishing the equivalence of phason strain and approximant crystals experimentally. The variety of electron diffraction patterns from quasicrystals was elucidated (Fig. 22.2). In the research studies on quasicrystals, an important discovery was the identification of irrational twins in the quasicrystal and its link with the rational twins through the coincidence site lattice concepts. It was found that the negative temperature coefficient of resistivity increased with the free ordering reaction in the quasicrystals. Rapid solidification and mechanical alloying routes were used to produce aluminium and titanium-based metallic glasses.

The development of specialized equipment was an important activity during this time. The cooperative growth in monotectic systems was investigated using a computer-controlled unidirectional solidification apparatus, which was specially designed and developed for this purpose. A unit for spray forming of metals and alloys was fabricated to produce very fine metallic powders and metal matrix composites.

In the area of mechanical metallurgy, the studies included impression creep of zinc, fracture behaviour of Mg alloys and near-threshold fatigue crack growth of high strength steel (Ravichandran et al. 1986). The moisture absorption characteristics of natural fibre composites were studied (Giridhar et al. 1986). Additionally, stress corrosion cracking studies in Al alloys containing Ti were carried out. The precipitate morphology in quenched and aged Al–2%Ge alloy was studied by transmission electron microscopy and the influence of molybdenum and silicon on the corrosion behaviour of Fe–Ni–B metallic glasses was assessed.

Prof. K. T. Jacob was appointed as Chairman of the Department in 1988, and Dr. M. K. Surappa (Ph.D., IISc) joined as Assistant Professor and Ms. H. Sukanya as Scientific Assistant. In the same year, the additional appointments comprised those of Mr. G. S. Avadhani as Scientific Officer and Mr. T. Raji as Scientific Assistant. In keeping with the policy changes made at the Institute to focus on post-graduate and research programmes, the 3-year B.E. degree programme was transformed to a 4-year M.E. degree in 1988, for students joining with a BSc degree. Prof K. I. Vasu retired from the services of the Institute in July 1989.

22.6 Recognition as a Centre of Advanced Study and March to the Golden Era

The nineties heralded several milestones in the history of the Department. In 1990, the Department was recognized by the University Grants Commission as a Centre of Advanced Study, the second Department of Metallurgy in the country to achieve the distinction. Dr. V. Jayaram (Ph.D., Stanford) was appointed as an Assistant Professor. Dr. S. Subramanian was promoted as lecturer, later to become Assistant Professor in 1993. During 1990–91, the retirements included those of Prof. M. Mohan Rao, Mr. C. R. Krishnamurthy and Prof. U. B. Nayak. In 1992, Prof. Y. V. R. K. Prasad was appointed as the Chairman of the Department. The first D.Sc. degree in Metallurgy in IISc was conferred in the same year to Prof. K. A. Natarajan. In 1993, the Institute suffered a profound loss in the sad demise of Mr. J. R. D. Tata, the President of the Court. The new faculty recruitments in 1993–94 included those of Dr. T. A. Abinandanan (Ph.D., CMU, USA) as Assistant Professor, Dr. Atul Chokshi (Ph.D., USC, USA) as Associate Professor and Mr. R. Ravi as a Scientific Officer. In 1994, Prof. C. N. R. Rao retired and Prof. G. Padmanaban of the Department of Biochemistry, IISc was appointed as the Director. The retirement of Prof. R. M. Mallya, also in 1994, was marked by the organization of the Conference on 'Reactivity of Solids'. In 1995, Prof. D. H. Sastry took charge as the Chairman of the Department. Mr. V. Babu was promoted as Scientific Assistant in the same year.

The research activities encompassed the entire gamut of metallurgy and materials engineering. Some examples are highlighted. In the area of mineral processing and bio-hydrometallurgy, an improved method was developed for the estimation of mineral-adhered biomass of *Thiobacillus ferrooxidans* by protein assay. The presence of a proteinaceous appendage on bacterial cells was found to aid in bacterial adhesion, resulting in enhanced leaching (Preston et al. 1993). Temperature-tolerant strains of *Thiobacillus ferrooxidans* were developed to improve bioleaching kinetics. The mechanisms of interaction of natural polysaccharides with various oxides, sulphides and hydrophobic minerals were delineated (Rath et al. 1997). The stability of alumina suspensions was enhanced using polymeric additives over a wide range of pH (Santhiya et al. 1999).

In the field of extractive metallurgy, a new solid state EMF technique was developed for measuring the chemical potential of carbon in its various allotropic forms. A novel design of solid electrolytes for use with dissimilar gas electrodes was formulated. Important contributions to the area of interaction parameter formalisms were made in the development of thermodynamically consistent relations of the ternary and quaternary systems. Based on Butler's equation, the studies of the thermodynamics of surfaces and adsorption were pursued on various alloys involving surface active elements. Mathematical and physical models for cupola (Viswanathan et al. 1997), blast furnace COREX and submerged arc furnaces, simulation of phenomena in the tundish, continuous casting (Gupta and Lahiri 1996) and strip casting, solid flow behaviour in shaft reactors were developed and validated using experimental data. High-temperature solid-state galvanic sensors for control and automation in extraction and refining of metals, manufacture of refractories and monitoring environmental pollution were developed. Oxygen and sulphur sensors for liquid metals and gas sensors for SO_2 and H_2S and conceptual development of multi-element sensors were some of the significant achievements.

In mechanical metallurgy, the pioneering work on the development of processing maps for a wide variety of materials including pure metals, solid solutions, two-phase alloys, multi-component complex alloys, intermetallics and metal-matrix composites gained international recognition, and a comprehensive hot working guide was published (Prasad and Sasidhara 1997), Fig. 22.3a. The fracture toughness and fatigue crack growth behaviour of a number of aerospace alloys as well as Ti–Zr alloys were studied. Employing a fracture mechanics approach, the stress corrosion cracking behaviour of lithium-containing Mg alloys was examined. The fatigue properties of Al–7Si–0.3Mg alloy were studied using trace additions of Be and Mn (Murali et al. 1997). The superplastic deformation of yttria-stabilized zirconia in both tension and compression was characterized. Analytical models were developed for deformation and failure in nanocrystalline materials. An analysis of the high temperature deformation process in nanocrystalline materials suggested that cavity nucleation might be easier in such materials due to a reduction in the critical cavity nucleus. High temperature creep studies on steel weldments and investigations on the fracture mechanics aspects of different zones in the welded structure were carried out. Electroslag refining studies on En24 and MDN showed improvements in mechanical properties. A lab scale technology developed for making lightweight aluminium lithium alloys for aircraft applications was successfully scaled up to pilot plant level of 200 kg. The high temperature creep behaviour of titanium aluminide alloyed with niobium was investigated and the optimum composition, microstructure and heat treatment were identified.

In the research on advanced ceramics and composites, a new bulk breakdown device based on $ZnO–La_2O_3$ and Y_2O_3 composites was developed for low voltage surge suppression and a patent was filed. A novel technique called sinter-forging was used to study the densification and high temperature deformation of nanocrystalline ceramics and composites. Simple, cost-effective and energy-efficient processes for the processing of MMCs based on aluminium and magnesium alloy matrices were developed for industrial applications. Al alloys were infiltrated under nitrogen to

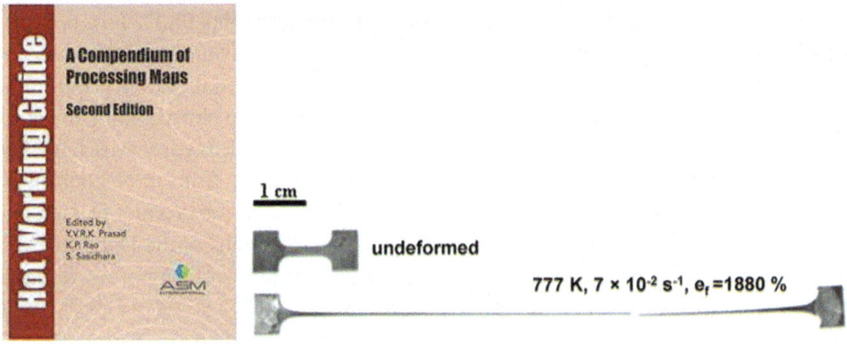

Fig. 22.3 **a** A compendium of processing maps containing 293 maps, published by ASM international (Prasad and Sasidhara 1997); **b** photographs of undeformed specimen with a gauge length of 5 mm and deformed specimen exhibiting maximum elongation of 1880% in nano-Nickel (Prasad and Chokshi 2010)

create Al/AlN composites. An aerodynamic levitation arrangement to levitate both spheres and discs inside an induction heated graphite furnace was designed and developed. Thermosetting matrix materials were reinforced with glass, carbon and aramid fibres and studies on interface modification involving both organic and inorganic materials were carried out. The modes of failure and impact response of elastomer-modified thermosets were evaluated. Intermetallic nanocomposites, soft nanodispersed immiscible alloys and intermetallic films and powders were synthesized using spray pyrolysis, rapid solidification, sol–gel and laser ablation routes.

The Department played a notable role in organizing several conferences, symposia and seminars. Special mention may be made of Indo-US workshops on 'Interfaces' (1989) and 'Nucleation and Growth in Solids' (1994), Discussion Meeting on 'Icosahedral Symmetry in Materials' (1993), International Conference on 'Inorganic Matrix Composites' (1995), 'Superplasticity in Advanced Materials' (1997) and 'Electrochemistry in Metallurgy and Materials Science' (1997).

The Golden Jubilee of the Department was marked by year-long events during 1996–97, and the grand finale was the successful organization of the International Conference on 'Recent Advances in Metallurgical Processes' in July 1997, with the participation of leading experts and alumni, both from India and abroad. The faculty recruitments in 1997 were those of Dr. Subodh Kumar (Ph.D., Imperial College, London), Dr. G. S. Gupta (Ph.D., Wollongong, Australia) and Dr. A. M. Raichur (Ph.D., Nevada, USA) as Assistant Professors. Prof. G. N. K. Iyengar retired in July 1997. In August 1998, Prof. G. Mehta from the University of Hyderabad took charge as the Director of IISc. An M.E. degree programme in Manufacturing Technology was started by the Department, with the curriculum involving courses from the sister Departments of IISc, namely the Centre for Product Design and Manufacturing as well the Computer Science and Automation, to which students with either B.E. in Metallurgical or Mechanical Engineering streams were admitted. The Institute approved the financial assistance of the international students on par with the scholarship of the Indian students, for both Master's and Ph.D. programmes, which encouraged international students, especially from Asian and African countries to pursue higher studies at the Department. A major collaborative programme on Aluminium Application Technology (ALTECH), with the Norwegian University of Science and Technology, Trondheim, Norway and Norsk Hydro, was launched in 1999 in the Department, with the participation of faculty from the Departments of Mechanical and Civil Engineering, as well as the Centre for Product Design and Development. In memory of Mr. Manish Narayan, an alumnus of the Department, a memorial fund was established in 1999 to provide partial travel support to research students to attend overseas international conferences.

22.7 Millennial Initiatives and Rechristening of the Department

At the threshold of the new millennium, the Institute took several initiatives promoting emerging and interdisciplinary areas through team efforts, developing active and close collaboration and networks with similar institutions worldwide and enabling enhanced interaction with industries through knowledge-wealth sharing process and promotion of entrepreneurship amongst its faculty. In 2000, Prof. K. A. Natarajan was appointed Chairman of the Department. An Indo-French Cell for Water Sciences was set up at the Institute jointly with the Institute of Research for Development (IRD), France, with the participation of faculty from the Departments of Metallurgy and Civil Engineering. In the same year, Prof. K. S. Raman retired and Dr. U. Ramamurty (Ph.D., Brown University, USA) joined as an Assistant Professor. In 2002, the Department received additional funding from the Department of Science and Technology through the first phase of the FIST programme which was extended for three phases up to 2020. The research activities were vigourously pursued and dealt with minerals, metals and alloys, polymers, ceramics and composites. A sample of these studies is detailed below.

The basic research carried out in the area of biohydrometallurgy for the past two decades, attracted national and international attention and culminated for the first time with the setting up of a pilot test bioreactor at Hutti Gold Mines Limited by the Department, with technical consultancy by Engineers India Limited. It was demonstrated that both gold and silver recoveries could be enhanced to >90% from about 50%. A kinetic model which accounted for the dissolution of the sulphide matrix due to both bacterial attachments onto the mineral surface and indirect leaching was proposed. A novel bio-leaching process to recover Cu, Co and Ni from Indian Ocean manganese nodules using organic reductants, at near neutral pH and ambient temperature, was developed and a patent filed. In bio-beneficiation, the utility of Bacillus polymyxa and Thiobacillus thiooxidans was explored for the selective separation of several sulphide and oxide minerals and the relevant surface chemistry was studied. The interaction of galena and sphalerite minerals with the metabolite obtained from

Bacillus polymyxa was examined through adsorption, electrokinetic, microflotation and flocculation tests. Galena could be selectively depressed or flocculated using the extra-cellular polymeric substances. A new method for characterizing the surface free energies of minerals by contact angle measurements was developed. The adhesion of the bacterium Bacillus polymyxa to three Indian coals, with different compositions, was investigated. Surface free energy calculations showed that the coal was hydrophobic, while the bacterium was hydrophilic. With respect to environmental remediation, the removal of fluoride and arsenic from solutions was achieved by using mixed rare earth oxides, and it was found that nearly 95–99% of both ions could be adsorbed (Raichur and Basu 2001). Biodegradation of organic flotation collectors was achieved using Bacillus polymyxa.

In extractive metallurgy, the fundamental research work carried out during the past twenty years in thermodynamics gained international recognition as a leading group for the development of new and accurate experimental techniques for thermodynamic measurements at high temperatures primarily using solid-state electrochemical techniques. These studies were complemented by innovative structural studies, in a highly sustained and fruitful collaboration with Japanese scientists. The insights from these studies have led to the development of models for predicting and extending data for un-investigated systems and domains. The data and models have found practical application in the development of phase diagrams of different types. The materials and systems of interest for energy generation (fuel cells, ceramic membranes) and catalysis (automotive exhaust, redox reactions) were the prime choice of investigations during the period.

A new technique to visualize fluid flow in porous media was developed. Process modelling and simulation studies were carried out with respect to silicon carbide and boron carbide manufacturing and carburization of steel. Some fundamental research work was to model the blast furnace raceway and unpacked bed dispersion. A new theory was proposed for packed and fluidized beds, which included the importance of frictional forces, and a patent was applied for based on the above theory. An experimental study was carried out to understand the void formation and breaking in a packed bed using two-dimensional glass models. The gas cavity size hysteresis in a packed bed was studied using DEM (Singh et al. 2007). A new X-ray visualization technique was developed to quantify the various local liquid holdups, hysteresis and dispersion in a packed bed (Basavaraj et al. 2005). A theory was developed to predict the raceway size in the blast furnace, and this work led to obtaining a worldwide patent.

Nanomaterials, another class of emerging new materials were a subject of intensive research. Different synthesis routines, their mechanical, electrical and magnetic properties and the melting behaviour of nanodispersions were studied extensively (Chinnasamay et al. 2001). Mesoporous materials based on titanium oxide and nitride were synthesized and characterized using small-angle X-ray diffraction and electron microscopic techniques. NiAl and FeAl coatings on single crystal superalloys were successfully carried out by laser ablation. A 3D model was developed for dissimilar laser joining. A process for creating laser clad of alloys containing soft dispersion for tribological application was developed that resulted in materials with low coefficient of friction. The superconducting behaviour of embedded nanoparticles was studied and a quantitative estimate of the effect of size on superconductivity was made. The mechanochemistry of replaced reactions in metal sulphates was studied using ball milling. Computer simulations using a Monte Carlo technique were used to study the simultaneous precipitation of two phases from a single parent phase. Elastic stress effects—in particular, the effect of elastic modulus mismatch—were studied in phase-separating systems using a diffuse interface model (Gururajan and Abinandanan 2007). Grain boundary (GB) effects on spinodal decomposition were studied using a phase field model (Ramanarayan and Abinandanan 2004). Using this model, a discontinuous spinodal decomposition regime was identified, and characterized in terms of GB mobility and atomic diffusivity at the GB. The structure and hard magnetic properties of barium hexaferrite with and without La_2O_3 prepared by ball milling were studied (Babu and Padaikathan 2002).

In the area of mechanical metallurgy, research studies on Portevin Le Chatelier (PLC) effect, spray forming, functionally graded materials and fracture toughness of aluminium alloys were carried out. New observations on PLC effect using LASER speckle technique were obtained (Shabadi et al. 2004). Spray deposition parameters were optimised to obtain good spray performance. Functionally graded materials from aluminium alloys were produced using a new method and the effect of specimen size on fracture toughness was also studied. Expert systems were developed for the design of a hot-forging process, based on material workability (Ravi et al. 2003). Studies on the development of techniques for the mechanical properties of thin hard coatings, development of ceramic–metal composites for electronic packaging applications in satellites, understanding of low temperature deformation in metastable, amorphous ceramics and the development of a combustion pyrolysis system for the deposition of thin films were carried out. The evolution of microstructure upon partial crystallization and its influence on the mechanical properties such as hardness, elastic modulus and viscosity in bulk metallic glasses were studied. Percolation theory was utilized in further substantiating the research outcomes. Energy absorption characteristics of Al foam were investigated using flat-end and spherical-end indenters with varying impact velocities. Reinforced syntactic foams were characterized using an ultrasonic imaging technique and the compressive properties were evaluated. The compression properties of fibre-free and fibre-bearing syntactic foams were compared. The influence of chopped fibres on the flexural behaviour of syntactic foam core system was evaluated (Karthikeyan et al. 2005). The temperature-dependent mechanical behaviour of passivated copper films was studied. Stresses in copper films of thickness ranging from 40 to 1000 nm, deposited on quartz or silicon substrates and passivated with silicon oxide were measured by using the substrate curvature method. The influence of martensite content and morphology on the tensile and impact properties of high martensite dual-phase (DP) steel was studied and it was found that DP steels with finely dispersed microstructures had excellent mechanical properties. Impression creep tests were used to study the superplastic behaviour of intermetallics such iron and titanium aluminides and the effect of alloying additions on the plastic flow characteristics was investigated.

The detailed fundamental and applied research studies carried out in the area of metal-matrix composites are highly cited and globally acclaimed (Surappa 2003; Saravanan and Surappa 2000). New knowledge was gained in the understanding of mechanisms of erosion in Al MMCs. AlN-Al matrix composites were fabricated by reactive infiltration. The effect of SiC particles on the size and morphology of eutectic silicon in cast A 356/SiCp composites was studied. Magnesium and magnesium alloy matrix composites were processed without the use of flux or inert gas. A process for the preparation of nano ceramic–metal matrix composites was developed and resulted in US and Indian patents (Raj et al. 2013). The sliding wear behavior of Al–Li–SiCp composites was studied.

The role of interfacial energy was examined with reference to prospects for superplasticity in ceramics and ceramic composites. Experimental studies revealed that alumina-spinel composites were likely to be superplastic, whereas alumina-YAG composites were unlikely to be superplastic, based on interfacial energy considerations. Studies on phase evolution and densification of spray pyrolysed ZrO_2–Al_2O_3 powders indicated the possibility of producing dense, amorphous pellets that could be heat treated further to produce nanocrystalline microstructures conducive for superplasticity. The mechanisms of pressureless infiltration of Al–Mg-based alloys into Al_2O_3 preforms were elucidated (Rao and Jayaram 2001). The indentation fracture toughness on ceramic and ceramic composites was studied using the acoustic emission technique. The occurrence of diffusion creep during high temperature deformation in alumina and yttria stabilized cubic zirconia, when microstructural changes accompanying deformation were correctly incorporated in the analysis, was conclusively established. Experimental evidence for diffusion creep in the superplastic 3 mol% yttria-stabilized tetragonal zirconia was obtained (Charit and Chokshi 2001).

The Department continued to actively pursue outreach activities by the organization of several Symposia, Seminars and Workshops, including those in honour of the retiring faculty. Mention may be made of the Indo-Norwegian Seminar on Aluminium Products-Design and Manufacture (2002), International Seminar on Mineral Processing Technology (2002), Eighth International Conference on Quasicrystals (2002), Conference on Light Materials: Science and Technology (2003), Discussion Meeting on Processing Maps and their Applications (2003), Seminar on Materials for the Future (2004), Symposium on Perspectives in Minerals, Metals and Materials Research (2004), Indo-US Workshop on Nanoscience and Technology (2004), Symposium on Perspectives in Metals and Materials Research (2005), Eighth International Conference on Nanostructured Materials (2006). An Advanced Facility for Microscopy and Microanalysis was established in 2004 by the institute, in which the Department played a leading role.

During the period 2003–2006, several retirements took place, considerably reducing the strength of the faculty of the Department. Prof. S. Ranganathan, Prof. E. S. Dwarakadasa, Prof. Y. V. R. K. Prasad and Dr. M. V. Bhat retired in 2003, Prof. J. P. Hajra, Prof. K. A. Natarajan and Mr. S. Sasidhara in 2004, Prof. D. H. Sastry and Prof. A K Lahiri in 2005 and Prof. K T Jacob and Prof. Kishore in 2006. Prof. K. Chattopadhyay was appointed Chairman of the Department in 2004. Prof.

P. Balaram took charge as the Director of IISc in 2005. The new faculty recruitments in 2005, involved those of Dr. Satyam Suwas (Ph.D., IIT Kanpur), Dr. Aloke Paul (Ph.D., Eindhoven) and Dr. S. Karthikeyan (Ph.D., Ohio, USA) as Assistant Professors. Mr. P. Padaikathan was promoted as Scientific Assistant in 2006. Taking into consideration the expansion of the research activities in the various domains of materials, encompassing minerals, metals, polymers, ceramics and composites, the Department was rechristened as Materials Engineering in 2006.

22.8 From Diamond to Platinum—In Pursuit of Excellence

In 2007, the Diamond Jubilee of the Department was marked by the organization of a Symposium. The other events organized included the Symposium on Bulk Metallic Glasses-Science and Technology, Mechanical Behaviour of Systems at Small Length Scales and Materials Design Workshop: An Indo-US School for Materials Discovery and Design. In the same year, Dr. Rajeev Ranjan (Ph.D., BHU) and Dr. Praveen C. Ramamurthy (Ph.D., Clemson, USA) were inducted as Assistant Professors.

The centenary of the Indian Institute of Science was celebrated during 2008–09, and the Centenary Conference was held in December 2008. At the Department, an International Conference on Multiscale Modelling and Simulation and a DST-SERC School on Texture and Microstructure were organized. The recognition of the Department by the University Grants Commission as the sole Networking Resource Centre for Materials (NRC-M) in the country in 2008, provided a fillip to the millennial research and outreach activities. The Department ran a highly successful UGC-NRC-M programme during 2008–2015 and took mentorship role in encouraging and strengthening research atmosphere in colleges/institutes/universities through collaborative research initiatives and organizing intensive workshops (including hands-on training on instruments). The Institutes which participated in this program included six different NITs (Calicut, Trichy, Rourkela, Tiruchirappalli, Durgapur, Surathkal, Jaipur), PSG College of Technology, Coimbatore, Bengal Engineering and Science University, Shibpur. Several Summer and Winter Workshops were organized covering a wide range of topics encompassing microstructural engineering, structural characterization techniques, mechanical property characterization, joining of materials, polymer synthesis, characterization and applications, finite element methods, principles and techniques of X-ray diffraction, integrated computational materials engineering, mineral biotechnology, Magnesium alloys processing, properties and applications, severe plastic deformation and bulk nano-structured materials, biomaterials, phase field modelling, molecular dynamics and Monte Carlo methods, corrosion engineering, to cite a few. In 2014, the Department, jointly with the Department of Mechanical Engineering, organized an International Conference on Friction Based Processes and an International Workshop on Iron and Steel Making.

The conception of NRC-M program facilitated the construction of a new building, enhancing the office and laboratory space of the Department. The research activities received support by many different governmental departments (DST, CSIR,

DRDO, DBT, UGC, ISRO, DAE, AICTE) and industries (Boeing, Alcoa, Cookson Electronics, General Motors, Rolls Royce, Tata Steel, TVS Motors). International research collaborations were forged with institutions and research laboratories in countries like Japan, China, Korea, Singapore, UK, France, Germany, Norway, Sweden, Mexico and the USA.

Prof. V. Jayaram was appointed Chairman in 2009. A Visiting Fellowship was instituted by Prof. K. Chattopadhyay in memory of Prof. Hubert Aaronson. Several Conferences were organized during the year namely, Seminar on Computational Materials, Interface Related Mechanical Behaviour of materials, Indo-French Workshop on Anthropogenic Impacts on Water resources and Soils sponsored by CEFIPRA, International Conference on the Advances in the Theory of Ironmaking and Steelmaking, to cite a few. The faculty appointments comprised of Dr. Chandan Srivastava (Ph.D., Alabama, USA) as Assistant Professor and Dr. Dipankar Banerjee (Ph.D., IISc) as Professor in 2010, Dr. Kaushik Chatterjee (Ph.D., Penn State, USA), Dr. Praveen Kumar (Ph.D., USC, USA), Dr. Suryasarathi Bose (Ph.D., IIT-B) in 2011, Dr. Abhik N. Choudhury (Ph.D., Karsruhe Institute of Technology, Germany) and Dr. Vijay A. Sethuraman (Ph.D., USC, USA) in 2013, as Assistant Professors. A 4-year undergraduate programme, with a flexible curriculum focusing on both science and engineering components, was commenced in August 2011 by the Institute, in which 'Materials' is one of the major subjects for specialization. The addition of this research-oriented 4-year UG program and diversification of the research portfolio of faculty-led to development of a comprehensive teaching curriculum, catering to the needs of the undergraduate, masters and doctorate students. In 2012, the Department hosted the XXIII Annual Meeting of the Electron Microscope Society of India, and the 16th International Conference on Strength of Materials (ICSMA) was organized.

Prof. Anurag Kumar of the Department of Electrical Communication Engineering of the institute took charge as the Director of IISc in 2014. Prof. T. A. Abinandanan was appointed Chairman in 2015. The faculty retirements included those of Prof. K. Chattopadhyay (2015), Dr. B. V. Narayana (2016), Prof. M. K. Surappa (2017), Prof. D. Banerjee (2017) and Prof. S. Subramanian (2019). Dr. Vijay A. Sethuraman resigned in 2017 and Prof. U. Ramamurty sought voluntary retirement in 2018. The new faculty induction comprised of Dr. Subho Dasgupta (Ph.D., Technical University Darmstadt, Germany) in 2016. The major international conferences organized during 2015–2019 were, International Conference on Metals Materials honouring Prof. S. Ranganthan on his 75th birthday (2016), International Conference on Texture, Micro-texture and Mechanical Behaviour (2017) and Advances in Process Metallurgy (2019) to mark the retirement of Prof. S. Subramanian. In addition, workshops on mechanical behaviour at small length scale, attracting participants from USA, Europe, Far East and India, became a regular triennial event.

In August 2020, Prof. G. Rangarajan of the Department of Mathematics, IISc, was appointed Director of the Institute and Prof. Satyam Suwas was appointed Chairman of the Department. Dr. S. K. Makineni (Ph.D., IISc) and Dr. Sai Gautam Gopalakrishnan (Sc.D., MIT, USA) joined the Department as Assistant Professors in 2020. To mark the birth centenary of Prof. K. P. Abraham, a Visiting Professorship was instituted in 2020.

From 2006 onwards, subsequent to the renaming of the Department as Materials Engineering, and appointment of new faculty members with background in polymers, biomaterials, electronic materials and devices, computational materials science, etc., the Department gradually developed an interdisciplinary flavour. While research on frontiers areas of metallic materials and metallurgical processes continued, this period is highlighted by strengthening of research in non-metallurgical materials (structural and functional polymers, ferroelectric and piezoceramics, tissue engineering, flexible printed electronics). The Department has strengthened its infrastructural facilities for carrying out cutting-edge fundamental and applied research. This period also witnessed a significant increase in the number of Ph.D. students. Currently, the department has strong research program with ~120 Ph.D. students. Faculty members are involved in international collaborations and in leading important projects of national strategic importance like the GTMAP (DRDO) and AUSC (DAE, IGCAR). In addition, the Department has attracted projects by private and public sector companies like, Tata Steel, GAIL, Boeing, Pratt & Whitney, General Electric, to cite a few.

The current research activities can be classified into six broad categories (i) Biomaterials, (ii) Functional materials, (iii) Mechanical behaviour and alloy design, (iv) Process metallurgy, (v) Materials processing and manufacturing and (vi) Physical metallurgy including computational materials science. In the field of biomaterials, the focus is on environmental biotechnology, bio-processing, tissue engineering and drug delivery. In the functional materials category diverse areas like electro-active polymers, organic photovoltaics, ferro/piezoelectric materials, polymer composites and printed electronics are being pursued. Research under the mechanical behaviour category includes study of creep, fatigue, fracture of complex alloy systems like high entropy alloys, Ni- and Co-based superalloys, Ti alloys and light metal alloys. New initiatives have been taken to develop capabilities for small scale testing, in-situ testing under SEM. In the same context, a new testing method has been invented based on digital image correlation and cantilever bending (Jalali et al. 2021), for examining creep response of the material in high throughput fashion, and successfully employed to study primary-cum-secondary creep response of model metals as well as complex engineering alloys such as steels and superalloys. There is ongoing focus on studies pertaining to microstructure and property control via high strain rate deformation, high temperature deformation and severe plastic deformation. An interdisciplinary research involving mechanical processing and biology tackles the role of shot-pinning in enhancing fatigue life of biomedical grade β titanium alloys (used as orthopaedic transplants) in corrosive environment as experienced in the human body (Bahl et al. 2017).

In process metallurgy, a mathematical model was developed and experimentally verified to describe slag-metal emulsion behaviour and its influence on heat and mass transfer. The raceway formation in a moving bed has been studied (Mojamdar et al. 2018). Related modelling efforts have been carried out in the new secondary steelmaking process, REDA and in granulation and sintering of iron ore. Another group worked systematically on bioremediation of lead, zinc and chromium from aqueous solutions, and role of the functional groups on the bacterial cell surface in

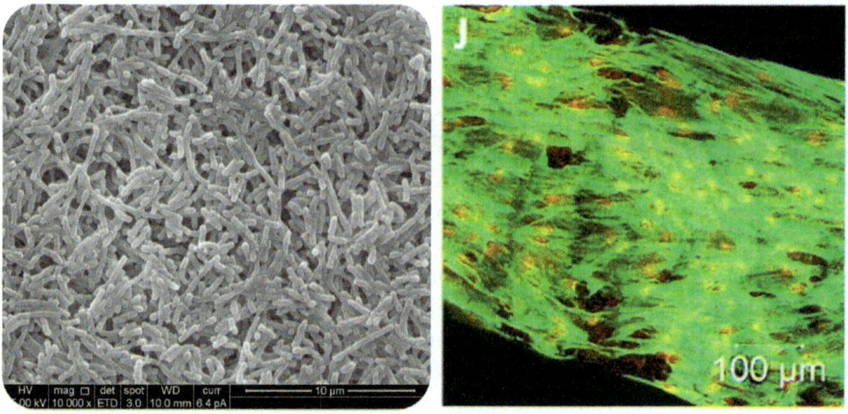

Fig. 22.4 **a** Scanning electron micrograph of *Pseudomonas fluorescens* used for bioremediation of lead **b** polymer scaffolds for tissue generation

metal complexation. For the first time, this group demonstrated a novel property of DNA as a flotation bio-reagent in the selective separation of sphalerite from galena, Fig. 22.4a (Vasanthakumar et al. 2012). The beneficiation of a low-grade siliceous copper ore from Tanzania was carried out by selective dispersion-flocculation and flotation techniques using polysaccharides.

In the field of materials processing, the department has established its leading role in the area of severe plastic deformation. Newer methods of severe plastic deformation have been invented and processing strategies developed for different materials (Bhattacharjee et al. 2013). The works on mechanical behaviour of severely deformed ultrafine grain materials have led to a deeper understanding of strength-ductility paradox, strength-conductivity paradox, contribution of different strengthening mechanisms, creep and superplasticity in the perspective of ultrafine grain materials (Mungole et al. 2014).

A major thrust area that has developed in the last decade and half pertains to the establishment of the state-of-the-art facilities for the investigation of texture development in materials with strong emphasis on processing-texture-property relationship studies. Some of the outstanding research in this area includes resolving the long-standing issue of texture transition in FCC metals with reference to stacking fault energy, introducing texture engineering concepts to produce ultra-fine grain sizes and desirable weak texture, required for many engineering applications (Madhavan et al. 2016). The demonstration of a novel deformation path to achieve sub-micron grain size and suitable texture in Mg-based light-metal alloys has shown the way to ductilise magnesium (Biswas et al. 2010). This has important implications for manufacturing of Mg-based engineering components.

The Department continues to hold its strong base in physical metallurgy with focus on alloy design, processing, and state-of-the-art characterization. In recent years, the emphasis has been on designing new materials for high-temperature structural applications, lightweight alloys for automotive, energy and thermoelectric materials. An important breakthrough in this field is the discovery of new low-density Co-based superalloys having γ/γ' microstructure that has the potential to compete with conventional Ni-based superalloys (Makineni et al. 2015, 2016). These alloys are based on Co–Al–Mo–Nb/Ta, Co–Al–V–Nb/Ta and Co–Al–Cr–Nb/Ta. A new strategy has been developed to raise the high-temperature capability of Al-based light metal alloys.

In recent years, the physical metallurgy discipline has got a new boost with the added capabilities in computational materials science. The focus here is on the numerico-experimental investigation of materials phenomena that involve a complimentary utilization of controlled experiments as well as modeling techniques for unravelling process-structure–property correlations. One of the principal focal points is the study of pattern formation during solidification using a combination of controlled directional solidification experiments and simulation studies using state-of-the-art phase-field models formulated in-house. The collection of experimental equipment is one of the few available set-ups in the country, which allow solidification experiments with an independent control of the thermal gradients and velocities, for alloys with a range of melting points. (Aramanda et al. 2020). Efforts are on to expand the work via collaboration with experimental groups dealing with multi-physics problems that allow a virtual assessment of both the process—structure correlations, relating the processing and chemical compositions to the bulk-hetero-junction microstructures and performance of organic photovoltaics, and coupled influence of composition gradient and coherency stresses in influencing failure mechanisms in electrical interconnects during material transport, caused by a combination of thermal and electric fields (Chakraborty et al. 2018). It was discovered that electric current through point contact probes led to the melting of and symmetrically circular flow of very thin metallic films, eventually leading to the removal or 'electro-etching' of the thin film from the region near the probe. The electro-etching phenomenon was harnessed to develop a lithography technique, christened as 'Electrolithography', which could be used to create patterns as narrow as 40 nm in the transferred material (Fig. 22.5) (Somaiah and Kumar 2018). In this context, a contactless scanning probe lithography tool, named as water-electrolithography, was also invented, with a significantly longer tip life, sharper patterns and little impact on environment, and the highest throughout amongst all scanning probe lithography tools. New discoveries in the area of material transport in both solid and liquid metals upon application of electric current, coupling between electromigration and thermo-migration and patterning of standalone structures using liquid electromigration were made, and their application space was explored (Jagtap et al. 2018).

Several groups pursued research on different aspects of the deformation behaviour of metals, alloys, intermetallics and composites. Studies included superplastic behaviour in nanocrystalline Ni and its alloys, Fig. 22.3b, deformation mechanisms in bulk metallic glasses, mechanisms for creep and quasistatic loading in titanium alloys, titanium aluminides and austenitic steels, high strain rate deformation of automotive steels and superplasticity in lightweight magnesium alloys (Prasad and Chokshi 2010;

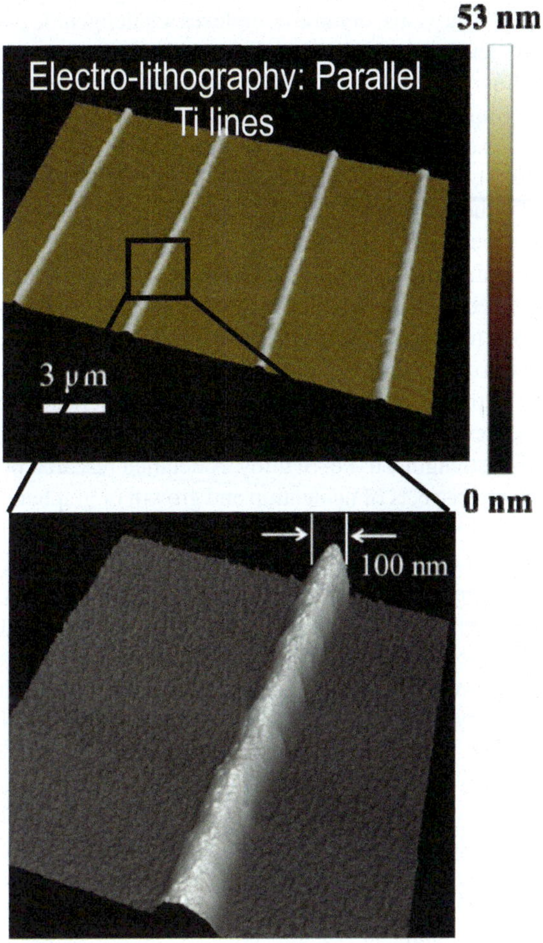

Fig. 22.5 Creation of metallic patterns upon passing electric current: A set of parallel 100 nm wide straight lines of Ti thin films, fabricated using electrolithography on SiO_2/Si substrate. The inset shows a magnified AFM topograph of one of the Ti lines

Kandalam et al. 2017). The important role of grain boundary sliding and identification of the rate-controlling diffusion species in ceramics, such as yttria doped zirconia and alumina was examined. The creep of complex alloy systems, such as high entropy alloys, Ni- and Co-based superalloys, Ti alloys and light metal alloys were systematically examined, and structure–property relationships were established, using mechanistic details, sophisticated characterization tools and first-principle calculations. The role of the interface between metal and ceramic in deformation and failure response of multilayered systems was examined.

The mechanical behaviour of polymer composites was investigated with respect to the finer details of architecture-mechanical property correlation and important recommendations pertaining to their airworthiness and impact-resistance were arrived at (Kundan et al. 2020). Pioneering contributions were made in the applicability of nano-indentation technique as a local probe of mechanical properties in materials as diverse as drug crystals, organic complexes, chalcogenide glasses and bulk metallic glass and for developing high throughput method to estimate the effects of composition, crystallographic orientation and temperature on mechanical response of material systems (Varughese et al. 2013). Extraordinary synergy was observed in the mechanical properties of polymer matrix composites reinforced with two nanocarbons (Prasad et al. 2009).

Studies on the processing of non-equilibrium alloy phases for exotic properties by chemical synthesis, high energy ball milling, injection and suction casting, melt spinning and laser processing were continued. Stable nanoparticle dispersions were created by ball milling of metal powders. A new quasi-chemical model was developed to explain stacking fault energies in Ni-base superalloys. Additionally, in these materials, atomistic and thermodynamic modelling tools were extended to design of alloys with desired composition and microstructure. Studies were carried out on the development of low mass density γ'-strengthened Co-based superalloys and a high strength Cu–Fe–Si alloy. A detailed research program was conceived on fundamental aspects of nucleation and growth of graphene and graphene oxide and the use of these materials as coating on various metallic surfaces was explored (Vamsi and Karthikeyan 2021a, b; Gupta and Srivastava 2020; Balachandran et al. 2017).

Experimental studies of diffusion-related issues in some of the new electronic packaging materials such as Cu–Sn, Ni(P)–Sn, Au–Sn systems, third-generation Ni-based alloys for gas turbine applications and high-performance Nb-based silicides were carried out. Classical interdiffusion studies were conducted to study the stability of phases and mixing in Nb–Si, Fe–Pt, Ni–Pt–Al, Co–Ni and electronic solders, and theories pertaining to the pseudo-binary and pseudo-ternary systems were proposed to study diffusion and microstructural evolution in complex alloy systems. Another development in recent years in this area is the formulation of a new theoretical framework for determining diffusion coefficients in inhomogeneous alloy systems with more than three components. New experimental methods have been developed for the estimation of intrinsic and tracer diffusion coefficients in multi-component systems, which enables a deeper understanding of microstructural evolution in complex material systems used in various high-temperature applications (Dash et al. 2020; Esakkiraja et al. 2020).

Recently, the department has initiated a major program in additive manufacturing. This emerging field requires fundamental understanding of microstructural evolution under different types of non-equilibrium processing of materials. With state-of-the-art characterization facilities and available expertise, the department is fully geared to undertake frontier research in this field. The role of post-processing on enhancing the mechanical behaviour of additively manufactured materials like 316L stainless steel, $(\alpha + \beta)$ titanium alloys and nickel base superalloys has been established (Bahl et al. 2019; Bawane et al. 2018).

Additionally, intensive research is also pursued in wide-ranging areas like developing strategies for new generation of drug delivery systems, design of organic materials for photovoltaics and pollution monitoring sensors, development of novel polymer processing methods for structural applications, water filtration, EMI shielding, tissue engineering, flexible electronics, piezoceramic sensors and actuators. A layer-by-layer technique of assembling pH and enzyme-responsive polyelectrolyte capsules was developed for the delivery of genes, anticancer drugs and probiotics for diagnostic and therapeutic applications as well as for environmental clean-up (Sreeranjini et al. 2016). In recent years, the group has developed expertise in the synthesis of encapsulated functionalized nanoparticles to target specific tissues and organs using external stimuli such as light and ultrasound. Another focus in the field of biomaterials research is to develop polymers and polymer nanocomposites for preparing 3D biodegradable scaffolds for engineering bone tissues, Fig. 22.4b. An important finding of this effort is the demonstration of the ability of carbonaceous and bio-ceramic nanoparticles in the scaffolds to impart multi-bio functional activity by augmenting tissue growth, while minimizing bacterial infections. An organotypic tissue model has been developed for breast tumor and cardiac muscle such that cells on these engineered biomaterials platforms can guide cells to organize into tissue-mimetic arrangements (Jaidev and Chatterjee 2019; Balachander et al. 2018). These tissue models are intended for investigating the molecular basis of diseases and for drug screening. Work is in progress to establish state-of-the-art facility for 3D bioprinting for fabricating tissue scaffolds.

The design and synthesis of organic molecules to tailor their properties for enhanced performance in the targeted applications areas like photovoltaics, sensors, EMI shielding and OFET's has been actively pursued in the Department (Ramamurthy and Mallya 2019; Sindhu et al. 2016). One of the ongoing efforts is to tailor molecules to enhance certain properties, which results in electrochemical sensors exhibiting exceptional selectivity towards analytes like volatile organic compounds, nitrates, heavy metals like lead, mercury and chromium ions as well as biological organisms such as *E. coli*. An important highlight in the area of polymer processing is the use of phase separation in polymeric blends as a technique for direct self-assembly and to control percolation of nanoparticles such as silver, CNTs and graphene in polymeric matrix. Such nanoengineering techniques were utilized to improve and tweak the separation properties of the membranes, with specific areas of interest such as antifouling membranes, membrane selective layer engineering and pore engineering for nanofiltration, reverse osmosis and forward osmosis, membranes as the next generation separation tools for industrial wastewater (Pawar et al. 2015; Bose et al. 2020).

Research activity in the field of functional ceramics is predominantly concerned with development of new ferroelectric compositions and strategies for enhanced electromechanical response. Notable contributions have been made in unraveling complex structure–property correlation issues in ferroelectric oxides-based piezoelectric materials, and innovative strategies in determining the nature of structural phases on the nanometric scale in $Na_{0.5}Bi_{0.5}TiO_3$-based Pb-free piezo-electrics, have been applied. A remarkable discovery of a polycrystalline piezoceramic exhibiting a (world) record high electro-strain of ~1% has been made (De and Ranjan 2020). The dual use of Raman and photoluminescence signals in rare-earth doped ferroelectrics for the design of an optical thermometer with significantly enhanced temperature sensitivity has been successfully demonstrated (Narayan et al. 2018).

Since 2016 the Department has initiated a major research program in the field of inkjet-printed electronics devices, wherein metal/semiconductor ink formulations, device architectures, TFTs, etc., are being investigated. Some of the outstanding issues related to network transport have been addressed through the performance of 2D electronic materials-based printed devices (Garlapati et al. 2018). Significant progress has been made in the area of low temperature curable TFTs, in enhancing mechanical reliability of devices and superior high frequency performances, when compared to the competing technologies (Cherukupally et al. 2020).

22.9 Accomplishments

Since its inception, the Department of Metallurgy at Indian Institute of Science, has made great contributions to the nation. It educated students who later received advanced degrees and became faculty members in new technical institutes and scientists in new research organizations, both in India and abroad. It produced the first Rhodes' scholar from India, and nine of its undergraduates became Members of the National Academy Engineering (USA). The faculty of the department has received many accolades for their research work, both at the national and international levels. These include Padmashri, Karnataka Rajyostava award, S. S. Bhatnagar Prize for engineering sciences, Fellowships of the World Academy of Sciences, Royal Society of Chemistry, Indian Academy of Sciences, Indian National Science Academy, Indian National Academy of Engineering, National Academy of Sciences and the Indian Institute of Metals, Swarnajayanti and J. C. Bose Fellowship; National Mineral Award of the Ministry of Mines; Metallurgist of the year, National Metallurgist of Ministry of Steel; Platinum Medal and G. D. Birla Gold Medal of the Indian Institute of Metals; MRSI Medal and Materials Scientist of the Year Award of the Material Research Society of India; INAE Professor Jai Krishna Memorial Award; Mineral Beneficiation Award of the Indian Institute of Mineral Engineers; Biotech Process Development and Commercialization Award and National Bioscience Award of DBT; Ramanujan Fellowship of DST. The International awards include International Hoffman Memorial Prize, Extraction and Processing Science Award (TMS), Willard Gibbs Equilibria Award, Friedrich Wilhelm Bessel Research award and Alexander von Humboldt Foundation Fellowship. The faculty members have several national and international patents to their credit. The prolific research accomplishments of the department could be gauged by the fact that over 200 papers have been published each year in the past decade. The alumni of the department continue to play important role in international academic arena as well as in major Indian institutions, R&D laboratories and industrial organizations.

22.10 IIM-IISc Linkage

The establishment of the Bangalore Chapter of the Indian Institute of Metals (IIM) is almost coeval with the establishment of the parent IIM body in 1946. Prof. Frank Adcock, the first Professor and HOD of the Department was a founder member of IIM. It is interesting to note that Prof. M. S. Thacker was Senior Vice-President of IIM in 1951, and later became President of IIM during 1954–56. Prof. Brahm Prakash was a member of the Council of IIM. The 7th Annual Meeting of the Indian Institute of Metals was held in Bangalore in February 1953 and the Local Chapter rendered much service in this connection. The 13th Annual Technical Meeting of the Indian institute of Metals was held at IISc, Bangalore on 1st and 2nd, December 1959, with the IIM Bangalore Chapter under the Chairmanship of Prof. A. A. Krishnan, who was also a Council member of IIM, playing a pivotal role. The Bangalore Chapter of the Indian Institute of Metals was given the responsibility to host the National Metallurgists' Day and The Annual Technical Meeting (NMD-ATM) of the Indian Institute of Metals during 1971, 1984, 1998 and 2010 with the active support of the Department of Metallurgy. Several faculty members of IISc and/or associated with the Bangalore Chapter have had the distinction of becoming the President of the Indian Institute of Metals namely, Prof. M. S. Thacker, Prof. Brahm Prakash, Prof. T. R. Anantharaman, Prof. E. G. Ramachandran, Dr. R. Krishnan, Dr. C. G. Krishnadas Nair, Prof. S. Ranganathan, Prof. D. Banerjee and Prof. K. Chattopadhyay. As a tribute to the first Indian Professor and Head of the Department of Metallurgy, IISc, the Prof. Brahm Prakash Memorial Lecture series was started in 1985. This lecture has been organized without interruption every year on 21st August, the birthday of Prof. Brahm Prakash and so far, 36 lectures have been delivered by eminent metallurgists and material scientists from the Departments of Atomic Energy, Space, Defence and Aerospace and from industry and academia. The Chapter also instituted a lecture series in commemoration of the Golden and Diamond Jubilee of the Indian Institute of Metals.

Acknowledgements The Department is grateful to all the academic, administrative, technical, scientific, support staff and students for their contributions to the growth and development of the Department. The Department is proud of the achievements of the alumni, who have brought honour to the Institute both by their work and the distinctions they have earned in academia, industry, R&D organizations and as entrepreneurs, both within and outside the country. The Department is highly indebted to the various Government and private funding organizations from India and abroad for their munificent support. The Department is beholden to the successive Directors and Registrars of the Institute for their guidance, encouragement and support.

References

Akila, R., Jacob, K.T.: Sensors Actuators **16**, 311 (1989)
Aramanda, S.K., Chattopadhyay, K., Choudhury, A.: Acta Mater. **197**, 108 (2020)
Babu, V., Padaikathan, P.: J. Magn. Magn. Mater. **241**, 85 (2002)

Bahl, S., Krishnamurthy, A.S., Suwas, S., Chatterjee, K.: Mater. Des. **126**, 226 (2017)
Bahl, S., Mishra, S., Yazar, K.U., Kola, I.R., Chatterjee, K., Suwas, S.: Addit. Manuf. **28**, 65 (2019)
Balachander, G.M., Talukdar, P., Debnath, M., Rangarajan, A., Chatterjee, K.: ACS Appl. Mater. Interfaces **10**, 33814 (2018)
Balachandran, S., Kumar, S., Banerjee, D.: Acta Mater. **131**, 423 (2017)
Basavaraj, M.G., Gupta, G.S., Naveen, K., Rudolph, V., Bali, R.: J. AIChE **51**, 2178 (2005)
Bawane, K.K., Srinivasan, D., Banerjee, D.: Metall. Mater. Trans. A **49**, 3793 (2018)
Bhat, M.V., Vasu, K.I.: Prakt. Metal. **20**, 130 (1983)
Bhattacharjee, D., Ray, R.K., Suwas, S., Biswas, S., Bhowmik, A., Singh, S.: A method to process interstitial free steels by multi-axial forging (MAF). TATA Steel and Indian Institute of Science, Bangalore, India, Pub. No.: 1831/KOL/08 dated 27.10.08, Int Appl No.: PCT/IN2009/000607. Granted on 13 March 2013
Biswas, S., Singh, D., Suwas, S.: Acta Mater. **58**, 3247 (2010)
Bose, S., et al.: Antimicrobial composite fabric and method thereof. Indian Patent Application No. 202041017738; Filing date: 25/04/2020
Chakraborty, S., Kumar, P., Choudhury, A.: Acta Mater. **153**, 377 (2018)
Charit, I., Chokshi, A.H.: Acta Mater. **49**, 2239 (2001)
Chattopadhyay, K., Ranganathan, S., Subbanna, G.N., Thangaraj, N.: Scr. Metall. **19**, 767 (1985)
Cherukupally, N., Divya, M., Dasgupta, S.: Adv. Electron. Mater. **6**, 2000788 (2020)
Chinnasamay, C.N., Narayanasamy, A., Ponpandian, N., Chattopadhyay, K., Shinoda, K., Jeyadevan, B., Tohji, K., Nakatsuka, K., Furubayashi, T., Nakatani, I.: Phys. Rev. B **63**, 18410–18411 (2001)
Dash, A., Esakkiraja, N., Paul, A.: Acta Mater. **193**, 163 (2020)
De, A., Ranjan, R.: Mater. Horiz. **7**, 1101 (2020)
Deb Roy, T., Abraham, K.P.: Metall. Trans. **5**, 349 (1974)
Deshpande, R.J., Nayak, U.B.: Trans. Indian Inst. Metals **42**, 109 (1989)
Dwarakadasa, E.S., Raman, K.S., Vasu, K.I.: Scr. Metall. **2**, 639 (1968)
Esakkiraja, N., Gupta, A., Jayaram, V., Hickel, T., Divinski, S.V., Paul, A.: Acta Mater. **195**, 35 (2020)
Garlapati, S.K., Divya, M., Breitung, B., Kruk, R., Hahn, H., Dasgupta, S.: Adv. Mater. **30**, 1707600 (2018)
Giridhar, J., Kishore, Rao, R.M.V.G.K.: J. Reinforced Plast. Compos. **5**, 141 (1986)
Gupta, D., Lahiri, A.K.: Metall. Mater. Trans. **27B**, 757 (1996)
Gupta, A., Srivastava, C.: J. Electroanal. Chem. **861**, 113964 (2020)
Gururajan, M.P., Abinandanan, T.A.: Acta Mater. **55**, 5015 (2007)
Hajra, J.P.: Metall. Trans. **11**, 215 (1980)
Hebsur, M.G., Abraham, K.P., Prasad, Y.V.R.K.: Eng. Fract. Mech. **13**, 851 (1980)
Jagtap, P., Narayan, P.R., Kumar, P.: J. Electron. Mater. **47**, 4177 (2018)
Jaidev, L.R., Chatterjee, K.: Mater. Des. **161**, 44 (2019)
Jalali, S.I.A., Jayaram, V., Kumar, P.: Acta Mater. **205**, 116535 (2021)
Kale, G.M., Pandit, S.S., Jacob, K.: Trans. Jpn. Inst. Metals **29**, 125 (1988)
Kandalam, S., Sabat, R.K., Bibhanshu, N., Avadhani, G.S., Kumar, S., Suwas, S.: Mater. Sci. Eng. A **687**, 85 (2017)
Karthikeyan, C.S., Sankaran, S., Kishore. Mater. Eng. **290**, 60 (2005)
Kishore, Vasu, K.I.: Scr. Metall. **5**, 997 (1971)
Kundan, K.V., Kumar, S., Bose, S.: Compos. Struct. **261**, 113272 (2020)
Kurny, A.S.W., Mallya, R.M., Mohan Rao, M.: Mater. Sci. Eng. **78**, 95 (1986)
Lahiri, A.K.: Trans. Indian Inst. Metals **30**(18), 29 (1977)
Madhavan, R., Ray, R.K., Suwas, S.: Philos. Mag. **96**, 3177 (2016)
Makineni, S.K., Samanta, A., Rojhirunsakool, T., Alam, T., Nithin, B., Singh, A.K., Banerjee, R., Chattopadhyay, K.: Acta Mater. **97**, 29 (2015)
Makineni, S.K., Nithin, B., Chattopadhyay, K.: γ-γ' Strengthened tungsten free cobalt-based superalloy. Patent No. IN201401981I4. Patent published 01.01.2016

Mojamdar, V., Gupta, G.S., Puthukkudi, A.: ISIJ Int. **58**, 1396 (2018)
Mungole, T., Kumar, P., Kawasaki, M., Langdon, T.G.: J. Mater. Res. **29**, 2534 (2014)
Murali, S., Arvind, T.S., Raman, K.S., Murthy, K.S.S.: Mater. Trans. JIM **38**, 28 (1997)
Narayan, B., Malhotra, J.S., Pandey, R., Yaddanapudi, K., Nukala, P., Dkhil, B., Senyshyn, A., Ranjan, R.: Nat. Mater. **17**, 427 (2018)
Pawar, S.P., Marathe, D., Kola, P., Bose, S.: J. Mater. Chem. A **3**, 656 (2015)
Prakash, B., Sundaram, C.V.: Proceedings of the International Conference on Peaceful Uses of Atomic Energy, Geneva, United Nations, New York, P/8768, 554 (1955)
Prasad, M.J.N.V., Chokshi, A.H.: Acta Mater. **58**, 5724 (2010)
Prasad, Y.V.R.K., Sastry, D.H., Vasu, K.I.: Scr. Metall. **3**, 771 (1969)
Prasad, K.E., Das, B., Maitra, U., Ramamurty, U., Rao, C.N.R.: Proc. Natl. Acad. Sci. **106**, 13186 (2009)
Prasad, Y.V.R.K., Sasidhara: Hot Working Guide—A Compendium of Processing Maps. ASM International (1997)
Preston, D., Natarajan, K.A., Sathyanarayana, D.N., Ramananda Rao, G.: Appl. Environ. Microbiol. **59**, 4051 (1993)
Raichur, A.M., Basu, M.J.: Sep. Purif. Technol. **34**, 121 (2001)
Raj, R., Surappa, M.K., Sudarshan: Process for preparation of nano ceramic-metal matrix composites and apparatus thereof. Patent No. US8540797B2 Filed 2011-05-12; Published 2013-09-24
Ramachandran, V., Krishnan, A.A.: Trans. Indian Inst. Metals **13**, 323 (1960)
Ramachandran, E.G., Sreenivasan, N.: Trans. Indian Inst. Metals **7**, 174 (1953)
Ramamurthy, P.C., Mallya, A.N.: Conducting polymer resistive based sensor for detection of nitrate ion in water Application No: 4639/CHE/2015. Date of filing: 02/09/2015, Patent No. 322412, Issued dated: 07/10/2019
Ramanarayan, H., Abinandanan, T.A.: Acta Mater. **52**, 921 (2004)
Ranganathan, S., Hajra, J.P.: Metall. Trans. **19B**, 649 (1988)
Rao, B.S., Jayaram, V.: Acta Mater. **49**, 2373 (2001)
Rath, R.K., Subramanian, S., Laskowski, J.S.: Langmuir **13**, 6260 (1997)
Ravi, R., Prasad, Y.V.R.K., Sarma, V.V.S.: J. Mater. Eng. Perform. **12**, 646 (2003)
Ravichandran, K.S., Dwarakadasa, E.S., Kishore: Mater. Sci. Eng. **83**, 11 (1986)
Sampath Kumar, T.S., Mallya, R.M., Hegde, M.S.: Appl. Surf. Sci. **35**, 63 (1988)
Santhiya, D., Subramanian, S., Natarajan, K.A., Malghan, S.G.: J. Colloid Interface Sci. **216**, 143 (1999)
Saravanan, R.A., Surappa, M.K.: Mater. Sci. Eng. A **276**, 108 (2000)
Seetharaman, S., Abraham, K.P.: Scr. Metall. **3**, 911 (1969)
Shabadi, R., Kumar, S., Roven, H.J., Dwarkadasa, E.S.: Mater. Sci. Eng. A **364**, 140 (2004)
Sindhu, S., Shishir, K., Krishna, B., Madras, G., Raghavan, S., Ramamurthy, P.C.: ACS Nano **10**, 6501 (2016)
Singh, V., Gupta, G.S., Sarkar, S.: Chem. Eng. Sci. **62**(22), 6102 (2007)
Sivan, V., Iyengar, G.N.K.: Trans. Indian Inst. Metals **29**, 83 (1976)
Somaiah, N., Kumar, P.: J. Appl. Phys. **124**, 185102 (2018)
Sreeranjini, P., Balaji, S., Rangarajan, A., Raichur, A.M.: ACS Appl. Mater. Interfaces **8**, 23437 (2016)
Srikanth, S., Jacob, K.T.: ISIJ Int. **29**, 171 (1989)
Surappa, M.K.: Sadhana **28**, 319 (2003)
Vamsi, K.V., Karthikeyan, S.: Intermetallics **132**, 107124 (2021a)
Vamsi, K.V., Karthikeyan, S.: J. Alloy. Compd. **860**, 158411 (2021b)
Varughese, S., Kiran, M.S.R.N., Ramamurty, U., Desiraju, G.R.: Angew. Chem. Int. Ed. **52**, 2701 (2013)
Vasanthakumar, B., Ravishankar, H., Subramanian, S.: PLoS One **7**, e39316 (2012)
Viswanathan, N.N., Srinivasan, M.N., Lahiri, A.K.: Ironmak. Steelmak. **24**, 476 (1997)
Yelloji Rao, M.K., Natarajan, K.A.: Int. J. Miner. Process. **27**, 95 (1989)

Chapter 23
Metallurgical Education in India: College of Engineering Pune's (COEP) Contribution

P. P. Deshpande

Abstract The College of Engineering Pune (COEP) is one of the oldest engineering colleges in India and was established in 1854 as the Poona Engineering Class and Mechanical School, with the primary objective to provide technical education to the subordinate officers in the Public Works Department. This article highlights the Department of Metallurgy and Material Science, which was established in 1948 under the leadership of Late Prof. G. K. Ogale. It was established to cater the needs of the country initially for steel manufacturing and development and has diversified itself into all domains of materials science and engineering. Since attaining autonomous status in 2003, COEP has been recognized by the World Bank-funded project as the best-governed institute in India, and has grown with its association with industry and professional bodies. The COEP is focussed on development of newer materials technologies while producing professionally skilled, knowledgeable human resources for the country.

Keywords COEP · Metallurgy · Materials science · Materials engieering · Pune · Maharashtra

---

P. P. Deshpande (✉)
Department of Metallurgy and Materials Engineering, COEP Technological University (COEP Tech), (A Unitary Public University of Government of Maharashtra), Formerly College of Engineering Pune, Pune, India
e-mail: ppd.meta@coeptech.ac.in

R. Divakar et al. (eds.), *Indian Metallurgy*, Indian Institute of Metals Series,
https://doi.org/10.1007/978-981-99-5060-7_23

23.1 College of Engineering Pune (COEP)

Prior to the nineteenth century, systematic investment in human resource development was not considered important in any country. Expenditures on education, training and other similar forms of investment were negligible. The scenario began to change during the last century with the application of science to the development of new products, first in Great Britain, and then gradually in other countries.

In such an era, the Poona Engineering Class and Mechanical School was established in July 1854 in Pune—cultural capital of Maharashtra state. It is the third oldest engineering college in India, after College of Engineering, Guindy, Chennai and IIT, Roorkee.

The objective was to provide technical education to the subordinate officers in the Public Works Department. The school became "Poona Civil Engineering College" during the course of time and later on the nomenclature was changed to the "College of Engineering, Poona". It was initially affiliated to the University of Bombay for a degree of Licentiate in Civil Engineering and later to the University of Pune. In the year 2003, the college got autonomous status, thus giving it the freedom to set its own curricula and manage its own finances. This has been the most significant change as far as pedagogy at COEP is concerned. COEP has been recognized by the World Bank-funded project as the best-governed institute in India. The institute is ranked 50th amongst the top 100 institutions under the category of Engineering, as per the National Institute Ranking Framework (NIRF). As of now, COEP has

11 academic departments and 2 Centres of Excellence namely Signal and Image Processing and Smart Renewal Energy Systems supported by TEQIP. The Institute offers 9 Under-graduate courses in engineering and 25 Post-graduate courses in engineering and 1 post-graduate programme in Business Administration, besides Ph.D. degree programmes in all the disciplines and One year PG Diploma course in ERP, Additive Manufacturing, Rail and Metro Technology, Electric Mobility, Data Science and Artificial Intelligence and Integrated Product Design and Development. COEP has more than 4500 students enrolled in its various courses. With 166 years of history in producing quality engineers, COEP has a rich Alumni base that has significantly contributed to the nation-building process. Alumni like Bharat Ratna Sir Mokshagundam Visvesvaraya; Prof. Thomas Kailath, Padmabhushan awardee; Prof. C. K. N. Patel, recipient of IEEE Medal of Honour; Dr. Vijay Kelkar, Padmavibhushan awardee; Late Shri. B. G. Shirke, Padmashree awardee; Mrs. Leela Poonawala, Padmashree awardee; Dr. Sanjay Dhande, Padmashree awardee; Mr. V. R. Katti, Padmashree awardee Prof. Ramesh Raskar, Lemelson awardee are few among those starward alumni of COEP who contributed immensely in the infrastructural development of the country. Life at COEP is not just limited to the classrooms; an extensive number of clubs and other social activities give students a flavor of what it is really like on this campus.

23.2 Department of Metallurgy and Material Science

The department of Metallurgical Engineering was established in 1948 under the leadership Late Prof. G. K. Ogale, who was a renowned metallurgist, a great visionary and institution builder of that time. It was renamed as department of Metallurgy and Material Science in the year 2004. Intake capacity of the department is 216 students in under graduate and 72 students in post graduate courses. Previously the department was offering an M. Tech. (Physical Metallurgy) course since 1957. The last two decades saw the remarkable development of new materials and technology. The department incorporated all these changes in the newly designed curriculum and designed M. Tech. Materials Engineering course to prepare the students for advances in technology. Based on the application made by the department, AICTE approved the change in nomenclature of M. Tech. Physical Metallurgy to M. Tech. Materials Engineering from the academic year 2016–17. The department was established to cater to the needs of the country initially for steel manufacturing and development. Now, it has diversified itself into all domains of materials science and engineering. The field of materials science seeks to explain and control the four basic elements: The structure and composition of a material as well as its synthesis and processing determine the resulting properties and performance. Merton C. Flemings and Praveen Chaudhari used these basic ideas to construct a tetrahedron (Fig. 23.1).

It will be noted that no particular material is mentioned in the materials tetrahedron. It applies equally well to ferrous and non-ferrous metals, ceramics, polymers and composites. Keeping this in mind, the department is actively involved in

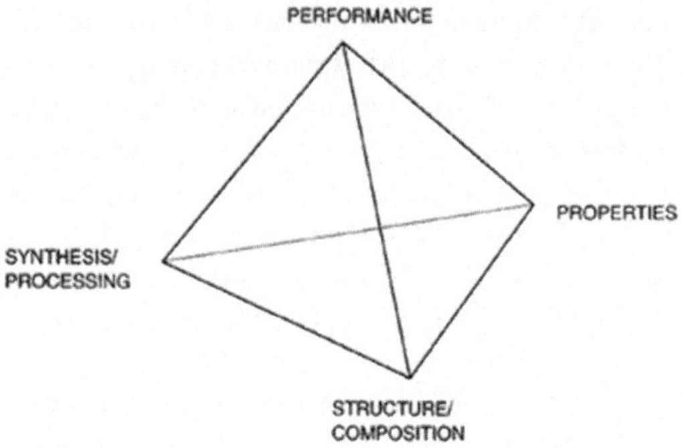

Fig. 23.1 Materials tetrahedron

the research and development activities in the areas of microwave dielectric, electroceramic, solid electrolytes, sodium-ion batteries, thermoelectric and electronic materials, glasses and glass ceramics, high entropy alloys, energy materials, inherently conducting polymers, aluminium base composite materials, polymer micro and nano-composites, thermomechanical processing, powder metallurgy, photo-catalysis and CO_2 absorption using ceramics, cryogenic treatment, tribology, alloy synthesis, metallurgical waste management, corrosion control and prevention, coating development, hot corrosion, surface engineering, laser material processing, materials joining, foundry technology, physical metallurgy of ferrous and non-ferrous alloys, metal working processes, light metal alloys, metal forming, iron and steel making, ferro alloys, flat rolling, galvanizing and integrated steel plant processing. The unique properties of materials due to nanoparticle size were recognized in the later half of the last century. The department has included a course on nanotechnology and has been working on nanotechnology-based research projects. The projects in all these research areas mentioned, received major funding from various government agencies such as DST, AICTE, BRNS, IGCAR, UGC, INSA, ISRO, DRDO and University of Pune. The department is a recognized Ph.D. centre affiliated to Savitribai Phule Pune University. Every year, on an average twenty research papers are being published by the department faculty in peer-reviewed national/international journals. The department faculty also received commercially important patents in the last few years. Students of the department excelled in all fields of materials science and engineering in India and aboard.

23.3 Professional Societies Activities

The department organizes various activities of the professional societies like Indian Institute of Metals IIM, ASM, PMAI, etc. The Indian Institute of Metals Pune chapter was established in the year 1947 and all staff members of the department are involved in the IIM Pune chapter. IIM NMD ATM 1976 and IIM NMD ATM 2014 were held in the department. Recently, EH-TACAG-17 was organized in association with Indian

Ceramic Society. The most recent was Asian Powder Metallurgy Association 2019 international conference. Apart from that PMAI "Powder Metallurgy Short Course", IIM "Dara P. Antia Memorial Lecture" and "Brahm Prakash Memorial Quiz", ASM "Materials Camp" are a few of the regular activities of the department. IIM Pune chapter was also co-organizer of IIM NMD ATM, organized by IIM Mumbai chapter and IIM Goa chapter, BITS, Pilani (Goa Campus) in Nov 2017.

23.4 Industrial Collaborations

Many industries including Bajaj Auto, John Deere, Eaton, Tata Motors regularly avail the department facilities for testing and consultancy. The department has signed a memorandum of understanding with various industries. EATON India engineering center, Pune established a state of art laboratory for advanced materials research in the department.

23.5 Way Ahead

Very recently, the department has received financial support to the tune of Rs. 5 crores from Firodiya Foundation and ALUCAST, the new setup is going to be named as N. K. Firodia School of Metallurgy and Materials Science. With the passage of time, educational wisdom, professional skills and acquisition of knowledge have become crucial factors in achieving productivity. One can even call the present century the "Age of Human Capital" in the sense that the determinant of a country's standard of living is how well it succeeds for the betterment of its population. The department is working in this direction. Its search for new and better materials technologies is never-ending.

Chapter 24
Metallurgy at PSG-Coimbatore: An Overview

J. Krishnamoorthi

Abstract PSG Tech is a renowned engineering institution established by the PSG & Sons' Charities Trust in 1951, with an effective industry-institute interaction. The Metallurgical Engineering program was established in 1968 as a part of the Department of Mechanical Engineering and subsequently as an independent department since 1974. PSG Tech's Department of Metallurgical Engineering is presently researching in the areas of Biomaterials, Ceramic Materials and Bainitic Steels. They have established a Centre of Excellence in Welding Engineering and Technology funded by the Department of Heavy Industry, Govt of India. The article discusses the history and progress of the Department of Metallurgical Engineering at PSG College of Technology in Coimbatore, India.

24.1 Introduction

PSG College of Technology, a renowned icon of engineering education, is one of the foremost institutions founded by the PSG & Sons' Charities Trust (1926) in the year 1951 at Coimbatore. The founder Principal Dr. G. R. Damodaran was instrumental in the planned growth of the institution from the humble beginnings to the present status of a world-renowned technological institution in about seven decades. An effective industry-institute interaction was ensured from the beginning, as the College is housed on the campus of PSG Industrial Institute, well known for PSG pumps.

J. Krishnamoorthi (✉)
Department of Metallurgical Engineering, PSG College of Technology, Coimbatore, TN 641004, India
e-mail: Krish@gmail.com; hod.metal@psgtech.ac.in

R. Divakar et al. (eds.), *Indian Metallurgy*, Indian Institute of Metals Series,
https://doi.org/10.1007/978-981-99-5060-7_24

The college, fondly known as PSG Tech, is extremely proud of its alumni, a considerable number of them being entrepreneurs or senior executives in industries both within India and abroad. Some of them are holding prestigious positions like Chief Executive and Managing Director and also as Chairman of various disciplines in universities abroad. A good number of alumni occupied the position of Vice Chancellor in various reputed universities in India. Several educational institutions have also been established by PSG Tech alumni.

24.2 Commencement of Metallurgical Engineering @ PSG Tech

Metallurgical Engineering was introduced at PSG Tech as an undergraduate programme, BE (Metallurgical Engineering), under the Department of Mechanical Engineering in the year 1968. The programme got its own identity when Department of Metallurgical Engineering was established in the year 1974 with Dr. N. K. Srinivasan as Head of the Department. PSG & Sons' Charities Metallurgy and Foundry division set up in the same year at Coimbatore ensured the industry-oriented academics in the Department. The foundry division caters the casting needs of engineering industries with its wide range of Gray cast iron, SG iron and Ni-hard castings.

From the year 2008, the student's intake for BE (Met Engg) was raised from 30 to 60 to cater to the industry recruitments. The BE Metallurgical Engineering programme of the Department has been accredited for 6 years by the National Board of Accreditation (NBA) in the year 2018. The postgraduate programme, ME (Industrial Metallurgy) was started in the year 1976 with an intake of 18. The Department has doctoral research (PhD in Metallurgical Engineering) in full time and part time basis. As of date, more than 50 research scholars have received their PhD degree through the Department and another 35 candidates are pursuing their research work.

Under the MoUs signed with BHEL—Educational Society, Tiruchirappalli, a PG certificate programme on "Welding and Quality Engineering" is run by the Department since 2009. A blend of theoretical learning on welding (at PSG Tech) with hands-on learning in welding shop floors (BHEL, WRI, Trichy) along with industry-related project work enables the students to be ready for industrial jobs.

24.3 Faculty and Visiting Professors, the Vigour of the Department

Department possesses excellent faculty with good academic & research records all over the years. A few faculty also have industrial experience. Apart from regular faculty, visiting professors were the major assets to the Department who contributed to the growth of the Department during their visiting tenure. Starting from Dr. Placid Rodriguez (IGCAR), Dr. Baldev Raj (IGCAR), Dr. P. C. Angelo (DRDO), Dr. Abhijit Dutta (DMRL), Dr. V. Ramaswamy (SAIL), Dr. K. Elayaperumal (BARC), Dr. S. L. Mannan (IGCAR) and Dr. R. Gopalan (ARCI Chennai) were instrumental in introducing latest trends in different areas of metallurgical and materials engineering to the students and faculty, initiating sponsored R&D projects and setting up of new facilities in laboratories. Ministry of Steel (MoS), Government of India had sanctioned MoS chair professor for the Department and MoS students' scholarship during 2013–2016.

24.4 Research Advancements

The tradition of working on sponsored projects was initiated in the Department with a project from Centre for Aeronautical Systems, Studies & Analysis (CASSA), Bangalore on 'Studies on failure of aluminium 7075 components used in aerospace applications' in the year 1999. From then to the present, about 50 projects have been received from sponsoring agencies like DRDO, DST, AICTE, UGC, BRNS, ISRO, SAIL, IGCAR, AR&DB, Tata Steel and NRB to the tune of to several crores. From these projects, major processing and characterization facilities were installed in the Department laboratories. The outcomes of the projects were published in reputed, high-impact factor journals.

Recently, a Centre of Excellence (COE) in Welding Engineering and Technology funded by Department of Heavy Industry, Govt of India was established in the Institution with a sanction of Rs 26.7 crores. This COE is aimed for three different technologies namely development of automated welding systems, development of waveform-controlled welding power sources and welding consumables development. The centre houses power sources for welding, welding robots, Gleeble thermo-mechanical simulation system, Sysweld welding simulation software and welding electrode extruder pilot plant.

Based on the faculty expertise and the facilities available in the Department, major research work being carried out are in the following areas:

- Biomaterials
- Ceramic materials
- Development of bainitic steels
- Development of welding consumables
- Diffusion bonding
- Foundry metallurgy
- Functionally graded ceramics
- High nitrogen stainless steels
- Heat treatment

- High strength steels
- High temperature ceramics
- High temperature coatings
- Materials modeling
- Metal matrix composites
- Mechanical alloying
- MIAB welding
- Nanomaterials
- Phase field modeling
- Powder metallurgy
- Semisolid casting
- Severe plastic deformation
- Surface modifications
- Super austenitic stainless steels
- Thermomechanical simulation
- Weldability studies
- Weld simulation

24.5 Industry—Institute Interaction

Apart from PSG industrial institute and PSG Sons' and Charities Metallurgy and Foundry division, the department is in active interaction with industries in terms of testing and consultancy activities. Several engineering manufacturing industries in and around Coimbatore, as well as in Tamil Nadu and in different parts of the country regularly visit the Department for their testing requirements. Recently, a major consultancy work from National High Speed Rail Corporation Ltd, Mumbai has been taken up for assessment of weld quality, review of WPS and welder qualification procedures.

24.6 Alumni of the Department

From the beginning of the Department, the students who have passion for working in metal industries or becoming an entrepreneur in metal industry opted for BE (Metallurgical Engg) and ME (Industry Metallurgy) at PSG TECH. Several of the alumni of the Department started their own foundry or other metal industry in Coimbatore as well as in different places in India. Several alumni rose to the top managerial positions in industries in public and private sectors. Many alumni also got positions in major research establishments namely VSSC, ISRO, IGCAR, BARC, NML, DMRL and DRDO. A few alumni who had chosen teaching as well as research are faculty members in IITs and NITs. The Department alumni are not only within the country but also around the world holding prestigious profession in industries, R&D organizations and academics.

24.7 The IIM Coimbatore Chapter

The Indian Institute of metals, Coimbatore chapter which is being nourished by the Dept of Metallurgical Engg is nearly five decades old. The chapter organizes various technical events for the benefit of its members and student friends. Wootz, a students' technical symposium is a flagship event organized in association with Metallurgical Engineering Association since 2009. In recent years, Coimbatore chapter has been playing definite role in the national council of IIM. IIM Coimbatore chapter bagged the best chapter award in the years 2011, 2012, 2013, 2014, 2015 and 2019 during NMD celebrations.

A crowning glory of the chapter was the successful conduction of IIM—NMD—ATM 2015 at Coimbatore. The 53rd National Metallurgists' Day (NMD) celebrations under the aegis of the Ministry of Steel and the 67th Annual Technical Meeting (ATM) of the Indian Institute of Metals were held at Hotel LeMeridien from 13 to 16th November 2015. Mr. Sajjan Jindal, Chairman of JSW Steel inaugurated the International Symposium Ms. Aruna Sundarrajan, Secretary and Mr. Sunil Barthwal, Joint Secretary, Ministry of Steel were also present for the symposium. The prestigious NMD awards were presented to the Awardees by the Hon'ble Union Minister for Steel & Mines Shri Narendra Singh Tomar. Mr. S. S. Mohanty, President of IIM and Dr. V. Ramaswamy, Professor were instrumental in organizing the programme. About 900 delegates participated in the event.

The students' affiliate chapters attached to the IIM Coimbatore chapter regularly organize guest lectures, webinars, workshops, quizzes to student community on topics of academic, R&D and industrial relevance to promote students' interest in taking up R&D, academic as well as entrepreneurship as their career.

24.8 Summary

PSG College of Technology, founded by the PSG & Sons' Charities Trust in 1951, is a renowned engineering institution with strong industry-institute interaction. The Department of Metallurgical Engineering was established in 1974 and has since produced numerous successful alumni who have established their own institutions or hold prestigious positions in various universities and industries. The research focus areas include biomaterials, metal matrix composites, ceramic materials, mechanical alloying, and nanomaterials, among others. Additionally, the department has received several sponsored projects and established a Centre of Excellence in Welding Engineering and Technology.

Chapter 25
Sixty Years of Metallurgy at IIT Madras

S. Sankaran, Uday Chakkingal, and Ravi Kumar

25.1 Introduction

The Department of Metallurgical & Materials Engineering is one of the oldest departments of IIT Madras, established in the same year as the Institute (1959). In the first few decades of its existence, then known as the Department of Metallurgy, it focused more on industrial metallurgy and contributed immensely to metal casting, forming, and welding. Over the past few decades, academic and research interest in non-metallic materials has grown globally. In tune with this change, the Department changed its name to the Department of Metallurgical and Materials Engineering to adapt to the transformations and expectations worldwide. The Department currently has a wide range of expertise in many areas of metallurgy and materials science & engineering. It continues to excel in all areas including the traditional ones.

From the early years of its existence, the department was internationally recognized as a center of learning and research in Metallurgy. Professor E. G. Ramachandran (EGR) was the first Head of the Department. Under his leadership, many research areas flourished, notably physical metallurgy, materials characterization, metal forming, and metal joining. Even to this day, EGR's legacy continues.

The following sections are devoted to the contributions of three individuals, namely, Prof. K. A. Padmanabhan (mechanical behavior), Prof. P. Venugopal (metal forming), and Prof. R. Vasudevan (X-ray diffraction) starting with a brief introduction of Prof. E. G. Ramachandran, the first head of the department.

S. Sankaran · U. Chakkingal · R. Kumar (✉)
Department of Metallurgical & Materials Engineering, Indian Institute of Technology-Madras (IIT Madras), Chennai 600 036, India
e-mail: nvrk@iitm.ac.in

R. Divakar et al. (eds.), *Indian Metallurgy*, Indian Institute of Metals Series,
https://doi.org/10.1007/978-981-99-5060-7_25

25.2 Professor E. G. Ramachandran (EGR)

EGR obtained his Ph.D. in 1947 at the very young age of 22 years from the University of Sheffield, England. Before joining IIT Madras, he served the Department of Metallurgy, IISc, Bangalore, as a faculty member for 9 years and National Metallurgical Laboratory, Jamshedpur as Assistant Director for 5 years. EGR joined IIT Madras in 1961 as the first Professor of the Department of Metallurgical Engineering and as the first head of the department. He was also the Director-in-Charge of IIT Madras for a brief period. He superannuated from IIT Madras in 1985. He was instrumental in establishing the industrial metallurgy division in the Indian Institute of Metals and later became its President in 1980. In recognition of his contributions, the EGR distinguished lecture series was started in 2013 with a generous financial contribution from one of his students Mr. Pukhraj and with efforts from Prof. B. S. Murty.

25.3 Professor K. A. Padmanabhan

Professor K. A. Padmanabhan received the "FORSCHUNGSPREIS" (Career Research Award) of the Humboldt Foundation, Germany and the higher 'Sc D' degree from the University of Cambridge, for his highly original research. His models for optimal superplasticity and inverse Hall–Petch effect in different classes of materials are described as "break-through from India" by "Nature Materials". His model for grain boundary sliding and its application to superplastic flow in microcrystalline metals and ceramics questioned the "central dogma" in high-temperature deformation modeling till that point that Grain Boundary Sliding is an inherently fast process, which at no stage could be controlling the rate of high-temperature deformation. In contrast, rate controlling boundary sliding process (see Fig. 25.1) could lead to mesoscopic boundary sliding/the formation of plane interfaces and give rise to superplasticity. With such an approach, this near-ubiquitous phenomenon, which is present in different classes of materials like pseudo-single phase and microduplex metals and alloys, ceramics, composites, dispersion strengthened materials, intermetallics, nanostructured materials, bulk metallic glasses, geological materials, etc. could be explained on a common basis.

His model for the inverse Hall–Petch effect in nanocrystalline and quasi-crystalline materials extends the ideas contained in the above model to explain why when the grain size goes below a value of ~10–15 nm, there is grain size softening, rather than the usual increase in hardness with decreasing grain size. His work for the Department of Atomic Energy was concerned mainly with deformation, fracture, stress corrosion cracking, fracture, and low cycle fatigue of AISI 304, 304 LN, 316,

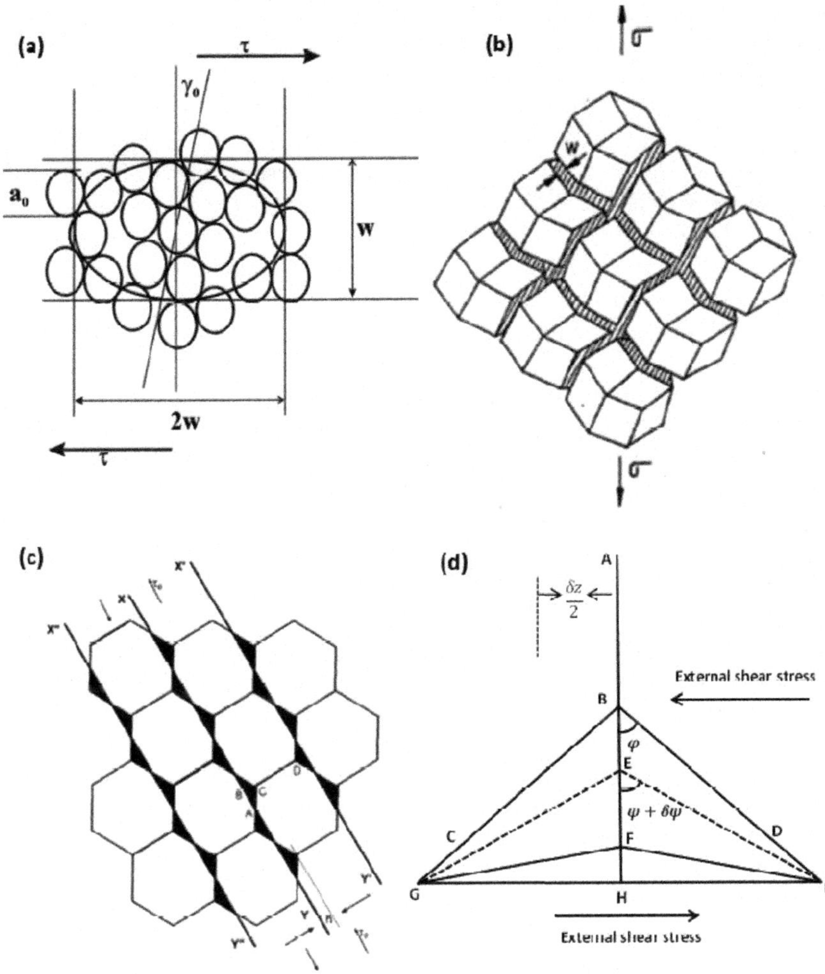

Fig. 25.1 Concept of superplasticity and models **a** Padmanabhan and Gleiter (2012), **b** Bhattacharya and Padmanabhan (1989), **c** Padmanabhan and Gleiter (2004), **d** Padmanabhan and Gleiter (2004)

and 316 LN austenitic stainless steels and Mo-containing grade D9, of vital interest as construction materials for the Fast Breeder Reactor program. He developed a procedure for the superplastic forming of hemispherical domes of Ti–6Al–4 V alloy of the minimum wall thickness of 4 mm for the Defence Research and Development Laboratory in the early years of the Integrated Missiles Development Program. Prof. Padmanabhan developed for the Liquid Propulsion Systems Centre, Indian Space Research Organization, an Al-alloy equivalent to the French alloy AFNOR 7020, needed for making water tankages. Production was done through an ordnance factory.

He also set up at Vikram Sarabhai Space Centre, Thiruvananthapuram a facility to superplastically form hemispheres of alloy Ti–6Al–4 V of 410 mm diameter.

He was a consultant to the R & D Centre for Iron and Steel, Steel Authority of India Limited (SAIL), when they produced for the first time in the country both Extra Deep Drawing (EDD) and Liquid Petroleum Gas (LPG) grades of steel. He developed a model for predicting the formability of steels under different stress states. He was a consultant to Tata Engineering and Locomotive Company (now known as Tata Motors) when they introduced the microalloyed ferrite–pearlite steel (first generation), 49MnVS3, for the first time in India for the forging of crankshafts, just a year or two after it was introduced by Daimler Benz of Germany in their vehicles. Most recently, he developed a multi-stage closed die forging technology for the production of undercarriage base plate fitting components for the airframe of a high-performance (supersonic) Aircraft for the Indian Aeronautical Development Agency. In all, ten technologies developed by him, and members of his research group are used in Indian industries.

25.4 Professor P. Venugopal

Professor P. Venugopal was associated with the Metal Forming Laboratory (now renamed as Materials Forming Laboratory) from 1968 till 2014. He was involved in setting up the lab in 1968 along with his guide and mentor, Prof. Hans Wilfried Wagener from the University of Kassel, Germany. (A photograph of the founder of the Metals Forming Lab, Prof. Wagener with Prof. Venugopal is shown in Fig. 25.2).

Professor Venugopal is an accomplished expert on the analysis of presses and hammers, machine selection, and metal forming tool design. His initial research on the efficiency of presses and hammers is considered groundbreaking and contributed immensely to industrial knowledge on press selection for different forming applications. A laboratory setup for determining the overall efficiency of a direct-driven friction screw press is shown in Fig. 25.3.

Prof. Venugopal is well known for his work on cold extrusion of titanium and sintered preforms of steel. A schematic of cold extrusion tooling designed by him is shown in Fig. 25.4.

He has published his research extensively in various metal forming and materials science journals. He has guided a large number of M.S., M. Tech., and Ph. D. students specializing in the area of metal forming besides numerable B.Tech. students. He is the recipient of several awards, like the FIE "FRI" Foundation National Special Achievement Award for Engineering and Technology besides many best paper awards. He has been Visiting Professor at Universities in Germany, Malaysia, Singapore, and Thailand. He has published more than 102 research papers on various aspects of metal-forming technology. He was invited to serve as a Member of the Advisory Board of the Annual Event of the Asia–Pacific Conference of J. Mechanical Working Technology for 12 years representing Asia. Professor Venugopal retired in 2006 and was invited by IIT Madras as Institute Professor Emeritus to join in 2007. He

Fig. 25.2 Prof. Hans Wilfried Wagener with Prof. Venugopal at Kassel, Germany in 1985

Fig. 25.3 Set up for determining overall efficiency of a direct driven friction screw press (1. Furnace to heat the job up to 1000^0C for hot upsetting, 2. Top hitting die, 3. Press ram, 4. Press frame, 5. Temperature recorder, 6. Temperature controller, 7. Carrier frequency amplifier, 8. 36 Channel moving coil galvanometer)

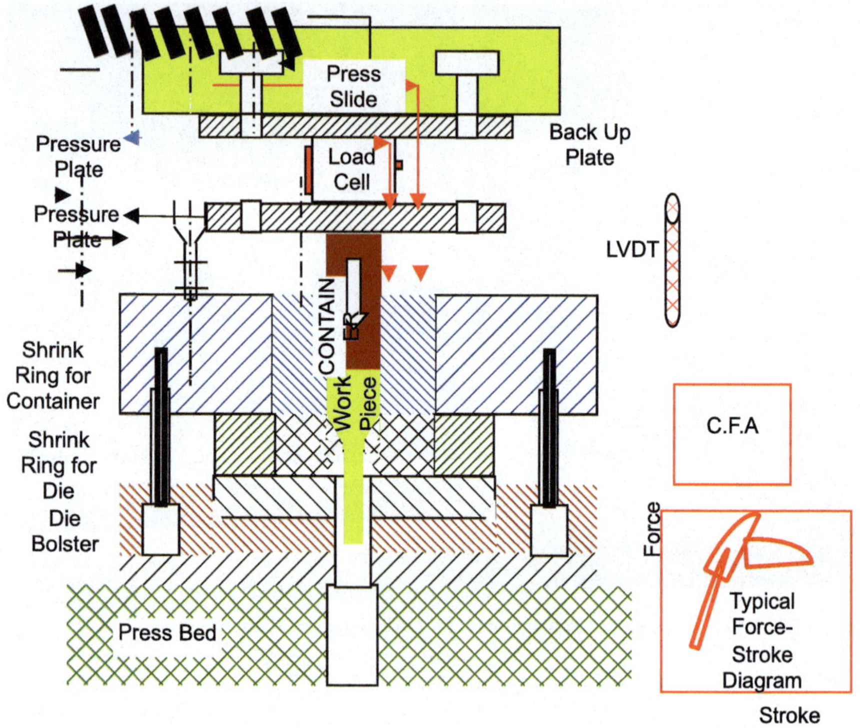

Fig. 25.4 Schematic of the experimental set up for cold extrusion process

served as Professor Emeritus till May 2014. He is an Honorary Technical Director of a Cold Extrusion Company in Chennai and has been a Retainer Consultant to reputed Metal Forming Companies. Professor Venugopal laid the foundations for metal-forming research at IIT Madras.

25.5 Professor Vasudevan

The Central X-ray Diffraction Laboratory (CXRD) in IIT Madras was established in the late 1960s. An unstabilized small X-ray unit and a 57.3 mm Debye–Scherrer camera were all that were available until 1969. Professor Dr. –Ing. R. Vasudevan was asked by Prof. Arcot Ramachandran in 1969, the then director of IIT Madras to compile a list of sophisticated XRD facilities that would help serve the needs of all the departments and the allotted budget was in the range of DM 200,000/- to 300,000/- (German Deutsche Marks). Two stabilized X-ray units, a powder diffractometer, D-S cameras of 114.6 mm diameter, a Weissenberg camera along with a Kratky small angle scattering unit were procured. With the excellent assistance provided by Philips

Fig. 25.5 Prof. Kesavan Nair (L) with Prof. R. Vasudevan (R) (picture taken in April 2013, courtesy Dr. Ravi Sankar Kottada). The image on the right hand side is an old residual stress analyser (Rigaku make, ~30 yrs old) which was used to measure residual stresses in central XRD laboratory

India, the laboratory was fully commissioned and functional within a month of the arrival of the equipment in India in the month of March 1973.

Professor Eckard Macherauch from the University of Karlsruhe, a recognized expert in the field of Residual Stress Measurements by XRD techniques was visiting India in 1973 and spent a period of one month in IIT Madras. During this time, the Central XRD Laboratory was introduced to the field of residual stress measurements using XRD and in due course added a Rigaku unit designed for this kind of study (see Fig. 25.5).

Kesavan Nair and Pathiraj who were then working with Prof. Vasudevan were pursuing their doctoral theses and carried out all the work coming to the laboratory in addition to attending to their own work. Dr. Kesavan Nair eventually joined the Department of Metallurgical and Materials Engineering, IIT Madras as a faculty member and continued to head the Central XRD Laboratory until his retirement in 2013. Dr. Pathiraj works now in the Faculty of Engineering Technology at the University of Twente, the Netherlands.

In addition to regular research and student work, the laboratory has assisted a large number of industries in the past in solving their technical problems, especially those connected with the analysis of residual stress and/or retained austenite measurements including phase analysis. Workshops and seminars were often conducted to assist the Indian industries in the areas of heat treatment, retained austenite, and residual stress measurements. Figure 25.6 exemplifies a seminar conducted in 1995 on the treatment of engineering components. The laboratory has also assisted medical doctors in their research work on the analysis of kidney stones of patients and has developed copyrighted software for the rapid analysis of kidney stone composition (Private communications with Prof. R Vasudevan; Private communications with Prof. Kesavan Nair).

Fig. 25.6 Prof. E. G. Ramachandran (left image) during a seminar on thermal treatment of engineering components in 1995

The Central XRD Laboratory continues to interact with a large number of industries within and as well as outside the country, offering consultancy services related to metallurgical failure analysis, phase analysis, and estimation of retained austenite and residual stresses using X-ray diffraction techniques.

25.6 Conclusions and Outlook

The Department is on an upward growth trajectory, with a progressively improved presence in all sectors, where metallurgical and materials research is essential. In a world that is racing toward becoming a knowledge economy, which also sees the need for converging technologies and flexible learning, we are acutely aware of the need to create human resources that can contribute to the most critical problems society faces. We aim to contribute to cutting-edge manufacturing, energy, sustainability, environmental protection, transport, electronics, healthcare, and many others, where materials skills will be directly relevant. The recent resurgence in data science and machine learning has created many opportunities for metallurgists and materials engineers to address technological challenges in a newer way.

With dynamic faculty members, dedicated staff, motivated students, and devoted alumni, we are poised for a period of extraordinary growth. The Department, with its commitment to continued growth and learning, and passion for contributing to society through education, training, and research, is poised for consistent growth in the foreseeable future.

References

Bhattacharya, S.S., Padmanabhan, K.A.: Trans. Indian Inst. Metals 42 (supplement) (1989)
Padmanabhan, K.A., Gleiter, H.: Mater. Sci. Eng. A 381 (2004).
Padmanabhan, K.A., Gleiter, H.: Curr. Opin. Solid State Mater. Sci. 16 (2012)

Private communications with Prof. R Vasudevan
Private communications with Prof. Kesavan Nair

Chapter 26
The Journey of the Department of Metallurgical and Materials Engineering, NIT Durgapur, Towards Achieving Academic Excellence

M. M. Ghosh, Satadipa Banerjee, Hrishikesh Kumar, and K. S. Ghosh

Abstract This article describes the Department of Metallurgical and Materials Engineering (MME) of the National Institute of Technology (NIT) Durgapur in India, which has been a significant center for metallurgical and materials teaching and research since its inception in 1960. The department currently offers B.Tech., M.Tech. and Ph.D. programs in Metallurgical and Materials Engineering, and has expertise in areas such as Extractive Metallurgy, Physical Metallurgy, Manufacturing Processes, Nano-Science and Nano-Technology, Modelling and Simulations, Powder Metallurgy, Advanced Materials, and Corrosion Engineering. The department has played an important role in metallurgy and materials-related research and development, and its faculty members have interacted with important academic institutes and organizations.

26.1 Introduction

> Engineering and Technology is All About Materials and Transforming Them, to Make Life Better for People—Anonymous

Ever since the inception of the Department of Metallurgical and Materials Engineering (MME) of the National Institute of Technology (NIT) Durgapur, it has remained a paramount centre for metallurgical and materials teaching and research in the eastern region, as well as the whole of India. The department started its journey from the inception of the Institute in the year 1960. Earlier, the department's name was Metallurgical Engineering of Regional Engineering College (REC) Durgapur under the University of Burdwan, West Bengal. The college was converted into a National Institute of Technology (NIT), Durgapur under the purview of the Ministry

M. M. Ghosh (✉) · S. Banerjee · H. Kumar · K. S. Ghosh
Department of Metallurgical and Materials Engineering, National Institute of Technology, Durgapur 713209, India
e-mail: madanmohan.ghosh@mme.nitdgp.ac.in

R. Divakar et al. (eds.), *Indian Metallurgy*, Indian Institute of Metals Series,
https://doi.org/10.1007/978-981-99-5060-7_26

of Human Resource Development (presently, Ministry of Education) of the Government of India. The department started initially by offering B.E. in Metallurgical Engineering. The postgraduate (PG) programme started in the year 1966. Now, the Ph.D. programme is in full swing after the Institute was converted to NIT from RE College status in 2003. The department has played indispensable roles in various metallurgy and materials-related research and developments. It has fostered and groomed generations of young metallurgists with the help of hardworking eminent faculty members.

The department has maintained high-level interactions right from its emergence, with important academic institutes, organizations of the stature of CSIR and other national laboratories, steel plants in India including SAIL, and some organizations in India and abroad. The department provides testing and consultancy services to different institutions and industries in India. This esteemed department has been accredited by the National Board of Accreditation (NBA) over the years and also had the feather of the STEEL CHAIR PROFESSER, honoured by the Ministry of Steel, Government of India. The Department was rated one of the best in India along with the Metallurgy Department of the Indian Institute of Technology (BHU), Varanasi. Our department got funding and commissioned a Transmission Electron Microscope (TEM) from UNESCO. The TEM served not only the research activities of the department but also nearby Institutes such as the Bengal Engineering College, now Indian Institute of Engineering Science and Technology (IIEST), Shibpur, Indian Institute of Technology (IIT) Kharagpur, and Indian School of Mines (ISM) Dhanbad. The machine ran for a quite long time, but due to technological development and due to ageing it became inactive and non-operative. The machine has been kept as an Institute heritage and can be seen along with the department picture. The Indian Institute of Metals (IIM) which is the largest professional body of the metallurgical community in India has recently opened a Students Chapter in the department. Under the banner of this newly formed Chapter, the students can organize various technical events and can interact with eminent persons of the metallurgical and materials fraternity.

The department has steadily progressed and today it provides B.Tech. (Metallurgical and Materials Engineering), M.Tech. (Metallurgical and Materials Technology) and Ph.D. programmes. The department is a participating department of the Centre of Excellence (CoE) on Advanced Materials at NIT Durgapur under TEQIP- II and III. It is a DST-FIST-sponsored department that is presently in Phase-II and also has received funding from TEQIP. The number of faculty members of the department is currently thirteen (13) with supporting staff members four (04). The faculty members are actively involved in teaching and research. Over the years, the department has developed excellent expertise in the areas of Extractive Metallurgy, Physical Metallurgy, Manufacturing Processes, Mechanical Behaviour of Materials, Nano-Science and Nano-Technology, Modelling and Simulations, Powder Metallurgy, Advanced Materials, Corrosion Engineering, etc. The department aims to incorporate a new dimension of understanding and evaluation of newer materials and to establish itself as a pioneer of materials research in order to effectively cater to the rising needs of industries and institutes in India and abroad. The department is now equipped with facilities like XRD, FESEM, UTM (Instron), wear testing, mechanical testing, high-speed computation facilities, DLS, DSC, reducibility testing unit, and many others. The department looks forward to achieving ultimate excellence in various fields to provide significant contributions to the nation in the coming era.

Present view of the entrance of the department.

26.2 Brief History of the Department and the Institute

Predominantly in the early 60 s, by the act of parliament, eight (08) Regional Engineering Colleges (RECs) were established in the major states of India to nurture engineering graduates of remarkable merit. As a part of the plan, the Government of India in partnership with the Government of West Bengal founded one such REC in Durgapur to regale the needs of several major ongoing industrial projects.

The college has proven to serve the nation as a significant education centre providing the highest standard of technical education. The college was rechristened as National Institute of Technology (NIT) Durgapur in 2003 along with other RECs, and it was uplifted as a deemed university with a fully funded eminent technological Institution status administered by an autonomous board of governors, under the Ministry of Human Resource Development, Government of India. NIT Durgapur is the first Institute in eastern India to be selected as a lead Institute under the Technical Education Quality Improvement Program (TEQIP) of the Government of India, funded by the World Bank. In 2007, the Institute received a glorious honour when the Union Government of India declared it an "Institute of National Importance".

The Institute's origin, the department of metallurgical engineering started its journey in 1960, since the inception of the Institute. The Institute debuted with only four branches of engineering, metallurgy being one of them, which is now known as Metallurgical and Materials Engineering (since 2004). Today, it has transformed from "regional" to "national" and from "college" to "institute". With generous funding from the government and immense effort from active faculty members, the department has registered a profuse growth and improved significantly in all possible ways.

26.3 Vision and Mission of the Department

26.3.1 Vision

To prove and be accepted globally as a leading department and as a quality establishment for gathering and disseminating knowledge in the field of Metallurgical and Materials Engineering.

26.3.2 Mission

- To produce highly educated, well-rounded and motivated graduates possessing fundamental understanding and knowledge in Metallurgical and Materials Engineering, who can provide service and leadership to the nation and the world
- To pursue creative research and new technologies in Metallurgical and Materials Engineering in order to cater to the needs of industry, government and society
- To accomplish visibility by active participation in conferences and technical activities.

26.4 Journey Towards Academic Excellence Over the years

The department has been working to achieve its mission with diligent and industrious faculties from its genesis. Industry-Institute-Interactions (I-I-I), research, academic growth, etc. were seen right from the early 60 s to the present day. The department has proved itself to be a place of tremendous significance for guiding numerous young engineers and researchers as well as serving aid to various organizations in real-time problems.

26.4.1 Academic Programmes

Over the years the intake capacity of UG students has been expanded from 40 (in 1960–2004) to 82 at present. The PG programmes along with research started in the year 1966. The PG students' capacity has been enlarged to 46 in 2009 compared to 30 (1966−2008) to be specific. The department has been fostering many full-time and part-time young researchers over the years.

26.4.2 Teaching and Research Activities

The teaching at the B.Tech. level focuses on the fundamental knowledge base development and hands-on training in various areas of metallurgical and materials engineering, such as extractive metallurgy, physical metallurgy, mechanical metallurgy, corrosion science and engineering, materials modelling and simulations. M.Tech. level teaching emphasizes on advanced knowledge input in new materials development and the latest techno-scientific progress in metallurgical industries. The research activities of the department are centred on B.Tech. and M.Tech. projects, Ph.D. programmes, sponsored research projects and consultancy projects. The faculties have been continuously updating themselves through various faculty development schemes, which they are regularly implementing in developing new curricula and revising the existing ones in order to meet the industrial/social demands. The rapid shift of focus to research works has also resulted in a steep rise in research output indicators (journal and conference papers) and sponsored projects. Many research papers have been published in reputed peer-reviewed journals. These all-round efforts of the department have been able to ensure that even in the time of severe recession in the economy, the student's placement has not been influenced.

26.4.3 Sponsored Research Projects

The faculty members have guided more than a hundred industry-oriented projects in the various fields in collaboration with different industries and research laboratories, since the initial days of the department. Each of these projects has been carried out under the guidance of a faculty member of the department and an expert from the concerned industry or research laboratory. The objectives of these projects have been the solution of real-time industrial problems utilizing the expertise and facilities of both the department and industry or research lab. Thus, guiding these projects has proved to be an effective means of industrial consultancy. In the 90 s MHRD, DAE, AICTE, etc. sponsored a number of high-valued projects on "Use of Rare Earth Elements in Iron and Steel", "AlN Thin films", "Thin Film based Sensors" etc.

Table 26.1 Recent sponsored projects

Project title	Funding agency	Amount (Lakhs)	Duration
An investigation on accelerated spheroidization and mechanical property evaluation of high carbon steel under cyclic forced air cooling	DST-SERB	31 0.0	2015–2017
Development of ceramic particle dispersed aluminum alloy composite	DST	19.2	2012–2016
Wear Behaviour of Al-Si Alloys at room temperature and at elevated temperature	SERB-DST	25.08	2015–2018
Development of self-lubricating nano-composite for wear-resistant applications	DST	4.55	2017–2020
IN-VITRO and IN-VIVO Study of Electrochemical Behaviour of Dental Amalgams of Various States in Oral Environments	DST	16.99	2017
DST-FIST program 2018	DST	165.0	2018–2023

The department was bestowed with a STEEL CHAIR PROFESSOR, from the Ministry of Steel, Government of India, which provided scholarship to five (05) students each year in the 3rd and 4th year of their study to pursue research work on ferrous metallurgy. Department of Science and Technology, ACRC division, Government of India has sanctioned Rs. 117 lakhs for strengthening the research and post-graduate teaching facilities of the department. Council of Scientific and Industrial Research (CSIR), New Delhi had also sanctioned funds for carrying out various projects. On 7th May 2013, after a yearlong effort, the Centre of Excellence (CoE) was formed by the department, jointly with the Physics department and under the CoE scheme of MHRD, the Government of India Rs. 5 crores was sanctioned for research purposes (Table 26.1).

26.4.4 Collaborative Research

Since the initiation of the Institute and the department, various collaborative programmes have been undertaken by the department. In the early 90 s, the department collaborated with Durgapur Steel Plant (DSP), Alloy Steels Plant (ASP), Mining and Allied Machinery Corporation, Durgapur (MANC), ABB-ABL Projects, Durgapur, Indian Tube Company Ltd., Jamshedpur, Rourkela Steel Plant (RSP) and many more. Also, the department indulged in consulting research laboratories like CMERI Durgapur, RDCIS Ranchi, etc. in order to identify and effectively solve real-time industry problems side by side with industry experts for the benefit of the students and society. A Memorandum of Understanding (MoU) was signed with DSP and other industries for activities and academic purposes in the early

2000s. From 2018 onwards, attempts have been made to strengthen the industry-institute-interaction (I–I–I) activities. Since 2019, the department has organized more than twenty technical events, viz. short-term courses, 1-day or 2 days lecture programs by industry experts, seminars, industrial visits, GIAN programs, webinars, etc., which provided effective platforms for interaction and network formation between the department and the industries/R&D laboratories/other academic institutes. MoU has been signed with industries like Tata Consultancy Services, Aranya Bioscience, Foundry Cluster Development Associations, etc. Efforts have been made to sign MoUs with Damodar Valley Corporation (DVC) and renew MoU with DSP. Experts from the industries are invited to deliver lectures at the Institute. At present, the department is also associated with academic Institutes such as IIT Kharagpur, Jadavpur University, National Institute of Foundry and Forge Technology (NIFFT) Ranchi and IIEST (Shibpur) for Joint Ph.D. guidance and research paper publication. Apart from CSIR-Central Mechanical Engineering Research Institute (CSIR-CMERI) Durgapur, the department collaborates with other research organizations such as National Metallurgical Laboratory (NML) Jamshedpur, Bhabha Atomic Research Centre (BARC), and Defence Metallurgical Research Laboratory (DMRL). and continue to fraternize with industries like SAIL (DSP, ASP), RDCIS (Ranchi), along with Alstom, Tata Iron and Steel Company (TISCO) R&D, Tata Research Development and Design Centre (TRDDC) Pune and many others. The department strongly encourages its students to carry out their work all over the world for which there is financial support from the Institute. The department is in continuous contact with different foreign organizations such as the European Organization for Nuclear Research (CERN) Switzerland, the National University of Singapore, Massachusetts Institute of Technology USA, the University of Michigan USA, Max Plank Institute Germany, ETH Zurich (Switzerland), Penn State USA, Rensselaer Polytechnic Institute, Troy (USA) to name a few. MoU has been set up between the department and Jindal Steels Limited (JSL) through which professionals from JSL will teach an elective course on "Stainless Steel" to the UG students and will provide placement to some competent students of the department.

26.5 Areas of Research at a Glance

The areas of research of the department have been presented by means of a pictorial representation, as below.

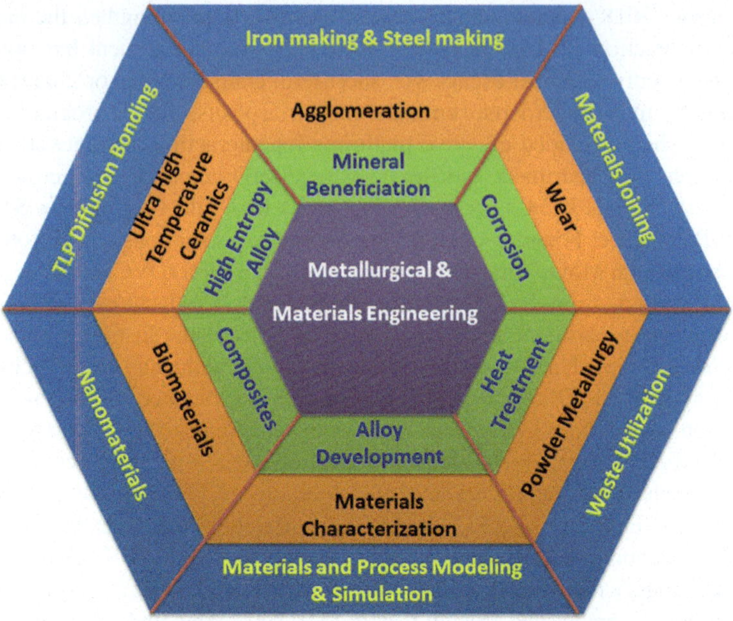

26.6 Research Facilities

The department takes pride in its wide spectrum of state-of-the-art laboratories equipped with both conventional and advanced instruments, which are used both for teaching and research purposes. These include Thermodynamics of Materials Laboratory, Metallography Laboratory, Mechanical Testing Laboratory, X-Ray Diffraction Laboratory, Manufacturing Processes Laboratory, Computation Laboratory, etc. to mention a few. Keeping its tradition of producing world-class metallurgists, the department has actively diversified to advanced materials for addressing today's technological challenges. The department has a reasonable number of instruments to engage the students in laboratory classes and research work. To mention a few are Slow Strain Rate Testing (SSRT) unit, Dynamic Light Scattering, Rolling Mill, 100 Ton hydraulic press, X-Ray Diffractometer, DSC, High-Temperature Wear Tester, Instron UTM, Leica Optical Microscope, FESEM, and Potentiostat/Galvanostat/FRA, etc. These laboratories and equipment of the department are often used by the researchers of many other Institutes as well as research labs. and industries in the region and all over India. Right from its juvenescence, the department offered technical assistance to various organizations, such as South Eastern Railways, Bharat Aluminium, BSC-RBM (Bidhan Nagar, Durgapur), CSIR-CMERI, Durgapur

Projects Ltd. (DPL), ABB-ABL Projects, and Durgapur Municipal Corporation (DMC).

The first Transmission Electron Microscope (TEM) of the eastern region was commissioned in the department under the support of UNESCO in 1967, adding glory to its pride. The TEM (Model: HS-7S, Hitachi, Japan) gave service to the Institute and all other Institutes in the region for generations.

With the help of the National Academic Depository (NAD) initiative of the Government of India facilitated by the Ministry of Human Resource Development, the department facilitates the issuance and storage of academic awards to the students in an online depository system. A large number of learning resources have been procured over time to improve the research–teaching–learning process. The networking has been revamped and the internet connectivity has been provided to all the wings and segments of the department. The Institute library has a large collection of text and reference books as well as many e-resources like electronic journals of national and international reputation in a variety of fields of materials research. The department takes pride along with the Institute for having a high-quality video conferencing facility, with 24 nodes—a facility among the best of its kind in the country. Moreover, the department has a departmental library with a collection of several books on metallurgy and materials.

26.7 Notable Contributions and Achievements Since the Past Decade

26.7.1 Faculties' Awards and Achievements

- Dr. Supriya Bera received DAAD summer Fellowship in 2018, Marie Curie Post-Doc Fellowship in 2015 and INAE Teacher Fellowship in 2013.
- Dr. A.K. Mandal received Outstanding Scientist Award in 2018 from the Organisation of Scientific Research and Development.
- Dr. A.K. Mandal received Outstanding Scientist in Metallurgical Engineering in 2018 from Venus International Foundation.
- Dr. D. Mandal received Prof. Shilowbhadra Banerjee Award for the best in-house project (NML Foundation Day Award, November 26, 2016) on "Development of Gas based Brass Melting Furnace for Rural Artisan" along with K.L. Sahoo, K.K. Paul and P. Poddar.
- Dr. Satadal Ghorai was awarded the second prize for poster presentation at the 67th Annual Technical Meeting of the Indian Institute of Metals held in Varanasi from 12 to 15th November 2013.
- Dr. Satadal Ghorai was awarded the second prize for oral presentation at the National Seminar on Exploration of Lean Grade Ore, Ore Fines and Urban Ores: Challenges, Problems and Solutions, LGO-2013, CSIR-NML, Jamshedpur, 22–23 January 2013.

- Dr. Manab Mallik was awarded second prize for oral presentation in the Metal Science group at the 67th Annual Technical Meeting of the Indian Institute of Metals held in Varanasi from 12 to 15th November 2013.
- Dr. Manab Mallik received the IIM-Swarna Jayanti Endowment Fund in 2013.
- Dr. Satadal Ghorai received the Shilowbhadra Banerjee Award for the best in-house project from NML for the project entitled, "Self-propagating high temperature sintering of LD sludge for recycling".
- Dr. Satadal Ghorai was awarded BOYSCAST fellowship 2007–08, Department of Science and Technology, Govt. of India.
- Dr. D. Mandal, received a BOYSCAST fellowship 2008–2009, to conduct advanced research on composite materials at the University of Alabama, Tuscaloosa, USA from August 2009 to August 2011.
- Dr. D. Mandal was awarded the 2nd best prize in poster presentation in the International Conference on Metals and Alloys: Past, Present and Future (METALLO-2007), IIT Kanpur, Dec 7th-10th, 2007.
- K.S. Ghosh: NACE (National Association of Corrosion Engineers, USA) International India Section (NIIS) CORROSION AWARENESS AWARD – 2004.

26.7.2 Research Publications and Patents

Our faculty members along with the students have been publishing their research works, in prestigious journals and conferences. Up to the early 2000s, 51 refereed journals, 57 conference proceedings, and 4 review articles were published by the department. The department has been publishing, on average, 40−50 papers per year in peer-reviewed journals over the past 5 years. Faculty members are actively engaged in the reviewing of technical papers of reputed international and national journals. Currently, the enhancement level with respect to research contributions has been very upright, and the numbers are overwhelmingly elevating.

26.8 Distinguished Alumni of the Department

The department has provided the required manpower for sustainable growth in India's backbone of the steel industry, aluminium industry and others. The performance of the students of the department has been excellent in all the areas from its early stages. Some of the ex-students of the department hold high positions in universities, research laboratories, and industries in India as well as abroad. The achievements of a few noteworthy alumni are mentioned below:

- Tarasankar DebRoy (1969) Professor, Materials Science and Engineering, Pennsylvania State University, USA.

- Kamanio Chattopadhyay (1971) INAE Distinguished Professor, IISC Bangalore (Recipient of S.S. Bhatnagar Award).
- Subhash Ranjan Pati (1971) Founder, Materials Technology Solutions, Ohio, USA.
- Partha Chattopadhyay (1982) Chief Operating Officer (COO), Tata Sponge Iron Ltd.
- Subrata Bhattacharya (1982) Director, Jindal Stainless Steel Ltd.
- T. Raghu (1984) Director (Management Services) & Head (Near Net Shape Group), Defence Metallurgical Research Lab., DRDO, Hyderabad.
- Bratin Saha (1989) Managing Director & Co-founder, AMM Enterprise GmbH, Germany.
- Sukanta Mitra (1991) CEO, Alpha Numero Technology Solutions, USA.
- Bikramjit Basu (1995) Professor, IISC Bangalore (Recipient of S.S. Bhatnagar Award).
- Aloke Paul (1996) Professor, IISC Bangalore (Recipient of S.S. Bhatnagar Award).
- Krishnan K. Sankaran (1997) Courtesy Faculty, University of North Texas, Adjunct Instructor, Washington University, St. Louis.

26.9 Summary

In conclusion, it has to be emphasized that the Department of Metallurgical and Materials Engineering at NIT Durgapur has played an important role in shaping Metallurgy and Materials Science in India through not only generating a vast pool of qualified human resources but also through yeomen fundamental research contributions and providing service to the metallurgical industry. NIT Durgapur has also supported the Indian Institute of Metals since the Department was established.

Acknowledgements The authors are thankful to the Head of the Department of Metallurgical and Materials Engineering, NIT Durgapur and all the faculty members and staff members of the Department for providing all the necessary data required to write this article. The authors are grateful to the Director, Dean (Academic Courses) and Dean (Academic Research), NIT Durgapur for extending all sorts of support needed for preparing the article.

Chapter 27
Metallurgical and Materials Engineering at the National Institute of Technology Karnataka: A Historical Overview

K. Narayan Prabhu

Keywords Karnataka Regional Engineering College · Surathkal · National Institute of Technology Karnataka · Metallurgical & Materials Engineering

27.1 Introduction

K. Narayan Prabhu (✉)
Department of Metallurgical and Materials Engineering, National Institute of Technology Karnataka, Srinivas Nagar, Mangalore 575025, India
e-mail: knprabhu@nitk.edu.in

R. Divakar et al. (eds.), *Indian Metallurgy*, Indian Institute of Metals Series,
https://doi.org/10.1007/978-981-99-5060-7_27

The Department of Metallurgical Engineering, one of the oldest Departments at the National Institute of Technology Karnataka (Formerly, Karnataka Regional Engineering College, Surathkal) was established in the year 1965 with an objective to impart education and training in Metallurgical & Materials Engineering. The Department owes a lot to the vision of its founder Head, Prof. T. Ramchandran. The Department is very active in research activities and keeps abreast with the state-of-the-art developments in Engineering and Technology. After almost four decades, it was renamed as the Department of Metallurgical & Materials Engineering. The Department offers one UG program (B.Tech. in Metallurgical & Materials Engineering, sanctioned strength: 64) and three PG programs (M. Tech. in Materials Process Technology with a sanctioned strength of 18, M.Tech. in Materials Engineering with a sanctioned strength of 33, and M.Tech. in Nanotechnology with a sanctioned strength of 18) as well as a masters program by research, and a full-time Ph.D. program. Students are given intensive training in conventional metallurgy, involving both practical and theory. Starting with basic subjects, students are taught application-oriented courses. The curriculum includes advance topics in Metallurgy and Materials Engineering. Apart from undergoing industrial training, students are required to carry out the project work as part of their curriculum. Our alumni are employed in various fields, such as industries, academia, R&D institutes, government, and public sector organizations. The Department has been active in carrying out funded R&D projects and many facilities for research have been added in recent years. The Department has shown a strong presence in the areas of corrosion, physical metallurgy, solidification, heat treatment, powder metallurgy, thin films, electrospinning of polymeric and ceramic nanofibers, metal joining, and forming. Some of the advanced facilities include TEM, SEM, XRD, FTIR spectrometer, contact angle meter, optical microscopes, and Jeol and Instron tensile testers. Metallurgical and Materials Engineering Association (MMEA) takes care of various students' activities in the Department and brings out an in-house magazine NITKAST. The Department library holds about 750 textbooks, back volumes, and journals. The Department has an excellent academic atmosphere with qualified and motivated young faculty and hard-working staff. The Surathkal Chapter of the Indian Institute of Metals (IIM) regularly conducts technical meetings, workshops, industrial tours, and visits to enrich the technical knowledge of students. During the past 57 years, about 1,500 graduates, 1,000 postgraduates, and 70 Ph.D. scholars have graduated from this department and have occupied positions of eminence in industries, R&D organizations, and academic institutions. The Department hosts the local chapter of the Indian Institute of Metals.

27.2 Historical Notes

Professor T. Ramchandran was the first Head of the Department and was ably assisted by renowned teachers. Born on 19th May 1929 at Thrissur (Kerala), Prof. T. Ramchandran graduated in Metallurgy from the Banaras Hindu University in 1950. He was awarded the Doctor Ingenieur degree by the Technical University of Clausthal in 1959. He spent the next 2 years as an IIT Madras trainee at the Institute for Physical Metallurgy and Metal Physics, T. U. Clausthal, attending specialized courses and conducting research in the field of textures under the guidance of Professor Wassermann. He was the first staff member to join the Department of Metallurgy at IIT Madras and worked there as Assistant Professor from 1961 to 1965. He moved over to the Karnataka Regional Engineering College, Surathkal on August 2, 1965, to start the Department of Metallurgical Engineering as its Professor and Head. He retired from the Karnataka Regional Engineering College in the year 1989 Building up the department at KREC, Surathkal over two decades of excellent teaching and developing infrastructure and a sound base for research, training, and development of a large number of distinguished students and teachers in metallurgy are the achievements for which Prof. T. Ramchandran can be truly proud of. The first batch of students graduated in 1970.

August 1st, 1972, saw the start of the first PG program in the Department, a Master's degree course in Process Metallurgy with specialization in Iron and Steel Technology. In the same month, Dr. TCM Pillay, a seasoned Scientist who had his education and training from the Imperial College, MIT, University of Wisconsin, and General Electric joined the Department. In the year 1973, the Department conducted a National Symposium on 'Electrometallurgy and Corrosion' in collaboration with the Electrochemical Society of India and the Indian National Science Academy.

27.3 Research, Projects and Events

Research leading to the award of Ph.D. was initiated in the mid-70s, and candidates were registered under Mysore University due to KREC's affiliation with the University of Mysore. Professor K. A. Natarajan spearheaded the research activity in the KREC setup. Funded projects from government agencies started coming to the Department. Dr. G. Premkumar's role in developing Extractive Metallurgy is unique and exemplary. The Department conducted several national conferences, seminars, workshops, and interactions with the R&D Departments of the Institutes of Higher Learning like IISc, IITs, and R&D Labs of National Importance like NAL, HAL, DMRL, MIDHANI, BARC, VSSC & RRL Trivandrum. A few students did their project work in these institutes and laboratories to fulfill the partial requirement of their degree (both UG & PG) program. In 1990, the Department started its second PG course, this time in Materials Engineering. The faculty members from other Departments of KREC also registered for Ph.D./MSc. Eng (Res) under the guidance of the faculty of the Department.

Research activity gained momentum in 1980 when KREC was first affiliated to Mangalore University, which then had only two engineering colleges under its jurisdiction. The first Ph.D. awarded in Engineering at Mangalore University was from the Department of Metallurgical Engineering in 1984. The testing and consultancy to neighboring industries were in full swing and continuing till now. In terms of generating internal revenue, the Department was second only to the Civil Engineering Department. In the initial years, the Department has benefitted immensely from the excellent faculty like Prof. K.G. Avadhani, Prof. V. Ramachandran, Prof. T.C.M. Pillai, Prof. N. Venkataraman, Prof. K.L. Bhat, Prof. H.V. Sudhaker Nayak, Prof. P.Prasad Rao, and Prof. K.R. Hebbar. Prof. Nayak and Prof. Rao carried out postdoctoral research work at prestigious universities in Germany, USA, and Canada. Between 1980 and 1985, Prof. K. Rajendra Udupa, Prof. P.L.N. Reddy, Prof. K. Srinivasan, and Prof. A.O. Surendranathan joined the Department and contributed in the fields of metal casting, electrometallurgy, metal forming, and corrosion engineering. In the year 1989, a national seminar on 'Metallurgy and Developments in Materials Science and Engineering in honor of Prof. T. Ramchandran. Post-1990, many faculty members joined and are very active in teaching, administration as well as research. The faculty members have also won national awards and fellowships from foreign professional bodies. The regular Ph.D. program was started in the year 2003 after the transformation of KREC to NITK. In the year 2005, M.Tech. (by research) the program was started in all PG branches. The year 2011 saw the start of a new PG program in Nanotechnology. In the same year, Dr. K.K. Prasad who was from RDCIS, SAIL, Ranchi joined the Department as Ministry of Steel Chair Professor. He served the Department for several years and both undergraduate and postgraduate students immensely benefitted from his rich experience. The placement statistics were good for the past few years with over 90% of the students getting placed in reputed industries. In 2015, the Department celebrated its Golden Jubilee and an international conference on 'Make in India—Role of Metals & Materials' was held to commemorate the event. A large number of workshops and seminars have been conducted by faculty members in the last decade for the benefit of metallurgists and materials engineers. The Department conducts every year Prof. T. Ramachandran Lecture Series in honor of Prof. T. Ramchandran, our Founder Director. The first lecture in the series was delivered by Prof. Krishnadas Nair, Former Managing Director of HAL, in the year 1990. The following photograph shows Dr. V. Ramchandran, Guest of Honour in one of the lectures in the series.

27.4 Recognition of Students and Faculty

National Institute of Technology Karnataka offers the following gold medals to the best outgoing undergraduate students: (1) Institute Medal, (2) Karthik Alloys Gold Medal, (3) Prof. H. V. Sudhaker Nayak Gold Medal, (4) SMIORE Gold Medal, and (5) 1986 Batch Gold Medal.

Postgraduate students securing the highest CGPA in the qualifying examination are given the following gold medals:

M.Tech. (Materials Process Technology): (1) Institute Medal, and (2) Smt. Sarojini Pillay Gold Medal,
M.Tech. (Materials Engineering): (1) Institute Medal, (2) Prof. K R Hebbar Gold Medal, and (3) Prof. K. L. Bhat & Prof. P. Prasad Rao Gold Medal,
M.Tech. (Nanotechnology): (1) Institute Medal.

Our students have secured several prizes and best paper awards including IIM students' prizes. We are proud of our students. The National Institute of Technology Karnataka is now ranked 12th among the top engineering institutions in India as per the National Institute Ranking Framework.

27.5 Summary

Looking forward, the Department of Metallurgical & Materials Engineering at NITK has implemented several changes in the curriculum as per the National Education Policy. The roadmap to graduation will be based on a curriculum designed on the processing-structure–property-performance paradigm. We look forward to guidance, support, and cooperation from all of our stakeholders.

Acknowledgements The author thanks Prof. H. V. S. Sudhaker Nayak, Professor (Retd.), Department of MME, NITK for his valuable inputs.

Chapter 28
Metallurgy at the National Institute of Technology Rourkela—a Polished अयस्

Archana Mallik and A. Basu

Abstract The article briefly describes the chronological progress of the Department of Metallurgical and Materials Engineering, National Institute of Technology (NIT) Rourkela from the 1960s till date. The rich history of the department along with the present status, particularly the focus on academia and research has been enumerated with milestones, facts, and figures.

☆अयस् (sanskrit): *n.* iron, steel, or in general, a metal.

A. Mallik · A. Basu (✉)
National Institute of Technology Rourkela, Department of Metallurgical and Materials Engineering, Rourkela, Odisha 769008, India
e-mail: basua@nitrkl.ac.in

R. Divakar et al. (eds.), *Indian Metallurgy*, Indian Institute of Metals Series,
https://doi.org/10.1007/978-981-99-5060-7_28

28.1 Introduction

Where the fields of supercomputers, ultrafast technology, megastructures, and super-advanced clinical research are at their peak, the field of Metallurgical and Materials Engineering has the herculean responsibility to serve as the backbone of the growing economy. India's rich Metallurgical alchemy heritage began before the third millennium and has evidence of the exchange of science to the rest of the world through the well-documented scripts by Nagarjuna (the Indian Alchemy and Metallurgy wizard) (http://cbseacademic.nic.in/web_material/Circulars/2012/68_KTPI/Module_8.pdf ; Rasopanisad 1928). However, five decades ago, India, as a unified and independent nation, was still in its infancy. It was the vision of Pandit Nehru and Dr. Humayum Kabir that led to the establishment of a set of unique institutions called regional engineering colleges (REC) in the country in the year 1959 (https://myvssut.wordpress.com/vssut-burla-story-of-a-magnificient/). Many such colleges were also established in different states in 1959 and 1960 with the central government's cooperation and the concerned state governments. Odisha (then Orissa) initially was not so lucky to get a REC. REC Rourkela came up as the second engineering college in Odisha in 1961, after the University College of Engineering Burla (now Veer Surendra Sai University of Technology, Burla), courtesy of the vision of Odisha's legendary chief minister, late Sri Biju Patnaik and the dynamic leadership of its first principal, Prof. Bhubaneswar Behera. The Department of Metallurgical and Materials Engineering (the then Metallurgical Engineering) at REC Rourkela was born in the year 1963 under such a positive and exciting climate to instill a sense of confidence and cooperation and provide high-quality technical manpower to our growing economy. The college appointed Prof. Sakya Singh Pani (Fig. 28.1a), a man with a plethora of experiences gained from different steel industries and national as well as overseas academic training, to establish the department. Prof. Pani, with his imagination, vision, hard work, and rigorous grooming embellished the department within a few years. Since then, this department has not only produced a large number of students who have gone on to occupy several top positions within the country and abroad but also served the needs of the industry. The following sections will briefly review the department's journey of struggle from an amateur teaching organization to one of an internationally visible academic education and research units.

28.2 The Journey of Struggle to Triumph: The REC Era

The first batch of undergraduate Metallurgists (B.Sc. Engineering in Metallurgical Engineering) came out in the year 1968 (Fig. 28.1b). Those were times when finding employment for all the graduates was not easy. However, the leadership of Prof. Pani and the tireless efforts of the faculty of high stature could lead the graduates to fetch class-1 jobs in Hindustan Aeronautics Limited, Bhabha Atomic Research Centre, and other public sectors including the Rourkela Steel Plant.

Fig. 28.1 **a** Picture of Prof. Sakya Singh Pani, **b** The pioneer batch

By the start of 1970, the department had taken a wonderful shape, not only in terms of well-accomplished students but a magnificent team of faculties and technical persons. A figure depicting the establishment of the different laboratories is depicted in Fig. 28.2a. At this point in time, the national and global call was to expand the teaching domain and initiate research in the field of extraction metallurgy. The department received its first international exposure through the United Nations Development Programme (UNDP) with the financial assistance of UNESCO in the year 1970 (http://web.undp.org/execbrd/archives/bluebooks/1960s/e-4451.pdf). A full-time postgraduate program "M.Sc. (Engg.) in Technology of Metallurgical Furnaces" with 16 seats was started under the supervision of Prof. Alkov (an expert from the Soviet Union). To strengthen the course further, faculty exchange programs were also formulated. During this decade, more faculty members joined the department mainly with a furnace technology background. In the mid-70 s, under the dynamic leadership of Prof. Somnath Mishra, Principal and a reputed metallurgist, the research field started nucleating. He was instrumental in getting financial aids from different government organizations. One such noteworthy research of the department is the extraction of vanadium from titanomagnetite vanadiferous deposits of Odisha financed by the Board of Scientific and Industrial Research, India. Through the next decade, the department witnessed a surge in teaching activities as well as research projects (Behera and Mohanty 1984; Sarnagi 1987).

More postgraduate courses as per the requirements of nearby industries were incorporated into the curriculum, thus initiating the industry-institute interface. Table 28.1 lists the various programmes offered until the year 2000. It is worth mentioning that the department simultaneously arranged for coaching of Associate Membership of Indian Institute of Metals (AMIIM) and Associate Membership of Institution of Engineers (AMIE) students on requests from their local chapters. During this period, laboratory equipment was slowly procured with different grants e.g. "IIT-REC Interaction Scheme" for the IIT interaction network and modernization of the laboratory. This is the time when the importance of teaching and research on materials was given importance. Hence, grants were offered to increase activities in the areas of Material Research. As a result, many high-end pieces of equipment were installed in the department (Table 28.1). A sponsored project on environmental

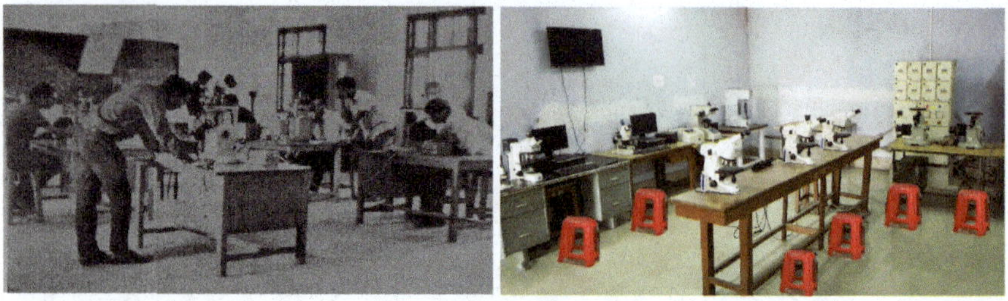

Fig. 28.2 **a** Metallography laboratory during the 1970s, **b** Part view of the metallography laboratory during the 2010s

pollution during Zn recovery was also handled by the department during this period. Subsequently, physical and mechanical metallurgy got a boost. The department started to make a mark in the global metallurgy and materials research community as the quantity and quality of technical papers in national and international journals improved. Research activities at all levels were initiated. Simultaneously sponsored and consultancy projects increased, which helped to get a higher accreditation grade, i.e., A grade by the Government. One such example is to help the Palaspanga sponge iron plant of Odisha (established in 1986) to get a quality product. The work was well-received by the Ministry of Science and Technology, Government of India. This was also the time when the department got a sizeable research share from the government through the Indo-UK collaborative project (in the year 1989). The department secured this project through tough competition with other contemporary RECs. The implementation tenure of the project was 1994–1999 with funding close to a million USD and an aim of procuring and setting up sophisticated equipment and training Indian experts in the UK. Under the supervision of six professors from UK universities, the "Materials Science and Engineering centre" was established with its own infrastructure and separate building. The list of major equipment procured through the scheme in the nineties is provided in Table 28.1. This not only benefited the department but also increased the face value of the college as well. The facility created under this program has helped the college in many instances, e.g., initiation of an MIS program, and coordination of the entire JEE Orissa program. This decade was a golden era as the department got another coveted and ambitious project, i.e., the INDO-US project, sanctioned by the US Naval Research Board. Department of Science and Technology (DST), Govt. of India monitored the grant through the Council of Scientific and Industrial Research. Cu Bearing HSLA steel was one of the research topics under this scheme (Ray et al. 2003). An Indo-Japanese collaborative project was also bagged by the department, leading to the training of many faculty members in Japan. Plasma-based extraction from ilmenite ore was another field of research during this decade.

In the early 2000s, in the light of the emergence of other materials, the curriculum of the programs run by the department was revised and subjects such as composite materials were introduced. Under Sambalpur University, the institute got an autonomous tag in 1993–94. Accordingly, the department changed its name to

Table 28.1 Academic programs and major facilities of the department of MME, NIT Rourkela (1963–2000)

Academic programs			
Sl	Program name	Starting year and initial intake	Remarks
1	B. Sc. Engineering in Metallurgical Engineering	1963, 40	• Changed to "B.E. in Metallurgical Engineering" in 1979 (graduated) • Duration decreased from 5 to 4 years in 1994 (graduated) • Changed to "B.E. Metallurgical and Materials Engineering" in 1994 • Changed to "B.Tech. in Metallurgical and Materials Engineering" in 2003 (graduated) • Continuing
2	M.Sc. (Engg.) in Technology of Metallurgical Furnaces	1970, 16	Discontinued in the late seventies
3	M.Sc. (Engg.) Ferrous Process Metallurgy	1981, 20	Discontinued from 2003
4	M.Sc. (Engg.) in Industrial Metallurgy	1981, 20	Discontinued from 2003
5	Ph.D	1980, NA	Continuing

Major equipment installed	
Other Govt. funding	Indo-UK project
DTA, DSC, TG, UTM (static INSTRON 1195), atomic absorption spectrometer, X-ray diffractometer (JEOL), scanning electron microscope with WDS (JEOL—T330) and high-temperature Microscope	X-ray diffractometer (Philips X'Pert MPD), Digital Ultrasonic Flow Dictator (Sonatest), Particle Size Analyzer (Malvern), Dilatometer (Netzsch), TG—DSC/ TG—DTA (Netzsch), Universal Testing Machines (Instron), Pin-On-Disc-Wear Tester, Pendulum Impact Testing Machine, Metallurgical Microscope, Planetary Ball Mill (Fritsch), Digital Melt Flow Index Tester, High-Temperature Tubular Furnace, Image Analyser, Heating Microscope (Leica), Raising Hearth Furnace, Universal Hydraulic Press, CHNS Analyser and Viscometer (Brook Field), Isostatic Cold Press

"Metallurgical and Materials Engineering Department". Students of the Department displayed excellent performance in competitive exams and secured very high ranks in the all-India GATE examination in the middle of the 1990s. During this period Prof. A. K. Mohanty (the then principal and a professor of the department) played a pivotal role in lifting the departmental activities to another level. His famous text-book entitled "Rate Processes in Metallurgy" (Prentice Hall of India) is still followed by many of the institutes even today. Many other textbooks have resulted from the acumen of the faculty members of the department.

28.3 The Change Over Period: The NIT Era

With time, technology has evolved; so has people's mindset. "Regional" has given way to "National", "College" to "Institute" and "Engineering" to "Technology"; in the year 2002, REC Rourkela was upgraded to a National Institute of Technology (NIT) like other RECs. The prospect and trend in academics, administration, resources, and research were going to be changed in the new century. A major change in the curriculum (mentioned in Table 28.2) and steady funding from the central government under the Ministry of Human Resource and Development (MHRD) was the critical mover during this change. Since the department already had a well-established research culture, teaching and research continued to flourish in this department, as reflected by publications of those days. Under the dynamic leadership of Prof. Sunil Kumar Sarangi, the first director of NIT Rourkela, the institute started taking the shape of a technical and cultural hub with several national and international initiatives.

In 2007, the institute got the tag of "Institute of National Importance" which resulted in a manifold increase in its infrastructure and varied research fields. Apart from core metallurgy fields, as discussed in earlier sections, research on composites materials, nanomaterials, thin-films, texture, computational materials science, surface engineering, etc. got a momentum as in other leading metallurgical/materials departments of this country. The changeover resulted in an increase of faculty as well as research scholars in the department. A representative figure of the 2010s undergraduate laboratory is given in Fig. 28.2b. Various master-level programs were initiated as per the needs of the time (refer to Table 28.2). The eligible academic programs of the department were continuously accredited by the All India Council for Technical Education (till 1994) or the National Board of Accreditation (after 1994). Steel Research Center was also established during the second decade of this century, with a memorandum of understanding signed with R&D Centre of SAIL to further enhance collaborative research. Major research facilities obtained by the department during this century are also shown in Table 28.2.

The developed research facilities are supported by the Technical Education Quality Improvement Programme (TEQIP) from the World Bank and individual extramural projects with different government and private funding managed by individual faculty members of the department. The department also got DST-FIST funding in the year 2014 in the Tier II category. Moreover, recently more than USD 140,000 have been sanctioned to the department by Higher Education Financing Agency (HEFA) and the procurement is under process. The summary of different activities and achievements of the department since 2000 is presented in Fig. 28.3. It can be seen from the figure that the department is geared for the 21st Century which has helped to breach the near-axiom class barrier between IITs and NITs by placing

Table 28.2 Academic programs* and major facilities of department, NIT Rourkela (2000-Till date)

Academic programs (*in continuation of Table 28.1)

Sl	Program name	Starting year and initial intake	Remarks
1	M. Tech. in Metallurgical and Materials Engineering	2003, 20	Continuing
2	Dual degree (B. Tech. and M. Tech.) in Metallurgical and Materials Engineering	2010, 10	Continuing
3	M. Tech. in Steel Technology	2013, 20	Discontinued from 2017
4	M. Tech. by Research	2004, NA	Continuing
5	Minor program (B. Tech.) in Metallurgical and Materials Engineering	2017	Continuing
6	Executive Ph.D	2019, NA	Continuing

Major equipment installed

Institute (Govt.) fund	TEQIP Fund	DST FIST Fund	Sponsored project fund
Potentiostat (Metrohm), Stylus Surface Profilometer (Veeco), Scanning Probe Microscope (Veeco), FTIR-Microscope (Shimadzu) (*partial project funding*), UTM (Instron), Dynamic Mechanical Thermal Analyser (Netzsch), Micro-Hardness Tester (Leco), Image analyser (ZEISS)	SEM (JEOL) DSC (Mettler), Dynamic Universal Testing System (BISS)	X-ray Diffractometer with Stress-Texture Cradle (Bruker)	High-Temperature Viscometer (Bahr), Ball Mills (Fritsch), Potentiostats (CorrTest, Ametek), Ultrasonic peening unit (Sonats ET), Ball-on-plate wear tester (Ducom), Dual Drive Planetary Mill (Gurpreet Engineering Works), Squeeze casting set up for Mg (SwamEquip), Impression creep testing machine (Spranktronics)

NIT Rourkela at the international level too. The glorious presence of its alumni members across the world at accomplished positions, student quality and research publications are few such indicators.

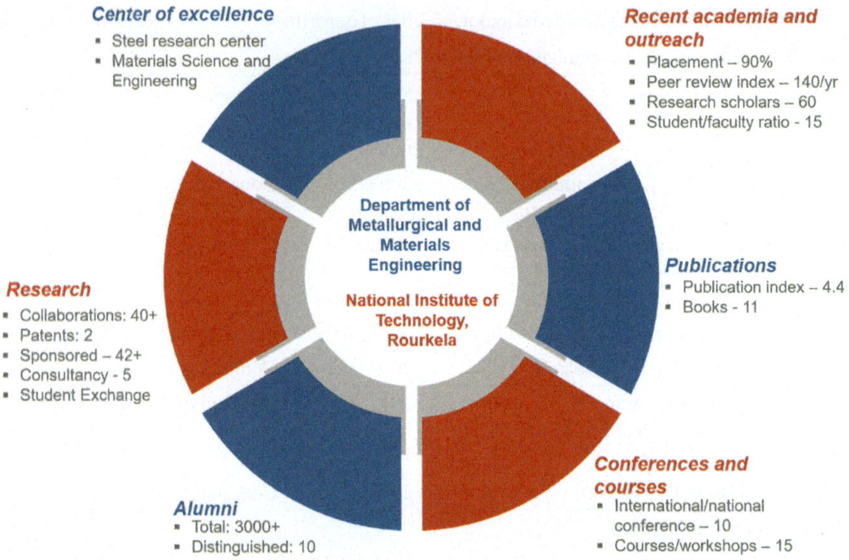

Fig. 28.3 Summary of departmental activities and achievements in the NIT Era

28.4 Concluding Remarks

Metallurgical and Materials engineering is perceived to be a predominantly experimental science and the department's research is at par with other metallurgical/materials departments of the nation and around the globe. However, over the past three decades, materials research came to the front, along with computational metallurgy. The aspect of the knowledge of associated equipment, design and development of capital equipment and analytical imaging that material scientists use in their routine research cannot be underestimated. It is time to give due importance to all aspects of materials engineering education. Furthermore, in the international scenario, this is a small department with suboptimal faculty strength and in need of a modern and liberal UG/PG curriculum. Despite the shortfalls, the department had and has all the potential to lead the way to make the country a leader in materials technology.

Acknowledgements The authors are thankful to the faculty and staff members of the department for their input and support. They are also thankful to the superannuated faculty members and alumni for providing information, without which it would not have been possible. Special mention goes to Prof. A. Sarangi, Prof. R. I. Ganguly, Prof. A. K. Panda, Dr. Amitava Ray, Dr. R. K. Paramguru and Mr. Deba Mohapatra.

References

Behera, R.C., Mohanty, A.K.: Alumino-Silicotheric reduction of sand ilmenite. Trans. Indian Inst. Metals **37**, 215–219 (1984)

http://cbseacademic.nic.in/web_material/Circulars/2012/68_KTPI/Module_8.pdf. Accessed on 10th May 2021

http://web.undp.org/execbrd/archives/bluebooks/1960s/e-4451.pdf. Accessed on 10th May 2021

https://myvssut.wordpress.com/vssut-burla-story-of-a-magnificient/. Accessed on 10th May 2021

Ray, P.K., Ganguly, R.I., Panda, A.K.: Optimization of mechanical properties of a Heat-treated Cu-bearing HSLA-80 Steel. Trans. Indian Inst. Metals **56**, 121–129 (2003)

Sambasivasastri, K., (ed.), Rasopanisad, printed by the superintendent, Govt. printing press, Trivandram, Trivandrm Sanskrit Series No. XCII (1928)

Sarnagi, A., Tripathy, K.K., Mishra, S.: Partial reduction of vanadiferrous titanomagnetite ores of orissa in rotary kiln, I. E. (India), **67**, 53 (1987)

Part IV
Individuals

Chapter 29
Six Decades of Academic Research in Materials Engineering: A Personal Perspective

K. Chattopadhyay

Abstract This article is a personal account of the evolution of the research ecosystem and academic world of metallurgical engineering in India, from the 1960s to the present day. The author emphasizes the efforts of academic leaders and political support in the establishment of newer steel plants and metallurgical engineering departments in post-independence India. The text reflects on the slow start of graduate education and research in the sixties, which only gained momentum in the seventies. The author notes the contributions of several visionaries, who initiated a transformation towards graduate education and research, leading to the growth of academic research clusters in India.

Keywords Metallurgy · Evolution of metallurgy education · Post-graduate education · Research ecosystem

29.1 Introduction

It is now 55 years since I walked into the portal of the Department of Metallurgical Engineering at Regional Engineering College (REC), Durgapur, and became part of the excitement that was ringing the metallurgical world with the setting up of newer steel plants and newer metallurgical engineering departments in the post-independence India of the fifties and sixties. My world widened further 5 years later when I had an opportunity to walk into the room of Professor T. R. Anantharaman at the Department of Metallurgy, Banaras Hindu University (BHU) and decided to pursue my graduate studies there. The decade of the 70s that I spent at BHU as a research student and later as a faculty provided a rare opportunity to absorb the prevailing dynamism of an increasingly confident profession. It was when the metallurgical industry was maturing while the academic world was striving to establish a global reputation. The four decades of association following this at the

K. Chattopadhyay (✉)
Department of Materials Engineering, Indian Institute of Science, Bangalore 560012, India
e-mail: kamanio@iisc.ac.in

R. Divakar et al. (eds.), *Indian Metallurgy*, Indian Institute of Metals Series,
https://doi.org/10.1007/978-981-99-5060-7_29

Indian Institute of Science (IISc), Bangalore, allowed me to observe my surroundings with newer insights flowing from the different roles donned by the undersigned. Our profession matured during this period with a continuous influx of newer people and opportunities. The generational changes and the confidence of newer people were striking. Following is a brief account of these observations from a personal perspective, emphasising the evolving research ecosystem and the academic world.

29.2 The Early Decades

Those who have their undergraduate studies in the 60s will remember the atmosphere of hope and expectation that used to prevail in every metallurgical engineering department. Most of them, except for a few like Banaras Hindu University (BHU), were new. When I first took a tour of the Department of Metallurgical Engineering at Durgapur as an undergraduate student, I was struck by the brand new and, as I now realise, highly well-equipped laboratories. The scenario was similar for most of the other newer metallurgical engineering departments of the country. It reflected a high degree of effort and commitment of the academic leaders that we had in our field at that time and political support. We need to appreciate their contributions. However, most of the efforts were geared towards undergraduate education, and the aim was to produce competent graduates for our emerging metallurgical industries. The visits to the new integrated steel plants were memorable occasions among students. A call for an interview for graduate trainees at Hindustan Steel Limited used to be an occasion for celebration. Graduate education started taking root rather slowly during this period and played only second fiddle to undergraduate education. The doctoral programme was mostly confined to a few teaching faculties aiming to obtain doctorate degrees by carrying out research in their spare time. However, this outstanding and committed group formed the core that initiated a transformation towards graduate education.

One can explore why academic research took time to take root in the sixties. The momentum started building only in the 70s. The oft-repeated statement of lack of facilities is not the correct explanation. Instead, the ecosystem that promotes independent thinking and new research was generally absent on the Indian soil. Most of those who returned after obtaining degrees from abroad were awed by their supervisors and the Ph.D. workplace. Hence, they only attempted an extension of their graduate research and ended up with what we can call incremental research. Like we got technology for setting up our steel plants, we probably hoped for a similar step for higher education and research. I often remember what was there at REC Durgapur. We had a transmission electron microscope of Hitachi make, the latest Siemens make X-ray diffractometer, a vacuum arc melter, a set of equipment for mechanical property measurements and an excellent chemical analysis facility. I could read all the major journals in my undergraduate days at our college library. Thus, the bottleneck was in our minds.

However, India and our generation were extremely fortunate that we had several exceptions. These doyens have seeded creative thinking in research and catalysed a culture of academic graduate research of world-class in the late sixties and early seventies. Many of us were influenced by people like Professor T. R. Anantharaman, Professor E. C. G. Subba Rao, Professor K. I. Vasu, Dr. R. Krishnan and Professor Ranganathan (those I knew personally, among several others). They mentored several of us, including the late Dr. Srikumar Banerjee and Professor P. Ramachandra Rao. They nurtured the critical period of transformation for higher education and research in our country and taught that higher education and research prepare us to do something new, walk into the unknown, and explore. They also taught that there is nothing national or local about innovation and creativity. It was also the time when research organisations in our strategic sectors took shape. Metallurgical engineering and materials profession played a key role in these developments with pioneering contributions from people like Prof. Brahm Prakash and Prof. C. V. Sundaram. The work under the leadership of these personalities brought confidence to the minds of many young research scholars who decided to stay in India and pursue graduate studies.

The 70s marked the consolidation of the Ph.D. programmes in the country and the growth of academic research clusters in both the older institutes like BHU and IISc and newer Indian Institutes of Technology (IITs) like Kanpur, Kharagpur and Madras. It also heralded a new generation of academic faculty who were primarily trained in India. During this period, the leadership of the Indian Institute of Metals (IIM) and academic institutions realised the importance of having a two-way exposure of scientists from India and abroad through collaborative research visits and conferences. Developing these contacts and the visits was not easy as the infrastructure available in our country for such visits was inadequate, and visitors often had reasonable grounds to fear about food, transport, and health. The Indian hosts had to make extraordinary efforts to make these visits a success. Two conferences in the early part of the 70s and one towards the end stand out in showcasing and internationalising the Indian research efforts. The first two occasions were the silver jubilee of the Indian Institute of Metals in 1971 and the golden jubilee celebration of the Department of Metallurgy, Banaras Hindu University, in 1973. Through these events, the Indian metallurgical community, particularly the younger ones, got to know the best in the world and succeeded in developing a lasting network.

An example is the present author's interaction with Professor Bruce Chalmers and his encouragement to publish the first paper from the master's thesis in *Scripta Metallurgica*. The other conference in 1978 was organised under the leadership of Prof T. R. Anantharaman and entitled 'Metal Science: The emerging frontier'. This event, attended by some of the outstanding world leaders, succeeded in bonding the entire physical metallurgy community of the country and the future generation. The continuing organisation of major conferences in the next three decades and their impact on globalising Indian metallurgy are often not appreciated enough. However, they played significant roles in shaping many young metallurgists and paved the way for major international collaborations.

29.3 The Period of Consolidation

The research carried out through the graduate programme at Indian educational institutes started making a global impact in the late 70s. The decade of the eighties and nineties saw several outstanding contributions from Indian academics and R&D laboratories soaring the nation's confidence. A few of these examples are: the role of vacancies in precipitation reaction at IISc, contributions towards the metallic glass and its thermodynamics at BHU, solid-state ordering reaction and its wave description at Bhabha Atomic Research Centre (BARC), the discovery of decagonal quasicrystal and its ordering at IISc, the significant contribution to Zirconium metallurgy at BARC, new titanium alloy development at Defence Metallurgical Research Laboratory (DMRL), research on superplasticity at IIT Madras and processing maps at IISc, discontinuous precipitation at IIT Kharagpur and early research on nanocrystalline materials at Tata Institute of Fundamental Research (TIFR) and IISc Bangalore.

The country also witnessed the capacity to develop indigenous technologies. It demanded an understanding of critical materials and particularly their mechanical properties and life predictions. Consequently, several clusters across academics and R&D laboratories emerged with associated infrastructures and inspiring leadership. Mentions can be made of high-temperature facilities at DMRL and National Metallurgical Laboratory (NML) and the creation of strong groups at IIT Madras and IISc, Bangalore.

29.4 Current Century

When the country entered the current century, metallurgical R&D was on a relatively firmer ground. One often used to hear a statement that metallurgical and materials engineers did not fail the country. There are many successes that others have also highlighted in this book. However, in the process, we may not have evaluated where we should have reached and the future steps needed at each decade to reach the desired next level. In the following, I shall present a personal view of what we missed and need to do to reach the elite levels focussing on advanced R&D and fundamental research.

One often hears debates about basic and applied research and requirements to cater to the country's needs. In this debate, we ignore that the only classification that one can do in research is excellent and impactful, which opens up new avenues. This kind of work contributes both nationally and globally. There is research that may sound either scholarly or tuned to the national need and give an impression of solving a national need. However, they contribute only incrementally and often stagnate the growth of science and technology. The distinction between these two must be made. Good research is always bold and creative and opens up innovation space. Thus they often cannot be anticipated, something that only visionary managers can anticipate.

However, our country has matured to accommodate all kinds of views and still make progress.

Thus present century saw the flowering of technologies like fast breeder reactors or aircraft or rockets where materials play a crucial role. The research on high entropy materials and newer high-temperature superalloys has attracted global attention. Many younger researchers are striving to achieve breakthroughs in their respective fields. However, are we satisfied with our innovativeness and rankings or academic standing? It is something we need to introspect.

29.5 Some Critical Issues from Personal Perspectives

The real difficulty in engineering disciplines like metallurgy and materials science is appreciating and bridging academic advancement with the industry's current requirements. In a country like India, industry, for the most part, is geared towards meeting the immediate demand of its populace. Thus, they may have limited bandwidth to invest in breakthrough academic discoveries and innovations and invest in new future-ready technology. It is only recently that some of our industries are projecting international leadership that requires such advancement. Thus, creative and advanced research cannot attract industry support in any significant way. Only the government and the strategic sectors have mainly sustained research efforts in our country and should get appreciation from the entire profession. In recent times, multinational companies with global reach started leveraging these capabilities. My experiences suggest a slow realisation in some Indian industries of the need to leverage the R&D base available in academic sectors. However, every single time funds need to be transferred, a chill sets in, and we cannot progress. Joint work at a smaller scale (even without money) needs to be initiated to build confidence. Unfortunately, with the increasing cost of research and time, this approach is not feasible in today's academic world.

There are multiple issues that need to be debated and addressed. The space constraint prevents flagging them. However, a final issue that I like to flag is the absence of what I like to term research support organisations or industry. If one works on something that can be applied to industry, the immediate question that is asked in our country is "Why do not you make a component and test"? The people who ask these questions often have struggled themselves to do the same in their organisation. The problem is not with the question but the lack of realisation of the ground situation. The absence of an intermediary organisation/industry that is mandated to support these activities represents one of our research ecosystem's most significant stumbling blocks. A simple example will suffice. In the process of validating the high-temperature properties of newly discovered alloys, the author needed materials in reasonable quantities (5–10 kg) that should be melted, cast and hipped. Of course, our country can produce 100 kg of such materials if mandated from the top but not routinely and in the quantity needed for intermediate development. Even if it is possible, it can only be obtained after numerous requests. One cannot blame the other side since time is valuable for everybody. Fortunately, one of my well-wishers in a multinational company came to the rescue and order was placed by his company to a USA company specialising in doing just that, research support. I do not know whether setting up similar companies for research support is financially feasible. However, without a solution to this problem, our scientists and engineers may not be able to transition their ideas to real-world applications.

29.6 Epilogue

The occasion of the platinum jubilee is a perfect occasion to reflect on the past and look at the future. We must be proud of what we have achieved starting from scratch due to a set of outstanding professional leaders in the past. We achieved world-class in several discoveries. Some of these probably were more recognised by the outside world than in this country. But we in India should do better. We do have the capacity to be world leaders and should now look forward to fructifying this aspiration by developing a research ecosystem that promotes excellence and appreciates the new and unconventional paths.

Chapter 30
Accelerated Discovery, Development, Manufacturing, and Deployment of New Materials Using Integrated Computational Materials Engineering (ICME) Tools and Digital Platforms

B. P. Gautham

Abstract The transformative potential of the integrated computational materials engineering (ICME) tools and technologies in accelerating the discovery, design, development, manufacturing and deployment of novel materials and products, is briefly reviewed. A practical approach to the realization of the benefits of ICME through judicious use of computational tools based on physics and machine learning along with the utility and power of digital platforms, such as TCS PREMAP, in actual industrial practice for solving challenging problems facing the industry, is illustrated with the help of an example of finding the appropriate composition and processing conditions for achieving a given set of target properties of a dual phase steel grade sheet for its application in the auto sector.

30.1 Introduction

It is well recognised that access to and self-sufficiency in terms of competence in the discovery, design, development and deployment of novel materials for specific applications in engineered products is a critical element of any nation's strategic plan for growth, since the availability of critical materials is one of the most important bottlenecks in various sectors of industry, in particular, for the strategic sectors of defence, aerospace and atomic energy in India. Immediate material challenges facing the industry in India span a wide range such as materials for light weight for automobiles, materials that can withstand elevated temperatures in the power sector and designing novel materials for medical prosthetics, preferably 3D printed. Conventional approaches of materials conceptualization to deployment, using empirical

B. P. Gautham (✉)
Gerald Tennyson, Sreedhar Reddy and Pradip, TCS Research, Tata Consultancy Services, Pune, India
e-mail: bp.gautham@tcs.com

R. Divakar et al. (eds.), *Indian Metallurgy*, Indian Institute of Metals Series,
https://doi.org/10.1007/978-981-99-5060-7_30

knowledge and trial-and-error-based experimentation, is known to take an enormous amount of time and effort, with typical timelines of 15–20 years being common (Council 2008). It is well acknowledged that an integrated computational materials engineering (ICME) approach on the other hand can accelerate this process, but there exist several challenges to overcome, in order to realize it in actual industrial practice (Council 2008).

The recent advances in computational tools and technologies pertaining to materials engineering, ability to deal with and leverage the vast amount of data through artificial intelligence (AI)/machine learning (ML) techniques and developing closed-loop iterations across these in silico technologies with data collected through laboratory experiments and manufacturing plants provides engineers an immense opportunity to accelerate materials development and deployment in the form of engineered industrial products. Such an approach also fits well with the current industry trends such as Industry 4.0 (Failed 2013) Engineering 4.0 (Foo and Ringel 2018), Materials 4.0 (Jose and Ramakrishna 2018), digital thread and digital twin (Mavris et al. 2018).

The concepts of Integrated Computational Materials Engineering (ICME) (Council 2008) and Materials Genome (OSTP 2011) have emerged over the past decade to address the above-mentioned challenges. The US National Academy of Sciences report on ICME is a landmark report that introduced the concept of ICME with an aim "to enable the optimization of materials, manufacturing processes, and component design, long before components are fabricated, by integrating the computational processes involved into a holistic system." A follow-up report commissioned by TMS identified specific ways by which ICME can be exploited in different sectors such as automobile, aerospace and maritime industries (TMS 2011). The report emphasised the need for "integration of personnel (for example plant engineers, designers, etc.), computational models, experiments, design and manufacturing processes across the product development cycle, for the purpose of accelerating and reducing the cost of development of a materials system or manufacturing process." Subsequently, several reports from TMS provide details on various challenges in realizing ICME, such as multiscale modelling (The Minerals Metals Materials Society (TMS) 2015), data infrastructure (Building a Materials Data Infrastructure: 2017) and validation and verification needs for materials modelling (Verification Validation of Computational Models Associated with the Mechanics of Materials 2019). While these reports predominantly focussed on structural materials, another parallel initiative in the US called the Materials Genome Initiative (OSTP 2011) focussed predominantly on functional materials with emphasis on the discovery of novel functional materials using data-driven technologies of AI/ML along with computational materials engineering tools. Similar efforts in other parts of the world, such as through process modelling (TPM) initiatives in Europe to accelerate industrial production of materials (Hirsch 2006) have provided further impetus to ICME and materials genome initiatives. The readers are also referred to a series of recent books and monographs on ICME by Horstemeyer (Horstemeyer 2012), Schmitz and Prahl (Schmitz and Prahl 2012) and (Koenraad 2007; Raabe 1998). Professional bodies such as NIST (NIST xxxx) and EMMC (EMMC xxxx) are playing a crucial role globally in developing standards and collaboration platforms for the realization of ICME in actual industrial practice.

Various initiatives in India such as the ICME National Hub at IIT Kanpur (Kanpur xxxx), ICME Lab at IIT Madras (Madras xxxx), and an interdisciplinary master's program at IIT Hyderabad have been launched to help create an appropriate ICME Eco-system in India. A good example is the recently launched Microsim open-source software for phase-field modelling by a consortium of Indian institutions, IISc Bangalore, IIT Bombay, IIT Madras and CDAC Pune (Microsim xxxx).

Tata Consultancy Services—TCS Research, the organization to which the authors belong, embarked on a strategic initiative more than a decade back and has been working on various fronts, to build appropriate tools, technologies and digital platforms to address the challenges of accelerated materials development through ICME (Janet Allen et al. 2013) centred concepts. This paper is based on the authors' own experience in solving challenging materials development related problems for our industry (Gautham and Reddy 2020) and their own insights based on this experience on how ICME, Materials Genome and other computational paradigms and recent advances in digital technologies can be gainfully leveraged for this purpose. A pragmatic ICME approach to materials development is presented in the following section. The need for digital platforms for industrial realization of such an approach along with a brief description of a digital platform developed by TCS called TCS PREMAP (Platform for Realization of Engineered Materials and Products) is presented in Sect. 30.3. A case study on the design and development of new grades of steel sheets, using TCS PREMAP is presented in Sect. 30.4 to illustrate the utility and power of digital platforms. A concluding section summarizes the immediate challenges and the efforts needed for industrial adaptation of ICME-based development of materials and products, in actual practice.

30.2 ICME for Industrial Problem Solving

Materials for engineering applications are developed for accomplishing a set of target properties and product-performance characteristics. Depending on the specific need and value to be delivered, the material is developed for a set of requirements for either a single product or a class of end-use applications. This section illustrates, with an example, the author's views on dealing with the challenges associated with arriving at a solution in the development of a new grade of steel sheet.

30.2.1 Traditional Process of Alloy Design to Production

Figure 30.1 depicts a chain of steps from design-to-industrial scale production, for a new grade of steel sheet for automotive applications. The starting point is to capture a set of requirements such as the strength, ductility, weldability and manufacturability

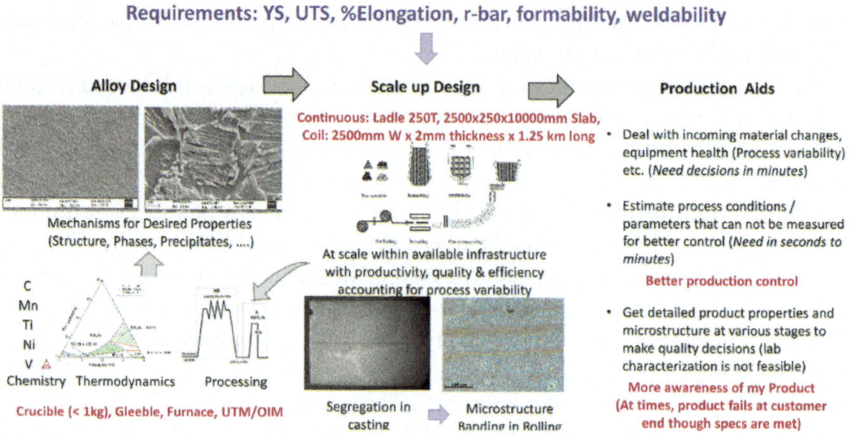

Fig. 30.1 Typical process from alloy design, scale up and production for new steel grades

based on the market needs for specific end-use applications. This information is then used to come up with the desired composition and microstructure of the manufactured steel sheet at its end stage of production along with appropriate solidification and thermomechanical processing conditions to achieve it. Typically, the alloy is designed at the laboratory scale, followed by engineering scale-up of the process and subsequently its production at the commercial scale by the industry. All these stages can be significantly accelerated using the computational paradigm of ICME. While there exist significant gaps in understanding and modelling and simulation of microstructure evolution across the process chain and its performance in actual application (The Minerals Metals Materials Society (TMS) 2015), these tools and techniques can be judiciously used for making good engineering judgements and significantly reduce the number of trials needed to accomplish the final goal.

Laboratory scale alloy design: Traditionally, the design of new steel grades involves experts using their tacit knowledge, phase diagrams and other related information to determine the appropriate chemistry and processing conditions. This knowledge guides the laboratory scale production of test specimens through the preparation of a cast bar and appropriate thermo-mechanical processing conducted in laboratory scale equipment such as Gleeble. Typically, the scale of production is in kilograms and the alloy is tested for the properties of these laboratory specimens. Several iterations are carried out until the targeted properties are achieved. However, translating the laboratory-scale results to industrial-scale production with accompanying constraints of available plant setup, is the next major challenge and often the key bottleneck in the industry.

Scale-up design for a given plant: While the laboratory scale testing is done to establish the alloy chemistry and the basic manufacturing process steps; during industrial scale production, where over 250 tonnes of steel per heat are cast into slabs of cross-section in the range of 2000 mm x 250 mm end up in a sheet about a millimetre in thickness and length in kilometres, one needs to consider various other critical aspects related to achieving the final properties of the product such as macrosegregation and its impact, variability of temperature and chemistry, non-uniformity of the temperature of the slab being rolled, microstructural banding that may get accentuated during large-scale production for some chemistries, etc. Very often, this challenge is dealt with, with the help and advice of a set of experts who based on their past experience and domain knowledge, come up with a set of experiments to confirm their suggestions and advice. Ultimately, the end result is achieved through extensive trial and error experimentation, either at laboratory scale, pilot plants or sometimes even in actual industrial production conditions on the shop floor (for example, test heats).

Production of end-product in the plant mills: In general, even after conducting trials on the shop floor, it requires a few batches of production before the plant engineers are able to stabilize the production of a new grade of steel and consistently meet the desired quality and production conditions at that scale. Further, even during normal production conditions, the plant engineers need to make multiple dynamic decisions such as dealing with any minor variations of chemistry or changes in process/equipment conditions that may adversely impact the product during production at scale and thus control the manufacturing process—again in the face of extremely limited data being available in the plant. This real-time decision-making of the manufacturing process again requires enormous experience and expertise and it varies from person to person and organization to organization.

30.2.2 Acceleration Through ICME Methodologies

As discussed earlier, ICME-based methods can be leveraged to derive significant benefits in accelerating the above process, reducing the number of experiments needed and thereby significantly reducing the time and cost of development (Cotton et al. 2012). Have illustrated the savings in time, cost and effort that can be achieved through ICME approaches at various stages of material product development. As shown in Fig. 30.2, various computational tools and digital technologies can be used to accelerate the product development process. While some of these tools are already in use in the industry, one requires new skill sets to be able to use more recent simulation tools based on either physics-based models and/or data-driven AI/ML techniques.

It is important to emphasise that the prudent approach to using such modelling tools lies in using them as decision-accelerators rather than as tools to make accurate predictions—often the debate over the accuracy of model prediction is not warranted (most of the time, the resistance to change in the current industrial practice stems from this argument) since the product developers or plant operators, need good-enough predictions to make decisions and finalize through experiments in product development or quality observations during production to achieve the targeted goals. It, therefore, becomes important to recognise the advantages and limitations of these

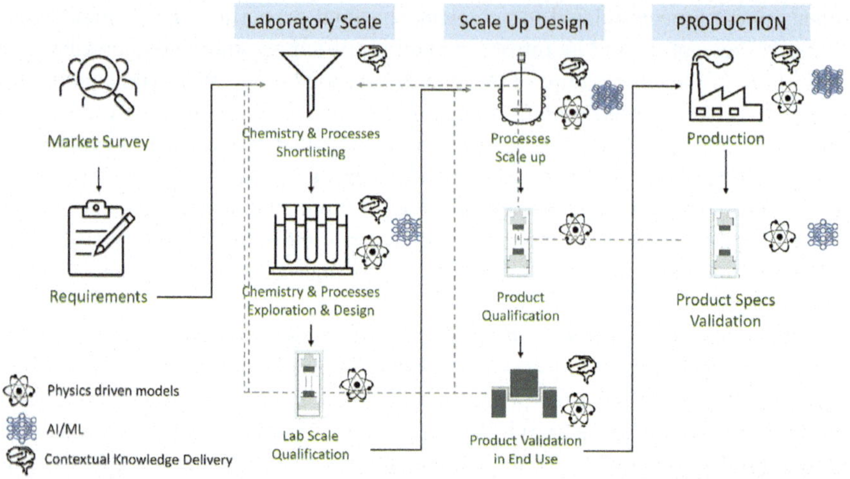

Fig. 30.2 Use of digital technologies to aid in accelerating product development

tools and come up with appropriate strategies to leverage them and while inarguably research is being carried out to continuously improve their predictive capabilities.

Alloy Design: Thermodynamic computations with CALPHAD-based tools and simulation studies in the kinetics of formation of different phases and precipitates can be used to perform an initial screening of alloy-chemistry and processing conditions to reduce the number of laboratory trials required to arrive at the final composition and processing conditions (Walle 2019). Further, microscale simulations at a representative volume element (RVE) can be carried out to predict the evolution of microstructure under different processing conditions using computational approaches such as phase field modelling, cellular automata, etc. These tools can be integrated with thermodynamic/kinetic modelling tools for acquiring relevant information, as needed thereby providing an enhanced view of microstructure evolution. Often it is prudent to use phenomenological or empirical models along with these sophisticated simulation tools to bridge the gaps in their ability to capture certain phenomena or to reduce the computational cost. Finally, micromechanics-based tools can be used to understand the deformation behaviour of the RVE (Khan et al. 2017). The readers are referred to publications describing excellent case studies on these lines—namely, the ICME-based design of aluminium alloys (Allison et al. 2006) and of steels (Farivar et al. 2017). Application of multi-objective design methods in conjunction with physics-based modelling tools is ideally suited for arriving at robust choices for alloy composition and processing conditions for alloy design as well as process scale-up (Khan et al. 2019).

Design for Scale-up: As discussed earlier, scaling up laboratory findings needs appropriate strategies to meet additional challenges. We need process models that simulate the transport (heat and mass) and deformation behaviour of the material along with its interactions with the equipment being used on the shop floor. A

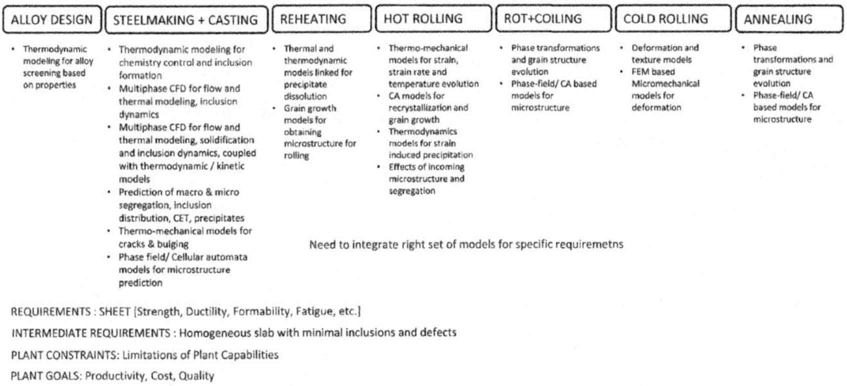

Fig. 30.3 Various phenomena to be modelled across the production value chain for steel sheets

variety of CFD/FEA-based tools are used for modelling manufacturing processes either linked with computational materials engineering tools and/or phenomenological models discussed above. These models need to be integrated across processes and scales, often called horizontal and vertical integration (Tennyson et al. 2015). For example, during the design of the continuous casting process, these tools are used to understand various phenomena such as macro-segregation (Mangal et al. 2018), columnar to equiaxed transition (Chaube 2015), etc. Figure 30.3 illustrates a broader view of various phenomena modelled across the process chain of steel sheet production.

Finally, it must be emphasised that in real-life scenarios, the challenge is to solve an inverse problem of what needs to be done for a stated final goal, whereas all the available tools provide forward predictions. For meeting this challenge, appropriate decision support tools are required, the most popular of these being, optimization tools used as inverse problem solvers (Nellippallil et al. 2020). Other techniques such as compromise decision support (cDSP) methodologies provide satisfactory but robust (need not be the most optimum but most stable under practical conditions of normal variability for different operating conditions) solutions, which will work even in the face of minor variations of process/equipment conditions in the plant (Shukla et al. 2015). Finally, while the production is underway, the time needed for decision-making will be extremely short and we require real-time predictive capabilities—either faster physics-based models or surrogate models or a set of empirical thumb rules or the more recent development of AI/ML-based models.

Use of AI/ML-based tools: AI/ML-based tools are increasingly being used to take advantage of the enormous amount of data available through simulations, experiments and/or data collected during plant scale production. AI/ML-based models can be used in a wide variety of contexts such as (a) bridging the gaps in physics-based understanding with the help of data-driven models—for example, estimation of the fatigue properties of a material from available data (Agrawal et al. 2014), which otherwise is difficult to model from physics-based methods (b) accelerating computation—for example, crystal plasticity computations accelerated using ML (Weber et al. 2020), (c) faster computation through building ML models to effectively act as simple transfer functions and (d) aid decision- making through classification or inverse inferencing (Sardeshmukh et al. 2019). The power of AI/ML techniques to build models from the available experimental or plant data has been demonstrated in actual operations and these models often provide a sufficiently good working model, which is extremely valuable in real-time decision-making.

Knowledge Engineering: While physics-based and AI/ML-based tools help aid decision-making to a significant extent, due to the complexity of the problem at hand across various length scales and over the complete chain of manufacturing operations, a large amount of formal or tacit knowledge is needed in making decisions at every stage. Typically, each industry possesses an enormous amount of usable- knowledge based on its own plant experience, often spread across various reports and documents within the company, in addition to traditional sources such as research publications, reference books and handbooks. It is important to emphasise that modern digital technologies can be leveraged to either institutionalize formal curation or even conduct deep domain specific search into these documents to provide the right contextual information to engineers as and when needed, thus reducing the dependence on their memory and accuracy of recall by those operating the plants and/or experts providing advice. This knowledge can be made available in a digitally processable format either for intelligent search leveraging natural language processing (Shah et al. 2018) or through modelling knowledge in a way that it can be delivered contextually (Yeddula et al. 2016).

Industrialization of ICME: The objective of a digital platform is to integrate various threads as well as sources of knowledge and after making use of all the above-mentioned computational and digital tools, to provide relevant information to assist decision-making, in an easy-to-use form by the users—which could be product designers, research scientists, plant operators, plant engineers, and/or plant managers (Gautham and Reddy 2020). A well-designed digital platform ensures that these different tools along with appropriate decision workflows are made available to a larger set of users who routinely make decisions of various kinds, at various stages of the manufacturing process in the plant, in an easy-to-use form, as and when needed. It is important to realize that the user need not be an expert. The platform provides the necessary expertise in appropriate form as and when needed tailor-made for the user. This aspect is further discussed in the next section.

30.3 Digital Platforms to Enable Industrialisation of ICME

30.3.1 Need for Digital Platforms for ICME

As discussed above, in addition to a wide variety of computational tools that need to be integrated appropriately, an industrial adaption of ICME or material genome paradigms necessarily requires digital platforms that can make the information available in an integrated form, for decision-making by a wide variety of users across the value chain. The digital platform should also facilitate the integration of new computational tools and methods being developed by researchers and make them available in a form appropriate for different users—for example, product development teams to carry out simulation and design activities, process engineers to optimise and enhance the manufacturing processes and plant operators for making routine decisions on the shop floor regarding different parameters to be controlled for a given steel grade, for instance. The key drivers for such a platform are described in greater detail in a prior publication (Gautham and Reddy 2020) and briefly summarized below.

- Standardization of terminology and digitalization of knowledge capture
- Library of master physics-based models for scenarios that are easily customizable for specific needs of users with wider applicability
- Easy deployment of such models developed by specialists to make them available for a larger set of users, who may not be experts
- Ability to construct and use integrated workflows involving knowledge and models for decision-making across manufacturing process operations
- Contextual delivery of knowledge both in terms of the appropriate use of tools, input parameters and tacit knowledge and thumb rules to enable decision-making
- Integrated models made available to appropriate research and plant engineers to make decisions with easy-to-use interfaces
- Enabling the building of new AI/ML models and the creation of new knowledge
- Connectivity with other digital platforms, already being used in the plant such as ERP and MES for information related to the raw materials (availability and cost), plant conditions, etc.
- Seamless improvement/upgrade/update of models/workflows with enhanced capabilities and expanding applications, for example, dealing with new process conditions
- Keep up with the latest technology trends such as cloud adoption and IIoT

We describe in the following section, one such platform TCS PREMAP, developed at TCS Research and currently being used by multiple end-user industries.

30.3.2 TCS PREMAP—Platform for Realisation of Engineered Materials and Products

TCS has developed a digital platform based on a state-of-the-art, model-driven software engineering paradigm to address the above-mentioned needs in the context of ICME. It is called TCS PREMAP, a platform for the realization of engineered materials and products (Gautham et al. 2013, 2017). This platform enables ontological definitions of various entities of interest such as a material, process, equipment, model, etc. using metamodels acting as templates (Reddy et al. 2017). These models become the core data and semantic models to express knowledge. Various forms of knowledge–e.g., a model with its applicability conditions, a right phase diagram applicable under specific conditions, a thumb rule that can be used to set the furnace temperature for a specific need—can be captured systematically using the ontology built into the platform. This enables easy storage and recall of the correct information as and when needed, appropriate for the context and intent of the user of that information (Yeddula et al. 2016). This knowledge, in executable or referential form, is delivered to users as they undertake the execution of workflows for decision-making.

A decision workflow builder allows for the construction of a series of steps such as execution of a simulation tool, showing specific outcomes to the user for the decision regarding the next step, carrying out the design of experiments (DoE) based simulations, optimization, model tuning, etc. These workflows can be constructed by experts and can be stored as part of the usable information in the platform. These workflows can then be used by a wider community of engineers to make decisions, as needed. The important point to emphasise is that the users of these workflows need not have expertise in modelling or familiarity with the modelling tools being used. These workflows are designed in such a manner that during the execution of these workflows, the platform automatically invokes and uses the appropriate simulation tool, fetches the required data from appropriate data sources and enables the decision maker to make appropriate decisions, appropriate for the problem at hand. The ontology base along with facilitators enables such a seamless data integration across various tools. It also allows for building machine learning (ML) models from past simulations or based on data from experiments for future use as and when required depending on the context. Finally, a set of decision support tools help enable the carrying out of studies such as design of experiments (DoE), robust design, etc. This platform can be used for offline decision-making purposes such as product development, process optimization and machine learning as well as online decision-making by plant operators for diagnostics and deviation management. The utility and power of TCS PREMAP are illustrated in the following section with the help of a case study, where ICME-based tools along with the digital platform TCS PREMAP were used successfully for the industrialization of ICME. The readers are also referred to another case study related to the development of new materials for gears, published earlier (John et al. 2017) (Fig. 30.4).

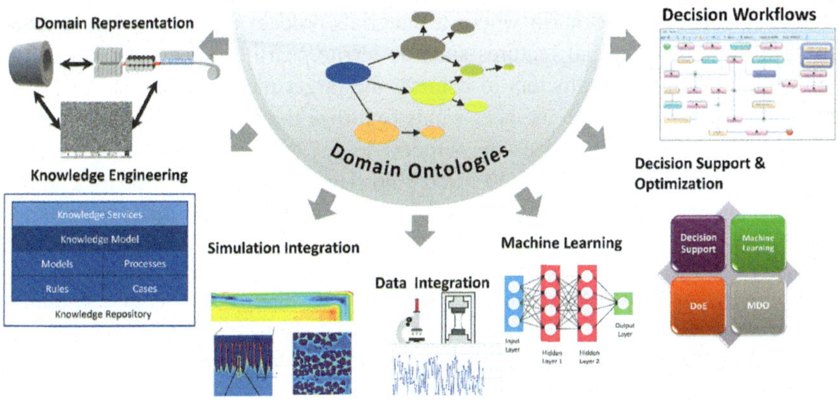

Fig. 30.4 TCS REMAP, an enabling platform for ICME

30.4 An Illustration—Design and Development of a Dual Phase Steel Sheet

Cold-rolled dual-phase steels are an important class of advanced high-strength steels (AHSS) that are being explored for automotive applications in order to achieve desired targets of light-weighting and yet meet all safety regulations, for example, having adequate crash resistance.

The challenge is to be able to find an appropriate combination of composition and processing conditions, based on the optimization of the properties and thus the microstructure of these steels to meet the ultimate requirements specified by the user industries. A phase-field model developed by (Zhu and Militzer 2015) is used to simulate microstructure evolution during inter-critical annealing and a FEM based micromechanics approach (Ramazani et al. 2013) is employed over the entire microstructure, to calculate the uniaxial flow curve for the resultant microstructure (Khan et al. 2019). This whole process of simulation and thus prediction of the final microstructure and the properties has been implemented on the integration platform TCS PREMAP as an easy-to-use workflow. A whole variety of different scenarios for a given combination of different processing conditions can thus be explored and final properties thus obtained can be evaluated. The workflow can utilize different models (simulation tools) and also a combination of different models which were built by different research groups. Thus the users, for example, product development and process optimization teams can simulate all kinds of possibilities using TCS PREMAP which not only provides the work-flow for the prediction of microstructure and properties but also provides all necessary help (for example on how to select various parameters) to conduct such optimization studies by providing appropriate knowledge needed for solving the problem based on the contextual search through the capabilities inbuilt in the TCS PREMAP digital platform.

Let us consider a scenario where a materials/process designer needs to come up with the desired initial features such as ferrite/pearlite ratios from hot rolling and optimal process parameters to achieve the target mechanical properties for a given alloy chemistry. The engineer wants to utilize the models/simulation tools to explore and evaluate the processing window to achieve the desired microstructure and properties as shown in Figs. 30.5 and 30.6. The engineer though being a domain expert, may not be fully aware of the models and their usage and is likely to encounter many challenges in the usage of the tools, some of which are listed below.

1. Guidance on the usage of the models/ choice of values of model and simulation parameters.
2. Guidance/suggestions on the initial guesstimate values of processing conditions to start with.

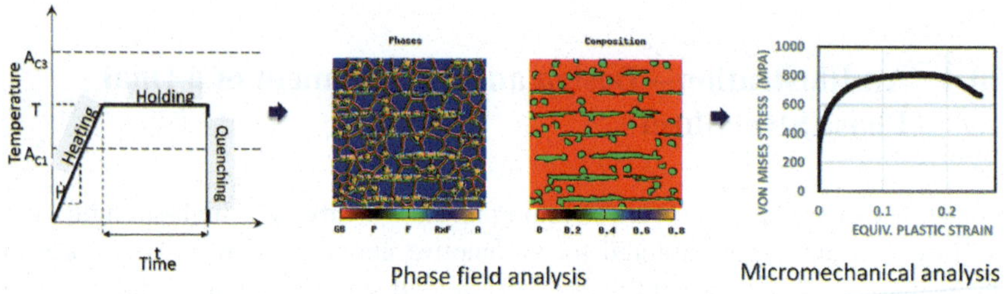

Fig. 30.5 Integrated workflow to simulate inter-critical annealing and properties

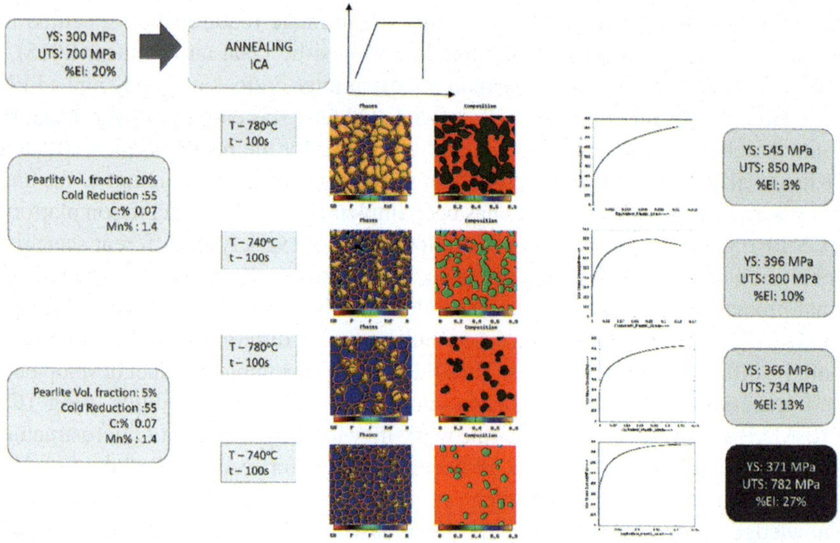

Fig. 30.6 Sequential execution of integrated simulation tools to achieve the desired target properties

3. For a given set of target properties required, whether an equivalent simulation has already been executed and the results available thereby saving time on redundant simulations.
4. Guidance on the results obtained that is, if the target requirements are not met, guidance is required on the parameters to be changed to obtain the desired results.

The integrated application is created on TCS PREMAP with appropriate ontologies, tools, knowledge elements, user interaction screens, and workflow as shown in Fig. 30.7. The implemented workflow is shown in Fig. 30.8. Once the application is created, the user can execute the workflow by providing the desired inputs as shown in Fig. 30.9. Knowledge elements built into the platform interact with the user to provide guidance on the choice of input parameters and also provide guidance to the user in making decisions with the results obtained. The data from all such executions are stored and can be reused. For example, the platform will guide the user whether the results are already available from a previous simulation with similar input conditions.

The workflow shown in Fig. 30.8 can be executed to arrive at the process conditions to achieve the required properties. A past simulation checker is used to extract the data

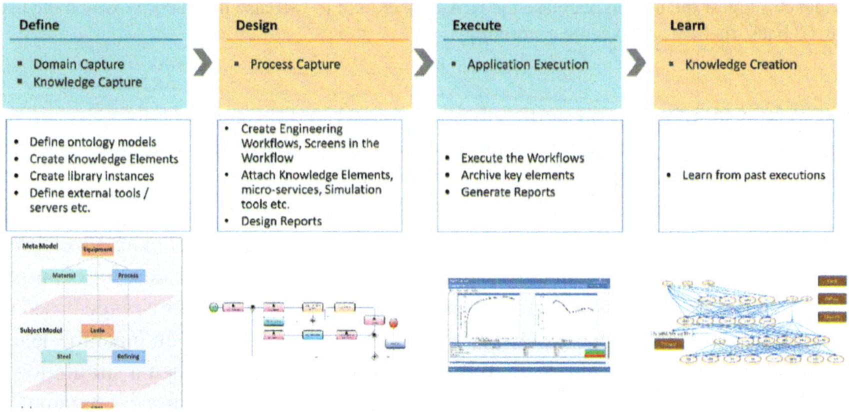

Fig. 30.7 Define-Design-Execute-Learn approach in TCS PREMAP

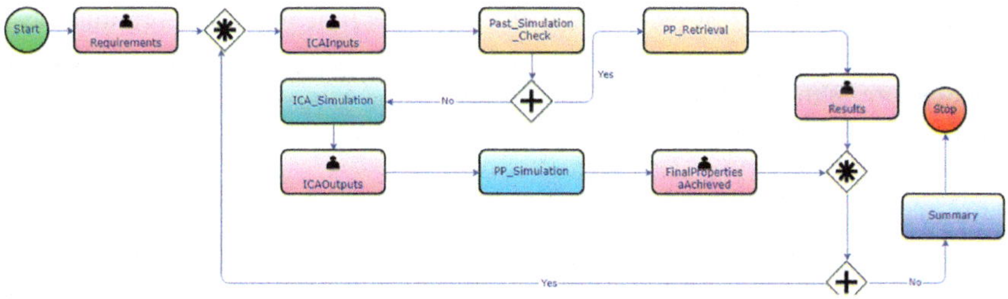

Fig. 30.8 An integrated workflow of intercritical annealing and property prediction

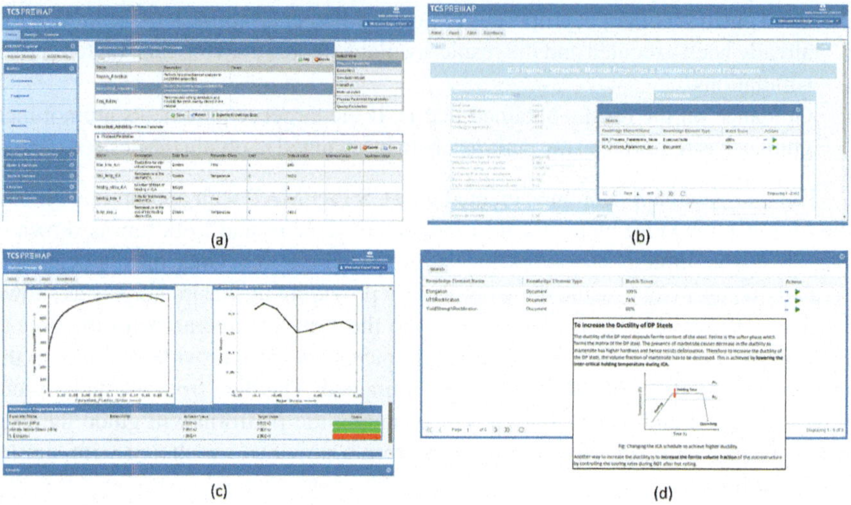

Fig. 30.9 **a** Ontology creation on TCS PREMAP, **b** During execution a knowledge element is guiding the user in selecting the inter-critical annealing parameters, **c** Outputs showing the flow curve and the FLC predicted with a low ductility. **d** Knowledge elements in the form of document evidence suggest the user to decrease the ICA temperature to increase the ductility which the user can go back, modify and execute accordingly

for a given set of target properties that were generated through previous executions thereby saving a significant amount of computational effort. The engineer proceeds further, if not satisfied by the results, in exploring a different set of process parameters. Here the engineer is guided in the choice of process parameters that may provide the desired outcome thus supporting the engineer to obtain the results in an accelerated manner. This knowledge can come from simple thermodynamic rules, for example, phase diagrams that can suggest the user the inter-critical temperature range based on the composition or simple expert-driven thumb rules or data-driven machine-learnt models. The engineer is also guided on the actions to be taken when a simulation is not providing the desired outcome, for example, on the choice of process parameters that can come from literature sources, expert-driven thumb rules or learnt models. With the help of TCS PREMAP, the user is thus able to screen different possibilities and arrive at a preferred set of operating conditions, best suited for achieving the desired final properties of the steel sheet. Enormous time and effort are thus saved in optimizing the operating parameters of various unit operations involved in the manufacture of the steel sheet.

30.5 Concluding Remarks

We have tried to illustrate the utility and power of an integrated computation materials engineering (ICME) approach to accelerate novel product development (including design, development and deployment of new materials and products) for a given set of target properties and thus reduce the cost associated with development. This is one of the immediate challenges facing the industry. We have also highlighted that a judicious use of these tools and methods will be of significant value to our industry and should be leveraged appropriately while the tools are of course continually being improved through research.

Digital platforms for ICME, for example, TCS PREMAP, as illustrated in this communication are needed to ease decision-making for engineers who are involved in the design, development and manufacturing of novel materials and products in the industry. A platform of this kind not only helps one to solve the immediate problem but the whole process of discovery and efforts to arrive at the best possible solution under various constraints and contexts is also captured by the platform. As data is generated and decisions get made using the platform, the platform also facilitates the storage of this knowledge for future use—basically to harness past experience to enhance and enrich the knowledge base for solving a similar problem in the future, thereby also providing a repository of knowledge—addressing another key problem of practical importance in the industry.

To summarize, we suggest the following, while adapting ICME-based computational approach for accelerating material design, development and deployment by our industry:

- Use an ICME-based approach to accelerate product and process development in terms of time and effort and reduced laboratory experiments/plant experiments or troubleshoot quickly, the production/process issues faced by plant operators on the shop floor
- It is not the accuracy of prediction that is paramount but the ability to harness whatever our past experience, capability and knowledge provide us through the use of digital platforms to arrive at a good enough prediction as inputs for solving the problem at hand.
- Democratize the use of models and make them available as decision-support tools for different stakeholders and users in the industry rather than all the knowledge and modelling tools remaining as research tools only.
- Ensure that all the models, data and knowledge in whatever form is available in the industry is stored in a platform-based knowledge repository. It is well-curated and properly maintained for sustained use and growth.
- The above-mentioned capability should be enabled through appropriate integration platforms like TCS PREMAP which will have immense value in the long term.

Acknowledgements The authors are grateful to TCS management and in particular, Mr. K Ananth Krishnan, CTO for his continuous support, help and encouragement to continue to work in this exciting area of immense value to our industry. The authors also gratefully acknowledge the kind help and support over the years from so many of our colleagues both within TCS as well as from those we have worked with in the industry and from our academic collaborators and advisers, who have contributed in different ways to reach where we are today in our efforts to make ICME-based accelerated materials and product development an industrial success story. We also acknowledge Dr. Appa Rao Chinta of Tata Steel India and Prof. Nilesh Gurao of IIT Kanpur for the microstructure images used in this manuscript.

References

Agrawal, A., Deshpande, P.D., Cecen, A., Basavarsu, G.P., Choudhary, A.N., Kalidindi, S.R.: Exploration of data science techniques to predict fatigue strength of steel from composition and processing parameters. Integr. Mater. Manuf. Innov. **3**(8), 90–108 (2014)

Allison, J., Li, M., Wolverton, C.: Virtual aluminum castings: An industrial application of ICME. J. Metals **58**, 28–35 (2006)

Chaube, S., T. G., Singh, A.: Modelling of Columnar-to-Equiaxed transition and inclusion distribution in continuously cast steel billets, Trans. Indian Inst. Met., **68**, pp. 1207–1213 (2015)

Cotton, J., Frohlich, C., Glamm, R.: What boeing wants from integrated computational materials engineering for metallic materials, In 53rd AIAA/ASME/ASCE/AHS/ASC Structures, Structural Dynamics and Materials, Hawaii, USA, (2012)

Council, N.R.: Integrated computational materials engineering: a transformational discipline for improved competitiveness and national security, Natl. Acad. Press. https://www.nap.edu/download/12199. (2008)

EMMC, EMMO: an ontology for applied sciences. Available: https://emmc.info/emmo-info/. Accessed 31 Dec 2021

Farivar, H., Rothenbucher, G., Prahl, U., Bernhardt, R.: ICME-Based process and alloy design for vacuum carburized steel components with high potential of reduced distortion. In Proceedings of the 4th World Congress on Integrated Computational Materials Engineering (ICME 2017), (2017)

Foo, E., Ringel, M.: The race toward engineering 4.0. Race Eng. 4.0, (2018)

Gautham, B., Reddy, S.: Materials design, development, and deployment in manufacturing industry: a digital paradigm, in advanced digital architectures for Model-Driven adaptive enterprises, IGI Global, pp. 251–265, (2020)

Gautham, B., Singh, A.K., Ghaisas, S.S., Reddy S.S., Mistree, F.: PREMAP: A platform for the realization of engineered materials and products, in ICoRD'13, Lect. Notes Mech. Eng. (2013)

Gautham, B., Reddy, S., Das P., Malhotra, C.: Facilitating ICME through platformization. In Proceedings of the 4th World Congress on Integrated Computational Materials Engineering (ICME 2017), (2017)

Hirsch, J.: Through process modelling. Mater. Sci. Forum **519–521**, 15–24 (2006)

Horstemeyer, M.F.: Integrated Computational Materials Engineering (ICME) for metals: using multiscale modeling to invigorate engineering design with science, Wiley, (2012)

IIT Kanpur: ICME national hub @ IIT Kanpur. Available: https://www.iitk.ac.in/ICME/. Accessed 31 Dec 2021

IIT Madras, Integrated Computational Materials Engineering (ICME) Laboratory. Available: https://mme.iitm.ac.in/icme. Accessed 31 Dec 2021

Janet Allen, F.M.J.P., Gautham, B., Singh, A., Reddy, S., Kulkarni, N., Kumar, P.: Identifying challenges in the integrated design of materials and products, In Proceedings of the 2nd World Congress on Integrated Computational Materials Engineering ICME'13, (2013)

John, D.M., Farivar, H., Rothenbucher, G., Kumar, R., Zagade, P., Khan, D., Babu, A., Gautham, B., Bernhardt, R., Phanikumar G., Prahl, U.: An attempt to integrate software tools at microscale and above towards an ICME approach for heat treatment of a DP steel gear with reduced distortion. In Proceedings of the 4th World Congress on Integrated Computational Materials Engineering (ICME 2017), (2017)

Jose, R., Ramakrishna, S.: Materials 4.0: Materials big data enabled materials discovery. Appl. Mater. Today **10**, 127–132 (2018)

Kagermann, H., Wahlster, W., Helbig, J.: Recommendations for implementing the strategic initiative INDUSTRIE 4.0—Securing the future of German manufacturing industry," M\"{u}nchen, (2013)

Khan, D., Shukla, R., Gautham, B.: In silico design of materials and processes: an application of ICME to carburizing steels. Trans. Indian Inst. Met. **72**, 2179–2185 (2019)

Khan, D., Suhane, A., Bhattacharjee, S.P,A., Tennyson, G., Pramod Zagade, P., Gautham, B.: Integrated microstructure based modelling of Process-Chain for cold rolled dual phase steels. In Proceedings of the 4th World Congress on Integrated Computational Materials Engineering (ICME 2017), (2017)

Koenraad, G.F. Janssens, Dierk Raabe, Britta Nestler.: Computational materials engineering, an introduction to microstructure evolution, Elsevier, (2007)

Mangal, S., Nirgudkar, H., Tennyson, G.: Macrosegregation of multicomponent steel in round billet continuous caster. In advanced high strength steel, Roy, T., Bhattacharya, B., Ghosh, C., Ajmani, S. Eds., Springer, pp. 195–203 (2018)

Mavris, D.N., Balchanos, M.G., Pinon-Fischer, O.J., Sung, W.J.: Towards a digital thread-enabled framework for the analysis and design of intelligent systems, In 2018 AIAA Inf. Syst.-AIAA Infotech @ Aerosp. (2018)

Microsim, First release of phase-field software 'Microsim'. Available: https://materials.iisc.ac.in/highlights/microsim-lab. Accessed 31 Dec 2021

Nellippallil, A.B., Allen, J.K., Gautham, B., Singh, A., Mistree, F.: Architecting robust Co-Design of materials, products, and manufacturing processes, Springer, (2020)

NIST, materials data repository, Available: https://www.nist.gov/programs-projects/materials-data-repository. Accessed 31 Dec 2021

OSTP, Materials genome initiative for global competitiveness, (2011)

Raabe, D.: Computational materials science: the simulation of materials. Wiley, Microstructures and Properties (1998)

Ramazani, A., Mukherjee, K., Quade, H., Prahl, U., Bleck, W.: Correlation between 2D and 3D flow curve modelling of DP steels using a microstructure-based RVE approach. Mater. Sci. Eng., A **560**, 129–139 (2013)

Reddy, S., Gautham, B., Das, P., Yeddula, R.R., Vale, S., Malhotra, C.: An ontological framework for integrated computational materials engineering. In Proceedings of the 4th World Congress on Integrated Computational Materials Engineering (ICME 2017), (2017)

Sardeshmukh, A., Reddy, S., Gautham, B.: Bayesian framework for inverse inference in manufacturing process Chains. Integr. Mater. Manuf. Innov. **8**, 95–106 (2019)

Schmitz, G.J., Prahl, U.: Integrative computational materials engineering: concepts and applications of a modular simulation platform, Wiley, (2012)

Shah, S., Vora, D., Gautham, B., Reddy, S.: A relation aware search engine for materials science. Integr. Mater. Manuf. Innov. **7**, 1–11 (2018)

Shukla, R., Kulkarni, N.H., Gautham, B., Singh, A.K., Farrokh Mistree: Design exploration of engineered materials, products, and associated manufacturing processes. J. Metals, **67**(1), pp. 94–107, (2015)

The Minerals Metals & Materials Society (TMS), Building a materials data infrastructure: opening new pathways to discovery and innovation in science and engineering, TMS, Pittsburgh, PA, USA, (2017)

The Minerals Metals & Materials Society (TMS), verification & validation of computational models associated with the mechanics of materials, TMS, Pittsburgh, PA, USA, (2019)

Tennyson, G., Shukla, R., Mangal, S., Sachi, S., Singh, A.K.: ICME for process Scale-Up: importance of vertical and horizontal integration of models. In Proceedings of the 3rd World Congress on Integrated Computational Materials Engineering (ICME 2015), (2015)
The Minerals Metals & Materials Society (TMS), Modeling across scales: a roadmapping study for connecting materials models and simulations across length and time scales, TMS, Warrendale, PA, USA, (2015)
TMS, Implementing Integrated Computational Materials Engineering (ICME) in the aerospace, automotive, and maritime industries, TMS, (2011)
Walle, A. a. d., Asta, M.: High-throughput calculations in the context of alloy design, MRS Bulletin, **44**(4), pp. 252–256 (2019)
Weber, G., Pinz, M., Ghosh, S.: Machine Learning-Aided Parametrically Homogenized Crystal Plasticity Model (PHCPM) for single crystal Ni-Based superalloys. J. Metals **72**, 4404–4419 (2020)
Yeddula, R.R., Vale, S., Reddy, S., Malhotra, C.P., Gautham, B., Zagade, P.: A knowledge modeling framework for computational materials engineering. In: The 28th International Conference on Software Engineering & Knowledge Engineering (SEKE 2016), (2016)
Zhu, B., Militzer, M.: Phase-Field modeling for intercritical annealing of a Dual-Phase steel. Metall. and Mater. Trans. a. **46**, 1073–1084 (2015)

Chapter 31
A Glimpse of the Growth of Mechanical Metallurgy Over the Past 50 years

Kalyan Kumar Ray

Abstract This report is an attempt to portray how development and growth in mechanical metallurgy have occurred in India over the past 50 years (~1970–2020). The report considers the period of growth in three phases as formative years (~1970–1980), pre-2000 (1980–2000) and post-2000 (>2000). The term 'Mechanical Metallurgy' (MM) appears to be a misnomer in the current era owing to rapid changes in the contents of this discipline. However, the report attempts to incorporate the contents with an emphasis related to its initial meaning for the sake of brevity; although its growth using the current broad implication of MM as Mechanics of Materials (MMs) is not completely neglected. The salient details of several activities related to the developments over the years are to be fetched only from the limited references cited here. A brief guideline is provided on the future activities for the growth of this discipline in India. The major focus is to acquaint the current generation of researchers with a short glimpse of the history of MM in the stipulated duration.

Keywords Mechanical metallurgy · Mechanics of materials · DCFF · Testing · Mechanism · Scale of study

31.1 Introduction

Mechanical metallurgy (MM) is the area of knowledge that deals with the science and technology of the behaviour of metals and alloys in response to imposed mechanical forces on these. Metallurgy is the domain of materials science and engineering that encompasses the understanding related to the chemical and physical behaviour of metals and alloys. The domain of this engineering discipline was broadly categorized in the initial years as chemical and physical metallurgy; mechanical metallurgy used to be considered as a constituent part of physical metallurgy. Establishment of the initial recognition of 'mechanical metallurgy' as a distinct discipline possibly can be credited to Dieter (Dieter 1961). The scope of the discipline is loosely defined.

K. K. Ray (✉)
Department of Metallurgical and Materials Engineering, Indian Institute of Technology Kharagpur, Kharagpur West Bengal, 721302, India

R. Divakar et al. (eds.), *Indian Metallurgy*, Indian Institute of Metals Series,
https://doi.org/10.1007/978-981-99-5060-7_31

However, one can possibly consider four major components within its scope: (a) experimental developments related to testing of metallic materials, (b) developments related to deformation, creep, fatigue and fracture (DCFF) of metallic materials, (c) developments related to mechanical processing like rolling, forging, etc. and (d) theoretical, numerical and computational simulations related to testing, DCFF and mechanical processing from nano to mega scale.

The primary aim of this report is to describe the nature of the development and growth of 'Mechanical Metallurgy' in India in the last fifty years (~1970–2020) depicting the state of this discipline in the formative years to its present state. The development and growth of any engineering discipline are commonly governed by its usability and applications for the need of social benefit of mankind; these are influenced primarily by a few factors like: human resources (manpower), physical capital, natural resources and existing technology and interactions between personnel from different organizations. This report would aim to illustrate the constituents of this sub-discipline together with the considerations of the factors influencing it to some extent. The period considered in this report is between the year (1970) when the first volume of the journal 'Metallurgical Transactions' appeared in which the first paper coincidentally is related to mechanical metallurgy (Hirth and Cohen 1970), and the hundredth year of the submission of the Griffiths paper related to his theory on brittle fracture (Griffith 1921). Interestingly several reports related to the history of metallurgy in India are available without almost any emphasis on attempts to highlight the development of Mechanical Metallurgy; possibly except one (Valluri et al. 1983) which deals with a part of MM.

The sketch of the usual definition of mechanical metallurgy as given in the first paragraph of this section has undergone significant deviations with the passage of time over the past fifty years, particularly with drastic changes in the curriculum of the concerned engineering discipline of 'Metallurgy'. In India, in particular, the discipline has transformed into 'Metallurgy and Materials' or simply 'Materials'; which encompasses the science and technology of metallic, ceramic, polymeric and all types of composite materials apart from several emerging materials based on the above as well as biomaterials. After the 1980s, this discipline has thus consequently demanded to consider dealing with the science and technology of the behaviour of all types of materials in response to mechanical forces imposed on these. These include mechanical processing, characterization of their mechanical properties, and the assurance of their structural integrity both experimentally as well as theoretically. Thus what used to be well-known as Mechanical Metallurgy 50 years ago is currently the domain of 'Science and technology of the behaviour of materials in response to imposed mechanical forces on these'; in simplified form this can be referred to as 'Mechanics of Materials' (MMs). This report emphasizes on summarizing issues related to the growth of 'mechanics of metallic materials' alias 'mechanical metallurgy'.

31.2 Factors Influencing the Growth and Development

31.2.1 Manpower

The primary manpower development for the domain of mechanical metallurgy (MM) in India originated from two major sources: Engineering and Technological Institutes or Universities and Government Research and Development (R&D) laboratories apart from a few industrial laboratories. In the first group, one can consider Indian Institutes of Technology (IITs) and National Institutes of Technology (NITs) apart from separate state government and private colleges. Currently, the number of IITs in India is 23 and that of NITs is 31 with a large number in the third category. However, interestingly there are only 47 institutes where Metallurgical or Materials Engineering Departments exist; this number is considerably less than 100th of the number of departments (each greater than 5000) pertaining to Mechanical, Civil, Computer Science and Electronics & communication engineering. Under the Government R&Ds, one can consider organizations like the Council of Scientific and Industrial Research (CSIR-38 laboratories, about one-fourth dealing with metals & materials), Defence Research and Development Organisation (DRDO), Department of Atomic Energy (DAE), Department of Space Research Organization (ISRO), Department of Civil Aviation and Railways (like RDSO). In addition to the Government manufacturing sectors like Steel Authority of India Limited (SAIL), Hindustan Aeronautics Limited (HAL), Bharat Heavy Electrical Limited (BHEL) or National Thermal Power Corporation (NTPC), etc. which are also contributing to the development of Mechanical Metallurgy.

31.2.2 Physical Capital

The economic growth of a country is primarily reflected in its physical capital related to construction and manufacturing industries, all types of transportation systems in land, water and air, power generation capabilities using thermal, nuclear and non-conventional methods, its defence capabilities and advanced vision for growth like space research. There has been a phenomenal growth over the last fifty years in India in all these ventures; just, for example, in the railways, passenger-kilometre and tonnes-kilometre (Fig. 31.1) has increased by approximately 400–500% (Indian Railways Facts 2017). The higher the amount of physical capital, the lower would be the cost of economic activity and better growth in Gross Domestic Product (GDP). Interestingly, attempts to enhance the physical capital provide the impetus to the growth and development of engineering research and *this is one of the primary driving forces for the rapid growth of mechanical metallurgy as it provides the assurance of structural integrity* of almost all types of engineering components in static or dynamic service.

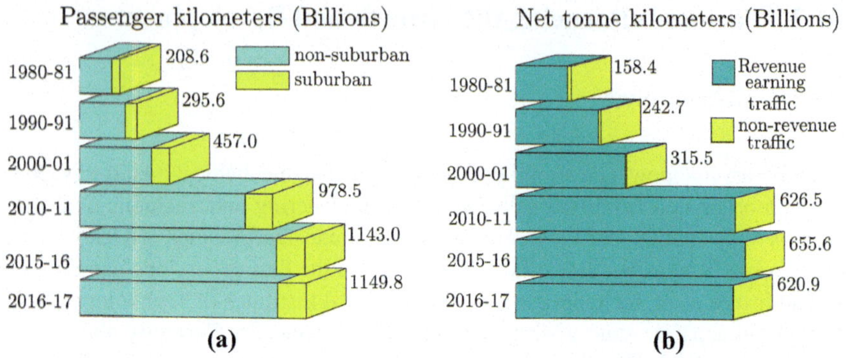

Fig. 31.1 A schematic illustration of India's capital growth in terms of increase in Passenger-Kilometer and Tonnes-Kilometer carried by Indian Railways from 1980 to 2016 (Indian Railways Facts 2017)

31.2.3 Technology and Investment

The existing technology in a country is another factor that is capable of accelerating its future improvement whether in research activities or in industrial production. In manufacturing, it provides higher productivity with improved quality but with a lower amount of labour, and reduces production costs to make it globally competitive. The driving force for the growth of mechanical metallurgy originates from this philosophy together with the demand for increasing the physical capital of a developing country. Specific emphasis has been laid on mechanical metallurgy as this discipline can lead to the awareness of the design and fabrication of engineering components resistant to failure. The failure can originate from inappropriate deformation, creep, fatigue, and fracture in ambient and severe environments.

31.2.4 Other Miscellaneous Factors

Natural resources in a country are a major factor in the growth and development of technological fronts to utilize the available resources for the country's economic growth. India is rich in several ferrous and non-ferrous minerals. However, the extraction of these metals primarily belongs to chemical metallurgy, with a minor role being played by MM. Interactions between different academic institutes, R&Ds and industry are also an important factor for the speed of growth for any engineering discipline; these interactions should also be considered inside the country as well as with institutes or organizations of other advanced or developing countries for better mutual benefits.

31.3 Formative Years in the Development of Mechanical Metallurgy in India

The developments (in research) in MM in the 1970s can be considered to be almost in an embryonic stage when the few existing IITs and the government R&Ds were not even two decades old, and this decade can be considered to be the formative years. The primary deficiency in this period was lack of facilities, lack of skilled manpower and lack of suitable interactions, and even lack of sufficient global technology related to this discipline compared to what exists in 2020s. The facilities for mechanical property determinations were based on universal testing machines (UTM) usually fitted with dial gauges, and the calculations of the results were based on simple calculators. In this decade, India could procure UTMs with suitable facilities for force–displacement results being obtained in chart recorders (e.g. screw driven Instron UTM machine). One involved in MM has to manually digitize the results and subject these to analyses. The author has remained witness to fracture toughness experiments in the mid-1970s being carried out using noting of dial gauge reading for the load and suitable load-line displacement (LLD) measurement using fabricated gauges using multi-meter readings with subsequent manual analysis (Pandey 1976). The basic concepts were well dealt with by persons engaged in MM research, but this is a good example of the lack of facility, or build-up of even global technology in this direction.

The evaluation of conventional mechanical properties specifically based on tension, compression, bend, hardness and impact and high cycle fatigue (primarily rotating bending) tests was possible with the standardized equipment available in the 1970s. In the academic arena and in R&D organizations, these properties were particularly used to evaluate the structure–property relations of varying thermo-mechanically treated alloys, and in industries, these tests were needed for quality control. The emphasis in metallurgical engineering during this period was more on chemical metallurgy, and on gaining an understanding of varied structure generations in different alloys in physical metallurgy which necessarily demanded property evaluations. Further, efforts were directed to reveal some of the fundamental directions of plastic deformation and to generate standardized creep, fatigue and fracture properties.

Typically, attempts to understand rate controlling mechanisms in deformation, deformation kinetics and thermal activation parameters for deformation, dynamic strain ageing, thermal activation strain rate analysis (TASRA) and Hall–Petch (H-P) analyses, and the nature and types of serrated flow in monotonic deformation in several polycrystalline materials are well cited in this period (Sastry et al. 1969; Sastry and Vasu 1972; Little et al. 1973; Rodriguez 1976; Mannan and Rodriguez 1972). A typical grain size dependence of activation parameters for plastic deformation in the 1970s is illustrated in Fig. 31.2 (Mannan and Rodriguez 1975). This was the period when Asia's one of the largest laboratories on creep testing was getting established in NML; a typical photograph of the laboratory is shown in Fig. 31.3. The objective of this laboratory was to provide creep data for the design of various power plant components. There were attempts to understand stacking fault energy

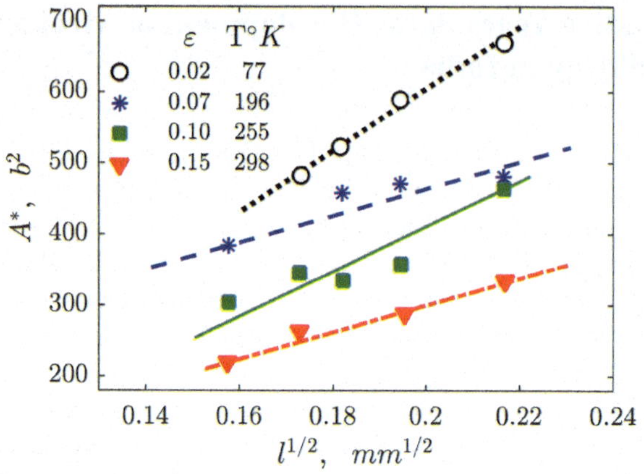

Fig. 31.2 Illustration of grain size dependence of activation area in cadmium at different temperatures (Mannan and Rodriguez 1975)

by X-ray diffraction line broadening (Wahi et al. 1971) and elevated temperature deformation analyses to understand creep and thereby even to construct deformation mechanism maps or to understand grain boundary migration during creep (Singh et al. 1973). However, limited or no electron microscopy facilities in the 1970s hindered in-depth analysis of these activities. In the field of fatigue and fracture, concerned mechanical metallurgists attempted to assess the fracture toughness behaviour of different materials; there were efforts directed to carry out tests using both linear elastic and elastic–plastic approaches, to understand the micro-mechanism of fracture and to reveal fracture mechanism maps (Pandey and Banerjee 1978; Paranjpe and Banerjee 1979; Rao et al. 1975), apart from some associated theoretical investigations and relatively limited work on fatigue (Singh et al. 1980).

The development of mechanical processing of materials like rolling, forging, extrusion and so on was usually limited to industrial developments with limited facilities in the academic institutes. The R&D activities in Steel Authority of India Limited (SAIL) initiated setting up rolling mills for studies to cater to industrial needs specifically for the development of different grades of steel, rather than for fundamental studies. Personal computers started entering Indian Institutes and other organizations in the latter period of 1970s and thus computational material mechanics was almost in its infancy.

31.4 Growth and Development in Pre-2000

The initiation of the different research activities in mechanical metallurgy in the formative years (1970s and earlier) led to the enhancement of manpower, instrumental facilities and technology build-up, and these with some increased funding

Fig. 31.3 Typical view of one of the largest creep laboratories of Asia developed in NML from the latter part of the 1960's

obtained from government and private sector agencies boosted its growth in a rapid manner during 1980 and 1990s. During this period, developments in transmission microscopy and computer facilities also helped enrich this field. These developments occurred both in academic Institutes and in the different R&D organizations. With reference to MM, the major facilities achieved for developments in mechanical property characterization were improved versions of the servo-hydraulic machines, the emergence of electro-mechanical/electro-dynamic UTMs, instrumented hardness testers and instrumented impact testers, apart from building up facilities for higher resolution scanning and transmission electron microscopes for examining fractured surfaces or sub-surfaces after deformation, creep, fatigue and fracture tests.

The major thrust for the development during this period originated from envisioning component and structural integrity to avoid failures arising out of plastic deformation, creep, fatigue and fracture (DCFF) in both ambient and aggressive environments. This also opened up activities in creep-fatigue, environment-assisted fatigue and fracture, together with attempts to understand the effect of different corrosive environments on plastic deformation (through slow strain rate tests). The nature of the work carried out in this period consisted of: (i) the development and setting up of different testing facilities with associated training/practices of the newer technology in the academic institutes, R&D organizations and industries, (ii) attempts to understand the role of microstructures on DCFF in different scale lengths with suitable interactions between different Institutes (both inside and outside the country); metallurgical community to a great extent usually lays more emphasis on this issue. Numerous conferences were held in this period for these interactions, but one of the major ones was the 'International Conference on Creep, Fatigue and Creep-fatigue Interaction' in India instituted by IGCAR in 1987 and supported by the Indian Institute of Metals, (iii) Progress in analytical and computational aspects in DCFF; the numerical and computational efforts got enhanced with the increased availability of computational facilities in different organizations, (iv) Structural integrity analysis with careful feedback from the failure of structural components and (v) Indian ventures to acquire different self-made assemblies, for example, Light Combat Aircraft (LCA), later named as 'Tejas'.

It would be natural to explore examples of the developmental activities in different institutions/organizations during this period. It is a difficult task to summarize the activities in this period in a short report; only some examples are being mentioned here. Test methodology development and analysis of test results occurred in many directions like tension, compression, fatigue and fracture, etc. (Sunder 1985; Ray et al. 1993; Sivaprasad et al. 2000a). In the domain of plastic deformation and creep, there was good thrust on work on super-plasticity (starting from 1970s) (Padmanabhan and Davies 1980; Chokshi et al. 1993), simple structure–property relations in several materials (Ray and Mondal 1991) elevated temperature deformation and constructing deformation mechanism/processing maps (Venugopal et al. 1996; Prasad and Seshacharyulu 1998; Chakravartty et al. 1995), creep of structural materials like pipelines, components of thermal and nuclear power plants and other metallic materials (Ghosh and McLean 1992; Swaminathan and Raghavan 1994; Sasikala et al. 2000; Malakondaiah and Rama Rao 1981), attempts to understand rate controlling mechanisms in deformation (Ray et al. 1994; Rao et al. 1986; Ray and Mallik 1983, 1984), understand dislocation characteristics (Sundararaman et al. 1988; Nandy and Banerjee 2000) with deformation in various classical and emerging materials. As an example, the strengthening of IN718 has been revealed by some authors (Sundararaman et al. 1988) due to precipitate shearing and precipitate-by-pass mechanisms in the alloy, making a distinct signature of how TEM work got involved with deformation characteristic analysis in this period.

During this period, a beginning was also made on standardizing fracture and fatigue work in different organizations. Besides standard fracture toughness estimations in mode-I for both linear elastic and elastic–plastic regimes (Srinivas et al. 1991; Singh et al. 1998; Sreenivasan et al. 1996), there were also attempts to understand mixed-mode fracture. For example, it is shown in Fig. 31.4 that increasing the mode III/I loading ratio causes a significant reduction in fracture toughness (Kamat et al. 1998; Kamat and Hirth 1994). The available facilities related to fatigue testing (specifically LCF) was in their early stage in the pre-2000 period primarily because of the lack of suitable facilities in several Institutes. This, however, does not appear to have deterred attempts to work in this area as evident from the investigations of different researchers (Sivaprasad et al. 2000b; Brahma et al. 1987; Tarafder et al. 1994; Choudhary et al. 2000).

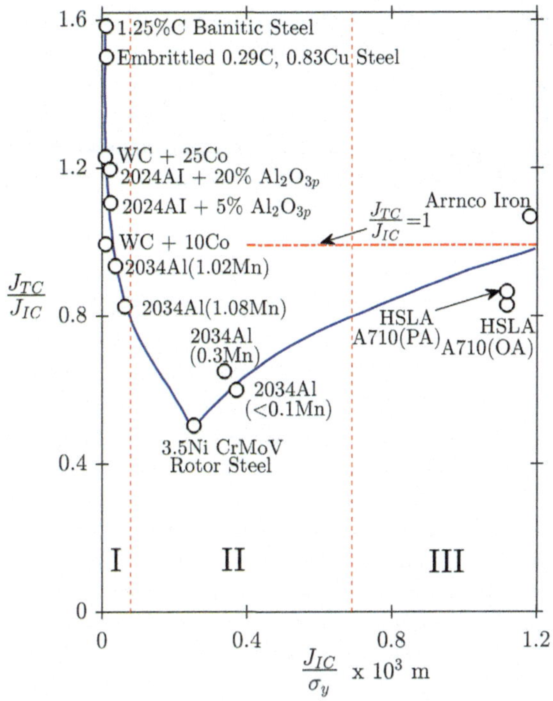

Fig. 31.4 The influence of mixed mode fracture (J_{TC}) on plane strain mode-I fracture (J_{IC}) relative to yield strength (σ_y) of materials (Kamat and Hirth 1994)

31.5 Growth and Development in Post-2000

The developmental activities in the formative years and in pre-2000 on different fronts of growth in various disciplines including MM led India to increase its educational facilities with the emergence of more number of IITs, NITs. This increase in the build-up of educational facilities is found to be more pronounced post-2000. The specific development that boosted enhanced growth in this period is the availability of improved later generation computer facilities and ease of procuring personal computers. The testing facilities which were available in pre-2000, emerged with higher data storage capacity and with in-built provisions for analysing the data by the machines to provide the required final output. For example, analysis of J-integral fracture toughness has to be done personally by the investigator in 1980s, but post-2000, one can get the value of the critical J-integral toughness directly from the machine after the test. This assisted specifically the R&D laboratories of the industries to acquire their property evaluation in a simpler way. Similarly, the data of creep experiments can be obtained directly as creep strain rate versus time/strain; improved testing facilities emerged for examining creep-fatigue interactions in materials. There were global improvements in sensor technology and the availability of non-contact sensors made some of the experiments to be performed with convenience; a specific example is the use of the digital image correlation (DIC) technique. Several of these globally available technological facilities were grasped by the Indian system and the growth of mechanical metallurgy got a boost. The major drive in this period possibly was on computational material mechanics, processing and characterization of smart and functional materials, high entropy alloys, intermetallics, nanomaterials, etc. apart from conventional experimental research with advanced facilities.

These are in addition to what started in the pre-2000 regime. In addition, one finds extensive work on the development of advanced ultra-supercritical (AUSC) power plants with the formation of a consortium incorporating BHEL, NTPC, DAE, DRDO, and several IITs and NITs. The work on the AUSC project is still ongoing for almost the entire decade of 2010–2020. This has led to tremendous advancement of engineering research and analysis on DCFF.

It is a difficult task to provide examples of research and development activities (in MM) in the post-2000 period in a satisfactory manner. One of the major reasons is that the growth in the classical MM in the pre-2000 period can be considered as standardization and setting up of facilities to a good extent with relatively less research outputs to the global scenario, in comparison to what emerged in the post-2000 period. The fruits of benefit started emerging in a burst in post-2000 era. This was accentuated by computational facilities and the availability of high-resolution electron microscopy. Some typical examples of these research outputs from different Institutions in the area of deformation, plasticity (Batra et al. 2001; Kashyap 2002; Rodriguez 2004; Chokshi 2005; Sahoo et al. 2007; Banerjee and Williams 2013; Sudhakar Rao et al. 2014; Suresh et al. 2014; Sinha et al. 2016; Laha et al. 2007; Raturi et al. 2021) and creep (Laha et al. 2009; Sakthivel et al. 2011; Bhaduri and Laha 2015; Shankar et al. 2017; Goyal et al. 2018; Sakthivel et al. 2020; Krovvidi et al. 2021) as well as fracture (Sivaprasad et al. 2000a; Singh et al. 2003; Bhattacharjee and Knott 2007; Roy et al. 2009; Ghosh et al. 2009a; Dutt et al. 2011; Chatterjee et al. 2015; Kumar et al. 2017; Krishnan et al. 2019; Shashank et al. 2011) and fatigue (Paul et al. 2010a, 2011; Sen et al. 2010; Dutta and Ray 2013; De et al. 2014; Prasad Reddy et al. 2015; Chowdhury et al. 2015; Saikrishna et al. 2016; Kumar and Ghosh 2018; Krishnan et al. 2018) are cited; with due apology from the author to several investigators whose important contributions could not be incorporated to keep the length of this article restricted. As an example, typical cyclic J–R curves for AISI 304LN base-metal under monotonic and cyclic fracture tests with imposed cycling loading at a constant plastic displacement of 0.1mm at different R ratios with the illustration of the mechanism of degradation of cyclic fracture toughness is illustrated in Fig. 31.5 with the help of the nature of Crack tip opening in monotonic and cyclic fracture tests for the steel (Roy et al. 2009).

There have been considerable developments in the domains of (i) texture analyses (Biswas et al. 2010; Samajdar et al. 2001) for understanding the anisotropic mechanical behaviour of materials, (ii) high-temperature materials like intermetallics (Mitra 2015; Germann et al. 2005; Chattopadhyay et al. 2006), (iii) understanding applicability of metallic glasses (Schuh et al. 2007; Narayan et al. 2012), (iv) micro and nanoscale testing practices (Jaya and Jayaram 2016; Ray 2012; Kumar et al. 2014; Acharya and Ray 2013) (examples of some in situ and ex situ micro specimens are shown in Fig. 31.6); (v) emergence of automated indentation tests at different scale

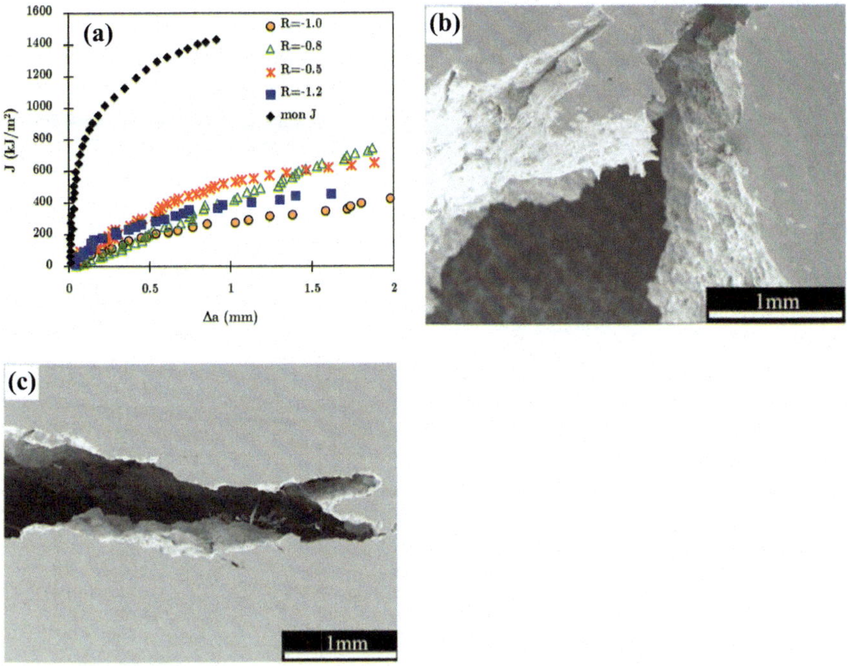

Fig. 31.5 **a** Typical cyclic J–R curves for AISI 304LN base metal under monotonic and cyclic fracture tests with imposed cycling loading at a constant plastic displacement of 0.1mm at different R ratios. Nature of Crack tip opening in **b** monotonic and **c** cyclic J-integral test for AISI 304LN stainless steel (Roy et al. 2009)

lengths for characterization of several emerging materials (Sujith Kumar et al. 2020); (vi) the development and mechanical characterization of high entropy alloys remain a new beginning in this period (Murty et al. 2019; Ray 2020; Ray et al. 2021; Sharma et al. 2018), (vii) tribology investigations appeared to achieve a new dimension (Roy et al. 2001; Das et al. 2010; Arun Prakash et al. 2012; Ray and Das 2017; Roy 2013), (iv) computational material mechanics flourished to a good level (Chakraborti 2004; Chakraborti et al. 2008; Ghosh et al. 2009b; Dutta 2017; Dutta et al. 2012; Pant et al. 2003; Paul 2016; Paul et al. 2010b; Arora et al. 2019) and so on. Some examples of computational work like finite element analysis of plastic zone size ahead of a crack with inclusion, and simulation of stress–strain behaviour of iron nanopillar with the simulated atomic positions at dislocations and twin boundaries by molecular dynamics are depicted in Figs. 31.7 and 31.8, respectively.

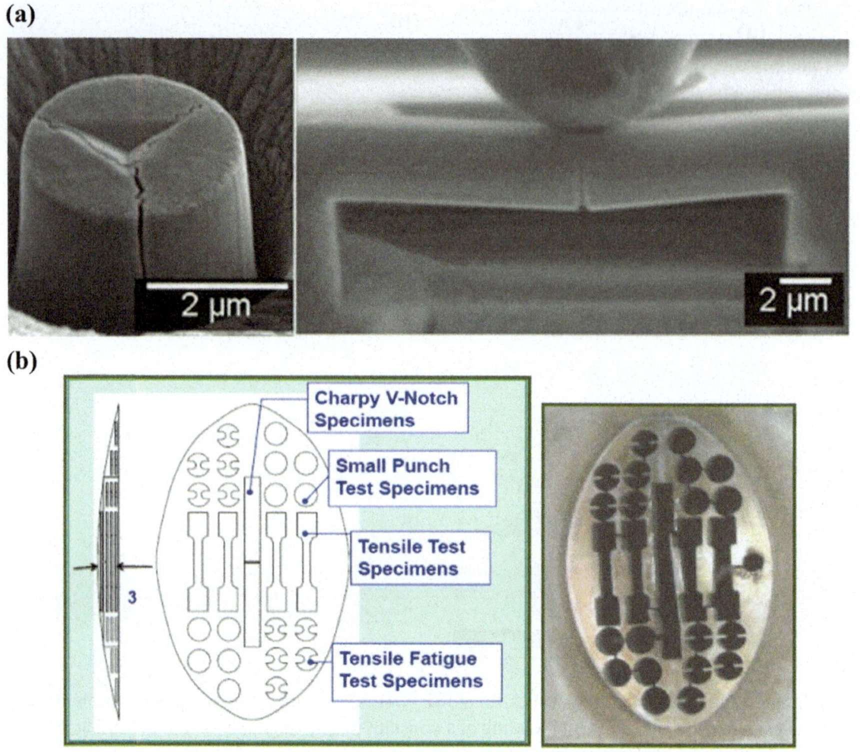

Fig. 31.6 Typical miniature micro specimens for different tests: **a** indentation and beam bending method (Jaya and Jayaram (2016)) and **b** scooped samples for different types of test (Kumar et al. (2014))

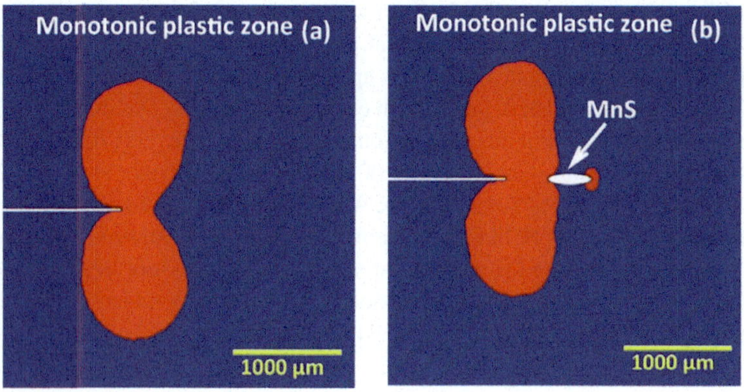

Fig. 31.7 Finite element simulation of monotonic plastic zone morphology ahead of **a** crack, and **b** crack with inclusion (Paul 2016)

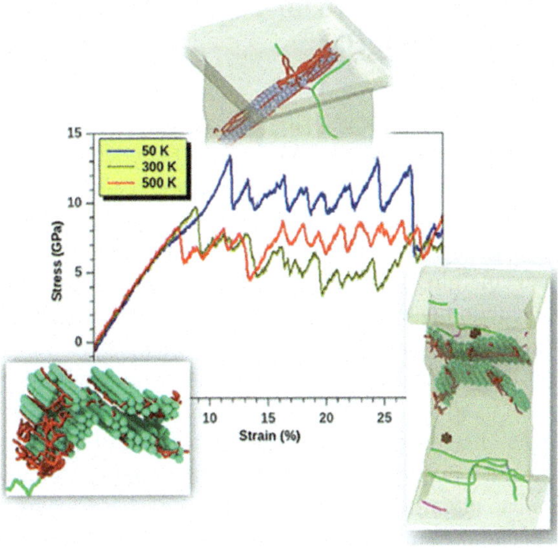

Fig. 31.8 Simulation of stress–strain diagram in bcc Fe- nanopillar with the simulated atomic positions at dislocations and twin boundaries (Dutta 2017)

31.6 Mechanics of Materials

The terminology mechanical metallurgy (MM) has started missing its inherent initial essence, and with the increased incorporation of accompanying studies on several composites, ceramics, polymers and numerous emerging materials of interest to all engineering branches along with metallic materials, the term MM is rapidly transforming towards the terminology like 'Mechanical Science and Technology of Materials' or simply 'Mechanics of Materials'. This is also echoed by the change in the name of the Departments from 'Metallurgy' to 'Metallurgy and Materials' or simply 'Materials' in several Universities/Institutes. With the introduction of courses like 'Introduction to Materials Engineering (Science)' in several engineering branches like mechanical, civil, aerospace, naval, etc., a large number of persons from these latter departments have also ventured to work on 'mechanics of materials' over the last several decades. So 'mechanical Metallurgy' or 'Mechanics of Materials' is not restricted to only the Department of Metallurgy or Materials; numerous advanced studies are emerging in this discipline from the investigations of concerned persons from different branches of Engineering.

31.7 Perspective for Future Research in Mechanics of Materials

The impetus for the growth in mechanical metallurgy has originated primarily to characterize the newly developed alloys to understand their structure–property relations, and to evaluate their potential for applications in varied engineering components. In the present era, the discipline has embraced several non-metal-based materials (like ceramics, or polymers/plastics), and composites apart from a variety of smart and emerging materials. The latter group incorporates nanomaterials, electronic and magnetic materials, biomaterials, or even the current emerging materials like high entropy alloys.

The demand thus necessitates understanding the response of forces on materials from nano to mega scale. The current developments in India can account to a good extent for the mechanical characterization of micro to macro-scale materials and are quite adaptable to follow the International specifications for these issues. However, as one attempt to reach the extreme of these scales, one finds the requirements for developing suitable techniques both for satisfactory specimen preparation, and their mechanical characterization as challenging. For example, one requires developing optical tweezer tests for bio-materials (Mills et al. 2004) and expanding the facilities for large component tests to ensure their component integrity.

The present report encompasses only developments related to metallic materials to restrict the content. However, there is a critical requirement to develop techniques and to characterize different types of non-metallic based composites (apart from metal matrix composites), which are rapidly replacing several alloys from their applications to acquire enhanced benefits for the components and assemblies (a suitable example is automobile industry). The micro and macro tests for these materials are to some extent developed having the benefit of similitude with the existing facilities for metallic materials, but there is a growing demand for alternate techniques. Just to provide some example, the characterization of nanofibres and carbon nanotubes requires specialized techniques and often demands experiments to be coupled with computational facilities for understanding their behaviour.

There has been phenomenal growth in surface engineering in the past 2–3 decades. One of the major problems faced by engineers/scientists working in this area is the assessment of adherence of the surface layer apart from its wear, erosion, environmental degradation and integrity. Although developments in the mechanics of materials in this direction have started, more thrust is required with enhanced development for materials being processed by 3D printing and laser surface alloying.

One of the growing demands in structural integrity analysis is to carefully detect the weakest part of an assembly and to carry out the design and fabrication of these with the mechanical properties of the weakest region. In general, the weakest part with respect to fatigue and fracture in large metallic structures is the welded joints. Over the past three decades, there have been several ventures to understand characteristics of the joined parts. This direction requires more attention with newer ventures like dissimilar material joining practices.

31.8 Summary

An attempt has been made in this report to portray the growth of Mechanical Metallurgy in India during the last fifty years (1970–2020). The stages of growth have been broadly categorized in the time scale as formative years (~1970–1980), pre-2000 (implying 1980–2000) and post-2000. In the formative years, possibly the primary driving force in the growth of MM was in manpower building for this discipline with the initiation of setting up numerous experimental facilities. In pre-2000, a good amount of experimental facilities for investigating deformation, creep, fatigue and fracture (DCFF) together with facilities for forming of materials were established. This was the period when more emphasis was being laid on computational material mechanics, mechanical characterization of different types of composites, development of newer types of facilities like creep-fatigue interactions, environmental fatigue and fracture, mechanical processing and characterization of several advanced structural materials based on iron, aluminium, titanium, copper and super-alloys, with the initiation for the development of numerous mechanical forming processes. The period 'post-2000' witnessed phenomenal growth in MM/MMs with feedback from the outputs of the earlier years; the amalgamation of sophisticated experimental facilities with analytical, numerical and computational facilities together with improved microstructural analyses, has opened up almost the latest state of the art of mechanics of materials in India. In the past decade, the success of the LCA project, cryo-engine ventures and work on advanced ultra-supercritical materials has boosted the facilities and other aspects of mechanical metallurgy/mechanics of materials.

Acknowledgements The author remains grateful to many personnel for their kind feedback while drafting this report. Specifically, the assistance rendered by Prof. Satyam Suwas, IISc Bangalore, Dr. G. Sasikala, IGCAR Kalpakkam, Prof. P. Chakraborti, Jadavpur University, Prof. K. Chattopadhyay, IISc Bangalore, Dr. K. Barat, NAL, Bangalore, Dr. R. Sarkar, DMRL Hyderabad and many others has enriched this report. In addition, the authors would like to thank Dr. Atri Nath for his assistance in the preparation of the figures and in the overall arrangement of the manuscript. The author solicits the excuses from several active persons in mechanical metallurgy if their work references cannot be accommodated. In general, the space limitation has also restricted the author from providing even the salient details of many interesting works in this period, but these can be accessed from the cited references.

References

Acharya, S., Ray, K.K.: Mater. Sci. Eng. A **565**, 405 (2013)
Arora, P., Gupta, S.K., Samal, M.K., Chattopadhyay, J.: Int. J. Fatigue **129**, 105209 (2019)
Arun Prakash, N., Gnanamoorthy, R., Kamaraj, M.: Surf. Coat. Technol. **207**, 450 (2012)
Banerjee, D., Williams, J.C.: Acta Mater. **61**, 844 (2013)
Batra, I.S., Dey, G.K., Kulkarni, U.D., Banerjee, S.: J. Nucl. Mater. **299**, 91 (2001)
Bhaduri, A.K., Laha, K.: Procedia Eng. **130**, 598 (2015)
Bhattacharjee, D., Knott, J.F.: Mater. Sci. Technol. **23**, 1269 (2007)
Biswas, S., Singh Dhinwal, S., Suwas, S.: Acta Mater. **58**, 3247 (2010)

Brahma, K., Sunder, R., Dattaguru, B.: Int. J. Fatigue **9**, 51 (1987)
Chakraborti, N.: Int. Mater. Rev. **49**, 246 (2004)
Chakraborti, N., Siva Kumar, B., Satish Babu, V., Moitra, S., Mukhopadhyay, A.: Appl. Math. Model. **32**, 1781 (2008)
Chakravartty, J.K., Dey, G.K., Banerjee, S., Prasad, Y.V.R.K.: J. Nucl. Mater. **218**, 247 (1995)
Chatterjee, A., Moitra, A., Bhaduri, A.K., Mitra, R., Chakrabarti, D.: Eng. Fract. Mech. **149**, 74 (2015)
Chattopadhyay, K., Balachandran, G., Mitra, R., Ray, K.K.: Intermetallics **14**, 1452 (2006)
Chokshi, A.H.: Mater. Sci. Eng. A **410–411**, 95 (2005)
Chokshi, A., Mukherjee, A., Langdon, T.: Mater. Sci. Eng. R. Rep. **10**, 237 (1993)
Choudhary, B., Rao, K., Mannan, S.: ISOMALM 2000. In: International Symposium on Materials Ageing and Life Management, pp. 85–92 (2000)
Chowdhury, T., Sivaprasad, S., Bar, H.N., Tarafder, S., Bandyopadhyay, N.R.: Eng. Fract. Mech. **148**, 60 (2015)
Das, D., Dutta, A.K., Ray, K.K.: Mater. Sci. Eng. A **527**, 2182 (2010)
De, P.S., Kundu, A., Chakraborti, P.C.: Mater. Des. **57**, 87 (2014)
Dieter, G.E.: Mechanical Metallurgy. McGraw Hill Book Company, New York (1961)
Dutt, B.S., Sasikala, G., Shanthi, G., Venugopal, S., Babu, M.N., Parida, P.K., Bhaduri, A.K.: Procedia Eng. **10**, 2725 (2011)
Dutta, A.: Acta Mater. **125**, 219 (2017)
Dutta, K., Ray, K.K.: Mater. Sci. Eng. A **575**, 127 (2013)
Dutta, A., Bhattacharya, M., Gayathri, N., Das, G.C., Barat, P.: Acta Mater. **60**, 3789 (2012)
Germann, L., Banerjee, D., Guédou, J.Y., Strudel, J.-L.: Intermetallics **13**, 920 (2005)
Ghosh, R.N., McLean, M.: Acta Metall. Mater. **40**, 3075 (1992)
Ghosh, S., Kain, V., Ray, A., Roy, H., Sivaprasad, S., Tarafder, S., Ray, K.K.: Metall. Mat. Trans. A **40**, 2938 (2009a)
Ghosh, S., Gabane, P., Bose, A., Chakraborti, N.: Comput. Mater. Sci. **45**, 96 (2009b)
Goyal, S., Mariappan, K., Shankar, V., Sandhya, R., Laha, K., Bhaduri, A.K.: Mater. Sci. Eng. A **730**, 16 (2018)
Griffith, A.A.: Philosophical transactions of the royal society A. Math. Phys. Eng. Sci. **221**, 163 (1921)
Hirth, J., Cohen, M.: Metall. Trans. **1**, 3 (1970)
Indian Railways Facts & Figures (2016–17). Ministry of Railways Railway Board, India (2017)
Jaya, B.N., Jayaram, V.: JOM **68**, 94 (2016)
Kamat, S.V., Hirth, J.P.: Scr. Metall. Mater. **30**, 145 (1994)
Kamat, S.V., Srinivas, M., Rama Rao, P.: Acta Mater. **46**, 4985 (1998)
Kashyap, B.P.: Acta Mater. **50**, 2413 (2002)
Krishnan, S.A., Sasikala, G., Moitra, A., Albert, S.K., Bhaduri, A.K.: IJSI **9**, 153 (2018)
Krishnan, S.A., Nikhil, R., Sasikala, G., Moitra, A., Albert, S.K., Bhaduri, A.K., Lakshmana Rao, C., Vishnuvardhan, S., Saravanan, M., Gandhi, P., Raghava, G.: Int. J. Press. Vessels Pip. **178**, 104008 (2019)
Krovvidi, S.C.S.P.K., Goyal, S., Bhaduri, A.K.: J. Mater. Eng. Perform. **30**, 2742 (2021)
Kumar, S., Ghosh, P.K.: Int. J. Fatigue **116**, 306 (2018)
Kumar, K., Pooleery, A., Madhusoodanan, K., Singh, R.N., Chakravartty, J.K., Dutta, B.K., Sinha, R.K.: Procedia Eng. **86**, 899 (2014)
Kumar, H., Ramakrishnan, V., Albert, S.K., Bhaduri, A.K., Ray, K.K.: J. Nucl. Mater. **484**, 141 (2017)
Laha, K., Chandravathi, K.S., Parameswaran, P., Rao, K.B.S., Mannan, S.L.: Metall. Mat. Trans. A **38**, 58 (2007)
Laha, K., Chandravathi, K.S., Parameswaran, P., Bhanu Sankara Rao, K.: Metall. Mat. Trans. A **40**, 386 (2009)
Little, E.A., Sastry, D.H., Vasu, K.I.: Scr. Metall. **7**, 131 (1973)
Malakondaiah, G., Rama Rao, P.: Acta Metall. **29**, 1263 (1981)

Mannan, S.L., Rodriguez, P.: Phil. Mag. **25**, 673 (1972)
Mannan, S.L., Rodriguez, P.: Acta Metall. **23**, 221 (1975)
Mills, J., Qie, L., Dao, M., Lim, C., Suresh, S.: Mol. Cell. Biomech. **1**, 169 (2004)
Mitra, R.: Structural Intermetallics and Intermetallic Matrix Composites. CRC Press (2015)
Murty, B.S., Yeh, J.-W., Ranganathan, S., Bhattacharjee, P.: High-Entropy Alloys. Elsevier (2019)
Nandy, T.K., Banerjee, D.: Intermetallics **8**, 1269 (2000)
Narayan, R.L., Singh, P.S., Hofmann, D.C., Hutchinson, N., Flores, K.M., Ramamurty, U.: Acta Mater. **60**, 5089 (2012)
Padmanabhan, K.A., Davies, G.J.: Superplasticity. Springer (1980)
Pandey, R.K.: Studies on fracture mechanics parameters in low strength steels. Ph.D. thesis, Indian Institute of Technology, Bombay (1976)
Pandey, R.K., Banerjee, S.: Eng. Fract. Mech. **10**, 817 (1978)
Pant, P., Schwarz, K.W., Baker, S.P.: Acta Mater. **51**, 3243 (2003)
Paranjpe, S.A., Banerjee, S.: Eng. Fract. Mech. **11**, 43 (1979)
Paul, S.K.: Eng. Fract. Mech. **159**, 90 (2016)
Paul, S.K., Sivaprasad, S., Dhar, S., Tarafder, S.: J. Nucl. Mater. **401**, 17 (2010a)
Paul, S.K., Sivaprasad, S., Dhar, S., Tarafder, M., Tarafder, S.: Comput. Mater. Sci. **48**, 662 (2010b)
Paul, S.K., Sivaprasad, S., Dhar, S., Tarafder, S.: Mech. Mater. **43**, 705 (2011)
Prasad Reddy, G.V., Kannan, R., Mariappan, K., Sandhya, R., Sankaran, S., Bhanu Sankara Rao, K.: Int. J. Fatigue **81**, 299 (2015)
Prasad, Y.V.R.K., Seshacharyulu, T.: Int. Mater. Rev. **43**, 243 (1998)
Rao, V.K., Taplin, D.M.R., Rao, P.R.: MTA **6**, 77 (1975)
Rao, K.B.S., Seetharaman, V., Mannan, S.L., Rodriguez, P.: High Temp. Mater. Process. (Lond.) **7**, 63 (1986)
Raturi, A., Biswas, K., Gurao, N.P.: Scripta Mater. **197**, 113809 (2021)
Ray, K.K.: MSF **736**, 192 (2012)
Ray, K.K., Das, D.: Mater. Sci. Technol. **33**, 340 (2017)
Ray, K.K., Mallik, A.K.: Mater. Sci. Eng. **59**, 59 (1983)
Ray, K.K., Mallik, A.K.: Trans. Indian Inst. Met. **37**, 426 (1984)
Ray, K.K., Mondal, D.: Acta Metall. Mater. **39**, 2201 (1991)
Ray, S.K., Bhaduri, A.K., Rodriguez, P.: J. Nucl. Mater. **200**, 70 (1993)
Ray, S.K., Sasikala, G., Rodriguez, P.: J. Mech. Behav. Mater. **5**, 377 (1994)
Ray, K.K., Roy, U., Ray, A., Kumari, M., Nath, A.: Theoret. Appl. Fract. Mech. **113**, 102948 (2021)
Ray, K.K.: High entropy alloys, 1st edn. In: Srivatsan, T.S., Gupta, M. (eds.) pp. 547–598. CRC Press, Boca Raton (2020)
Rodriguez, P.: Low-temperature deformation in hard HCP metals. Ph.D. thesis, Indian Institute of Science (IISc), Bangalore (1976)
Rodriguez, P.: Metall. Mat. Trans. A **35**, 2697 (2004)
Roy, M., Ray, K.K., Sundararajan, G.: Metall. Mat. Trans. A **32**, 1431 (2001)
Roy, H., Sivaprasad, S., Tarafder, S., Ray, K.K.: Eng. Fract. Mech. **76**, 1822 (2009)
Roy, M.: Surface Engineering for Enhanced Performance against Wear. Springer (2013)
Sahoo, S., Hiwarkar, V., Samajdar, I., Dey, G., Srivastav, D., Tiwari, R., Banerjee, S.: Scripta Mater. **56**, 963 (2007)
Saikrishna, C.N., Ramaiah, K.V., Paul, D., Bhaumik, S.K.: Acta Mater. **102**, 385 (2016)
Sakthivel, T., Vasudevan, M., Laha, K., Parameswaran, P., Chandravathi, K.S., Mathew, M.D., Bhaduri, A.K.: Mater. Sci. Eng. A **528**, 6971 (2011)
Sakthivel, T., Sasikala, G., Vasudevan, M.: Mater. Charact. **159**, 109988 (2020)
Samajdar, I., Ratchev, P., Verlinden, B., Aernoudt, E.: Acta Mater. **49**, 1759 (2001)
Sasikala, G., Mannan, S., Mathew, M., Rao, K.B.: Metall. Mater. Trans. A **31**, 1175 (2000)
Sastry, D.H., Vasu, K.I.: Acta Metall. **20**, 399 (1972)
Sastry, D.H., Prasad, Y.V.R.K., Vasu, K.I.: Acta Metall. **17**, 1453 (1969)
Schuh, C., Hufnagel, T., Ramamurty, U.: Acta Mater. **55**, 4067 (2007)
Sen, I., Gopinath, K., Datta, R., Ramamurty, U.: Acta Mater. **58**, 6799 (2010)

Shankar, V., Mariappan, K., Sandhya, R., Laha, K., Bhaduri, A.K.: Int. J. Fatigue **98**, 259 (2017)
Sharma, A.S., Yadav, S., Biswas, K., Basu, B.: Mater. Sci. Eng. R. Rep. **131**, 1 (2018)
Shashank Dutt, B., Nani Babu, M., Venugopal, S., Sasikala, G., Bhaduri, A.K.: Mater. Sci. Technol. **27**, 1527 (2011)
Singh, V., Rao, M., Rao, P.: Int. J. Fatigue **2**, 12 (1980)
Singh, P.K., Chattopadhyay, J., Kushwaha, H.S., Tarafder, S., Ranganath, V.R.: Int. J. Press. Vessels Pip. **75**, 271 (1998)
Singh, P.K., Ranganath, V.R., Tarafder, S., Prasad, P., Bhasin, V., Vaze, K.K., Kushwaha, H.S.: Int. J. Press. Vessels Pip. **80**, 745 (2003)
Singh, V., Rama Rao, P., Taplin, D.M.R.: J. Mater. Sci. **8**, 373 (1973)
Sinha, S., De, P.S., Mukherjee, S., Kundu, A., Chakraborti, P.C., Bhattacharya, B., Shome, M., Bhattacharjee, D.: Fatigue Fract. Engng. Mater. Struct. **39**, 110 (2016)
Sivaprasad, S., Tarafder, S., Tarafder, M., Ray, K.K.: Int. J. Fatigue **22**, 593 (2000a)
Sivaprasad, S., Tarafder, S., Ranganath, V., Ray, K.K.: Mater. Sci. Eng. A **284**, 195 (2000b)
Sreenivasan, P.R., Ray, S.K., Mannan, S.L., Rodriguez, P.: J. Nucl. Mater. **228**, 338 (1996)
Srinivas, M., Malakondaiah, G., Armstrong, R.W., Rama Rao, P.: Acta Metall. Mater. **39**, 807 (1991)
Sudhakar Rao, G., Chakravartty, J.K., Saibaba, N., Mahobia, G.S., Chattopadhyay, K., Santhi Srinivas, N.C., Singh, V.: Mater. Sci. Eng. A **603**, 114 (2014)
Sujith Kumar, S., Kumar, I.A., Marandi, L., Sen, I.: Acta Mater. **201**, 303 (2020)
Sundararaman, M., Mukhopadhyay, P., Banerjee, S.: Acta Metall. **36**, 847 (1988)
Sunder, R.: Int. J. Fatigue **7**, 3 (1985)
Suresh, K., Rao, K.P., Prasad, Y.V.R.K., Hort, N., Kainer, K.U.: Prod. Manuf. Res. **2**, 241 (2014)
Swaminathan, J., Raghavan, S.: High Temp. Mater. Process. (Lond.) **13**, 277 (1994)
Tarafder, S., Tarafder, M., Ranganath, V.R.: Eng. Fract. Mech. **47**, 901 (1994)
Valluri, S.R., Raju, K.N., Rama Rao, P.: Progress in Fracture Mechanics, pp. 71–83. Elsevier (1983)
Venugopal, S., Mannan, S.L., Prasad, Y.V.R.K.: MMTA **27**, 119 (1996)
Wahi, R.P., Rao, P.R., Anantharaman, T.R.: Scr. Metall. **5**, 97 (1971)

Chapter 32
Texture Research in the Past 75 Years: Historical Perspective and a Personal Journey

Ranjit Kumar Ray

Abstract During the processing of materials such as solidification, deformation or annealing, they acquire crystallographic textures that are unique to those processing steps, which are characteristic of the given material. If the crystallographic orientation of each grain is different from all other grains, the material is said to be 'randomly oriented', and if one or more orientations are preferred by most of the grains, then the material is said to be 'textured'. As texture affects the properties of materials, it is essential to understand its role in the behaviour of materials under different processing conditions. Sustained research in India on the texture of materials, for the past five decades put the country on the Texture map of the world. This article is an attempt to reveal the fascinating story of texture research in India, with special reference to the journey taken by the author.

Keywords Texture · Research · Characterisation · Properties

32.1 Introduction

India has a rich tradition and heritage in the field of the metallurgical sciences. The famous 'Iron Pillar' in New Delhi, built sometime in the third century AD, during the Gupta period, stands testimony to ancient India's mastery over the science and engineering of metallurgy. The urge as well as the need to highlight the old metallurgical glory of the country, and also to spread the message to continue this rich tradition down to the younger generations, led to the establishment of the Indian Institute of Metals (IIM) in the city of Calcutta on the 29th of December 1947, the year India attained her independence from the British. Since its inception, the IIM has been contributing immensely towards promoting the activities of Indian metallurgists in industry, academia, research and development as well as in the government sector. When such an institute completes 75 years of its glorious existence, the event becomes

R. K. Ray (✉)
Formerly, Professor of Materials & Metallurgical Engineering IIT, Kanpur, India
e-mail: rkray@iitk.ac.in; rkray1942@gmail.com

R. Divakar et al. (eds.), *Indian Metallurgy*, Indian Institute of Metals Series,
https://doi.org/10.1007/978-981-99-5060-7_32

a matter of national pride and celebration. This becomes the appropriate time to look back and reminisce about the long and eventful journey of the Institute, which may help in charting out the way forward and the future of Indian metallurgy, in the years to come. It is in this context that the present article has been written, in order to recall the beginning of systematic research in the area of Crystallographic Texture in India, some 50 years ago, in a very humble way and which, with time, has grown significantly to place India, solidly and firmly, on the texture map of the world. This fascinating story will be depicted in the next few pages to give a historical perspective of research in texture worldwide, with special reference to the tryst of the author with texture research in India for the past five decades.

32.2 Texture: Definition Measurement and Representation

Any polycrystalline material is made up of a large number of grains, separated from one another by grain boundaries. Each grain, in its turn, is nothing but a single crystal. If in a piece of material, the crystallographic orientation of each grain is distinctly different from those of all the other grains, the material is called 'random', having isotropic properties. If, on the other hand, one or more orientations are preferred by most of the grains, then the material will be called 'textured'. It is very interesting to note that in this world, it is impossible to find any material which is perfectly random. In other words, all the materials, under the sun, are textured, with varying degrees of sharpness. The grain orientations in a highly 'random' and a highly 'textured' material have been illustrated in Fig. 32.1a and b, respectively. It must be emphasized here that when we talk about an orientation being preferred by a large number of grains in a material, it does not mean that all those grains have exactly the same orientation, but that the orientations are very similar, differing by a few degrees only. Whenever a material is subjected to any kind of processing, whether solidification, deformation or annealing, the material assumes textures that are unique for those processing steps, and also characteristic of that material. Now, the next question that comes to our mind is how to define an 'orientation'. In the context of this paper, 'orientation' will be defined in terms of a rolled sheet material, characterized by the three mutually perpendicular directions, the Rolling Direction (RD), the Transverse Direction (TD) and the Normal Direction (ND), which is perpendicular to the Rolling Plane. Since the majority of metallic materials are used in the form of sheets, this paper will be concerned with the texture of sheet materials only. Figure 32.2 illustrates how the orientation of a grain in a sheet material is defined. If the $(h\ k\ l)$ plane of a grain in the sheet lies parallel to the rolling plane and if the $[u\ v\ w]$ direction of the grain lies parallel to the rolling direction, then the orientation of the grain is written down as $(h\ k\ l)\ [u\ v\ w]$. Figure 32.2 shows the orientations of several grains in the sheet of a cubic material. Texture, as has been defined above, essentially refers to a statistical measure of the orientations of the constituent grains in a material, without any regard to the spatial location of any particular grain or an aggregate of grains. Therefore, this definition of texture essentially means the average texture

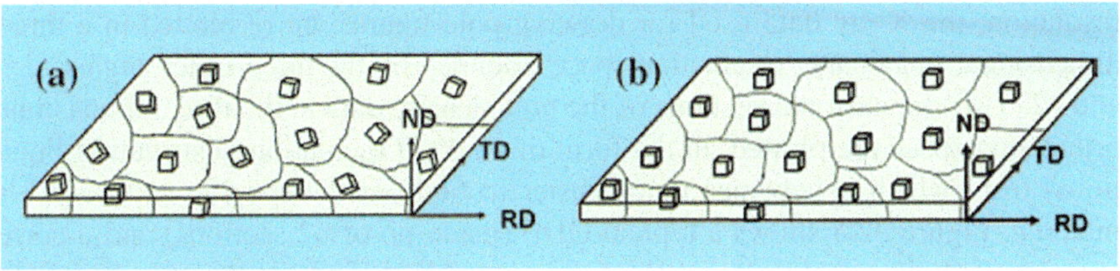

Fig. 32.1 Grain orientations in **a** a random material and **b** a highly textured material

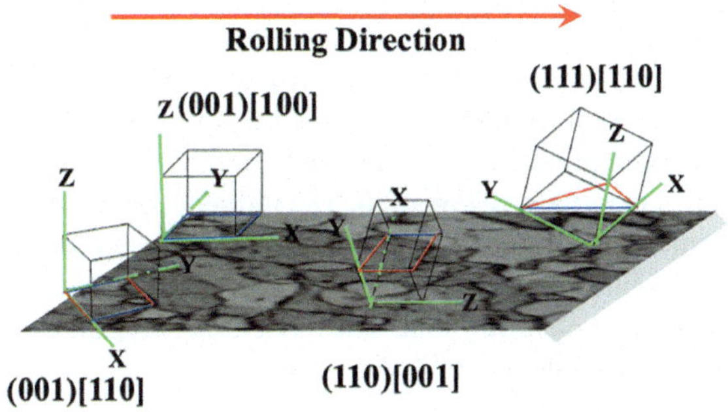

Fig. 32.2 Definition of an orientation in a rolled sheet material

of the material or the 'macrotexture' (or global texture). Sometimes it is necessary to measure texture from a much smaller population of grains in a polycrystalline material, in a localized region. The resultant measured texture pertains only to the chosen ensemble of grains and is not at all representative of the average texture of the extended material. This is known as 'micro-texture' (or local texture).

The mainstream techniques of texture measurement are shown in Fig. 32.3. Available techniques for both 'macrotexture' and 'micro-texture' measurements are indicated in this figure. Out of all these, the most popular method for 'macrotexture' measurement is X-ray diffraction, while that for 'micro-texture' measurement, it is the Scanning Electron Microscope (SEM), coupled with the Electron Back Scattered Diffraction (EBSD) attachment. The experimental arrangements for both these techniques are shown in Fig. 32.4. A detailed description of the measurement techniques can be found in Suwas and Ray (2014). In the X-ray diffraction method, the sample, in sheet form, is placed on the specimen holder, for diffraction to occur from a particular $(h\ k\ l)$ plane of the sample. Keeping the diffraction condition fixed, the sample is given a number of specific rotations to cover as many grains or crystals as possible, in the sample, to give rise to the total diffracted radiation. The intensities of these diffracted radiations, which are proportional to the density of the particular $(h\ k\ l)$ plane, are plotted in the form of diagrams, known as 'pole-figures'. A pole figure essentially shows a two- dimensional distribution of 'poles' or plane-normals to the $(h\ k\ l)$ planes in the different grains of the sample. In fact, it is nothing but a stereographic projection of the $(h\ k\ l)$ plane-normals, on which the sample parameters (RD, TD and ND, for a sheet sample) are superimposed. In order to have a better resolution, the X-ray data used for drawing pole figures are re-plotted in a three-dimensional imaginary 'orientation space', defined by the three Euler angles ø1, ø and ø2. Two-dimensional sections of the pole density data in the three-dimensional orientation space are plotted, in the form of ODFs (Orientation Distribution Functions) from which it becomes much easier to find out the specific texture of the material. Figure 32.5 shows a typical ODF (made up of ø2 sections) and a corresponding pole figure plot, from which it is possible to find out the specific texture a material possesses. All details regarding texture representation can be found in Suwas and Ray (2014).

The question that has very often been asked is why the study of textures of materials is so important. The reason is that texture affects all materials' properties, physical, chemical, mechanical, electrical, magnetic and whatnot. Just to give some idea, the Lankford parameter r, which is so important during deep-drawing operations, has been plotted against the angle to the rolling direction, θ, for differently textured sheet materials, in Fig. 32.6. The influence of texture on this parameter r, which is so important for mechanical engineers and materials scientists, is very apparent from this figure. Many such examples could have been cited here, but the limitation on space will not allow that. The reader may consult (Suwas and Ray 2014) to get more insight into these aspects.

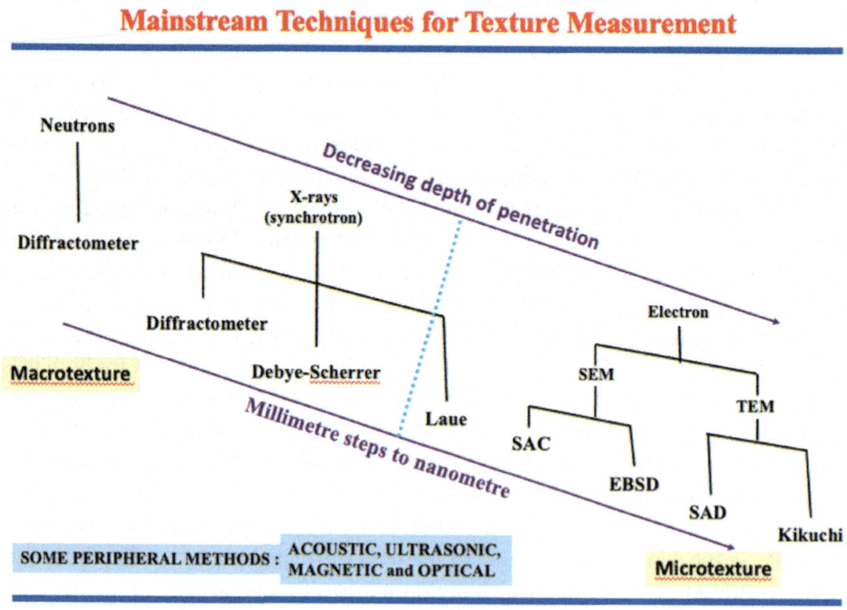

Fig. 32.3 Available techniques of texture measurement

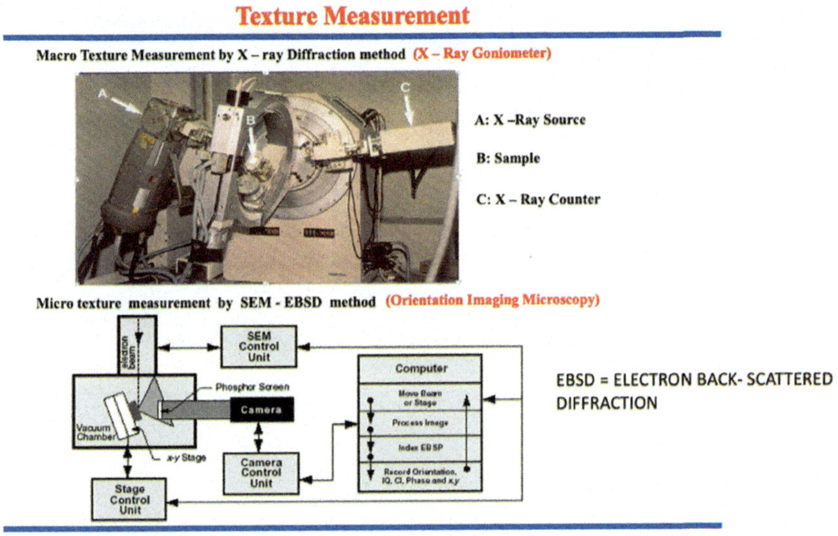

Fig. 32.4 Common techniques of macro- and micro-texture measurement

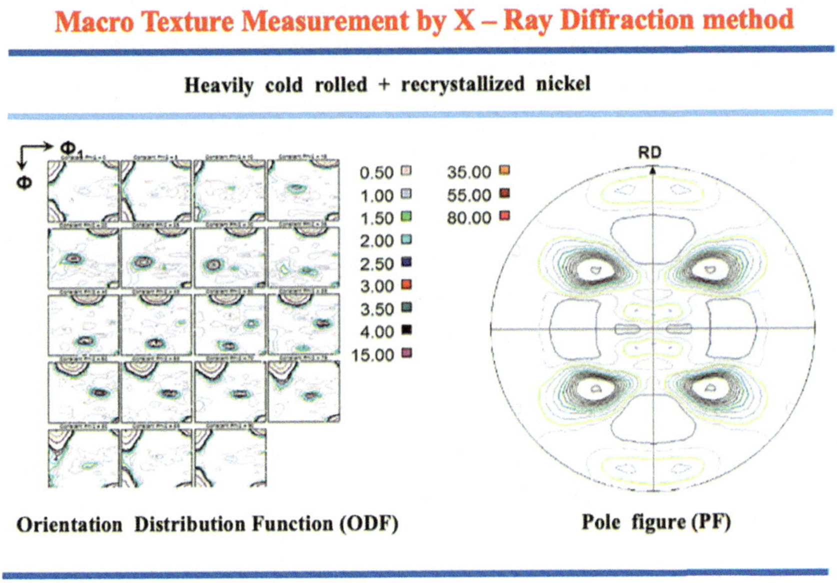

Fig. 32.5 ODF and pole figure plots for a heavily cold-rolled and recrystallized pure nickel

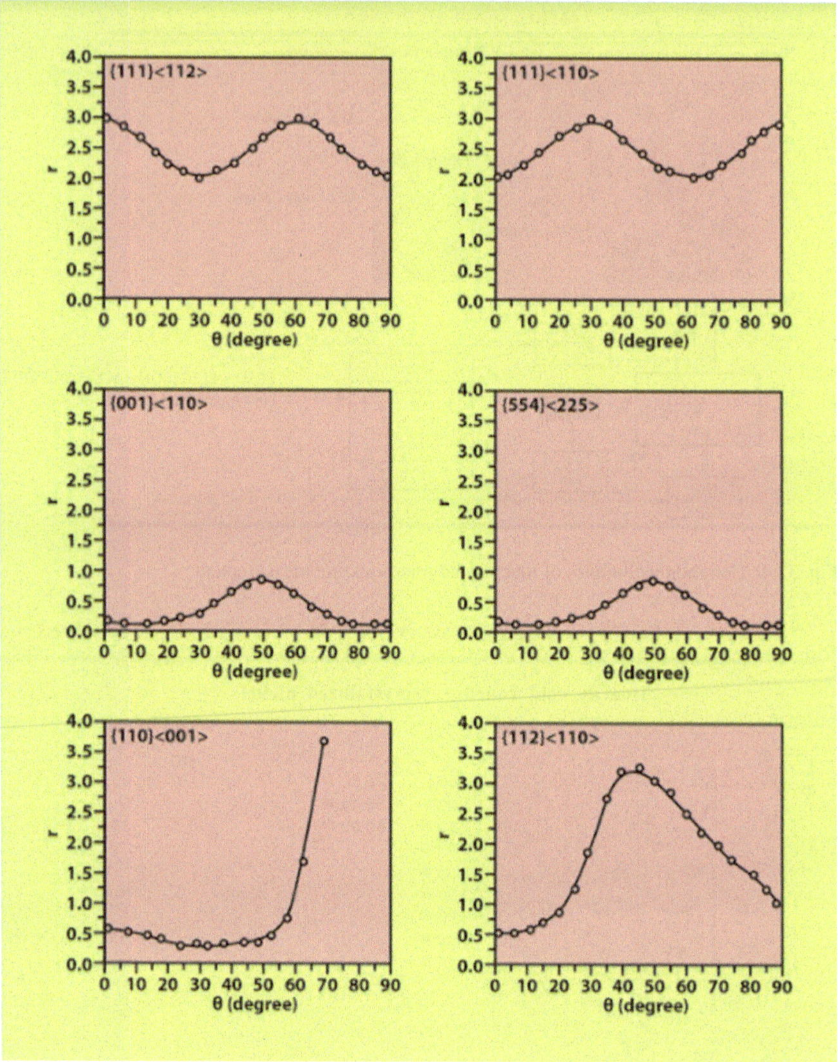

Fig. 32.6 Plot of r (Lankford parameter) versus angle to the rolling direction, θ, for differently textured sheet materials

32.3 Historical Perspective: The International Scene

It is generally believed that the idea of using crystallographic texture was first mooted by geologists, who studied the deformation of the fault-related crystalline rock minerals present along the 'San Andreas Fault' zone in Western California. They realized that the study of the texture of these rock minerals, which deform by continuous aseismic creep and micro-seismicity, could contribute to the understanding of the mechanism of their deformation, enabling the prediction of future earthquakes in the region. Mathematicians and Physicists contributed immensely to developing the theoretical basis of such measurements and now the materials scientists and engineers are the biggest beneficiaries of this characterization technique. Historically speaking, the first pole figure was generated by Wever in Berlin in 1924, from evaluating the inhomogeneous intensity distribution along the Debye–Scherrer rings in X-ray powder photography. The first texture goniometer was introduced by Decker et al. in 1948. The next year Schulz initiated the modern quantitative X-ray texture analysis method. Further development of pole figure determination by neutron diffraction started in the 1960s. The mathematical basis of ODF determination and related methods were introduced by R. J. Roe, R. O. Williams and H. J. Bunge during 1966–1969. Nowadays highly automated computer-controlled systems and corresponding measuring techniques are well established.

In the micro-texture measurement front, in 1928 Kikuchi first observed the 'Kikuchi lines' from thick single crystals during diffraction in a Transmission Electron Microscope (TEM). The Kikuchi line pattern, produced by diffraction from a particular grain, gives an idea of the exact orientation of that grain. The SEM-EBSD method is based on collecting all the Kikuchi line patterns from the grains in the selected area of the sample, from which the texture can be determined. This was followed by the observation of 'back-reflected Kikuchi lines' in a TEM by Alam. Between 1969 and 1979, 3 diffraction techniques, namely, Selected Area Channeling Pattern (SACP), Kossel diffraction and Electron Back Scattered Diffraction (EBSD) were developed. During the period 1982–1984, the first computer-assisted indexing of EBSD was introduced. Fully automated indexing of EBSD patterns became available from 1990 onwards. Very fast automated pattern analysis started in the year 2000.

Side by side, with the development of the experimental techniques, a lot of effort was directed towards theoretical texture modelling, starting from 1928. As a result, many satisfactory models are now available to explain the texture developed during the deformation of materials, although texture models of the recrystallization and phase transformation processes are still not that successful.

The systematic study of texture first started at the Kaiser Wilhelm Institute in Berlin in the 1920s. Several scientists, Polyani, Schmid, Burgers and Wassermann were associated with it. Professor H. J. Bunge, also known as the father of modern texture studies, established a world-renowned Texture laboratory in the Institute of Physical Metallurgy and Metal Physics at the Technical University of Clausthal in Germany during 1976–1997, when he was the Director of that institute. During almost the same time (1958–1989), Professor Kurt Luecke developed an excellent texture research centre, as Director of the Institute of Physical Metallurgy and Metal Physics, at the Technical University of Aachen, in Germany. Now, crystallographic texture is studied routinely in many academic and research institutes as well as industrial research laboratories throughout the world. The importance of crystallographic texture can be visualized from the fact that an International Conference on Textures

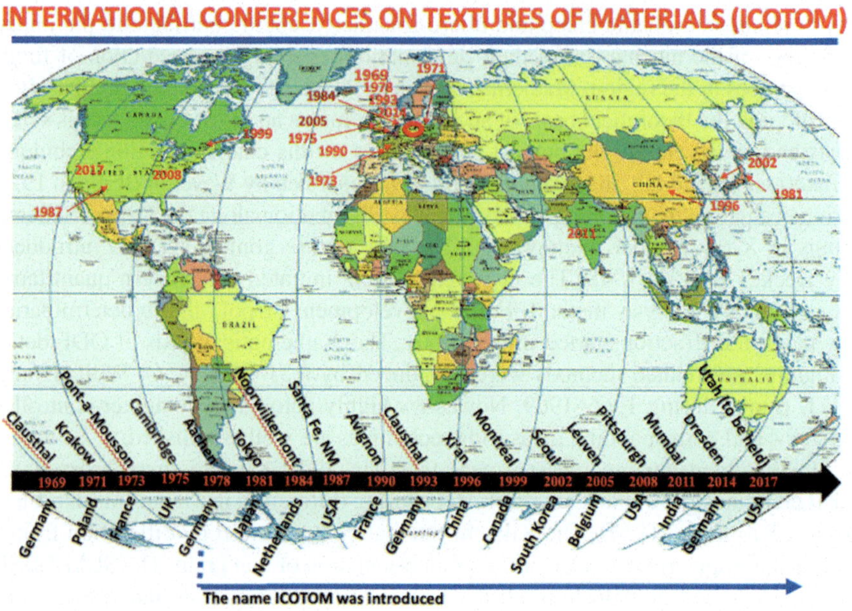

Fig. 32.7 Venues of ICOTOM conferences

of Materials (ICOTOM) is held once every three years, in different parts of the world. The ICOTOM conferences, held to date and their venues can be seen in Fig. 32.7.

32.4 My Tryst with Texture: The Indian Perspective

The first time when systematic texture studies and research started in India was in 1977, when the author joined IIT Kanpur as a faculty member in the Materials and Metallurgical Engineering Department. He had the opportunity of working on the deformation and recrystallization textures of copper and copper-base alloys during his Ph.D. days at the University of Birmingham during 1970–1973. In Kanpur, he started working with a few students, using a primitive texture goniometer that was available in the department. Many of his former students later distinguished themselves, by dint of their own merit, and set up Texture laboratories in a number of IITs, national institutes and industrial research laboratories. The years 1997–1998 witnessed two important developments that made a turnaround in Indian Texture research. The first-ever conference dedicated to Texture in India (NASAT-97) was organized at DMRL Hyderabad in 1997 by the author and a few of his professional colleagues. The following year, a 'National Facility on Texture & OIM' was established at IIT Bombay with generous support from the DST, DAE and DRDO. The first ICOTOM in India was held in IIT Bombay in 2011. Now, when we look back,

Fig. 32.8 Texture centres in India in 1977

it is very satisfying to note that what started as a single person's effort in 1977 at IIT Kanpur blossomed with time, culminating in the establishment of a large number of research centres in India, putting India firmly in the elite group of countries where high-quality Texture research, experimental as well as modelling, is carried out on a regular basis. The enormous stride in this direction, during the 40 years between 1977 and 2017, can be visualized in Figs. 32.8 and 32.9.

32.5 Conclusions

Now, 50 years after my tryst with texture began, and at the completion of glorious 75 years of the establishment of the Indian Institute of Metals, anyone will be curious to ask: what next? Where is Texture going from here? What is going to happen in the future in this area? To answer these questions, it must be emphasized that our primary job, as materials scientists, is to develop materials with specific properties. As we are very much aware, the property of a material is a function of four different parameters, namely, microstructure, texture, grain boundaries and internal stress. Therefore, in order to develop some desired property, one has to look into all these four aspects and their inter-relationships. Future research in this area must comprise a well-coordinated approach to develop holistic materials models, based on texture and the three other parameters, mentioned above. This will benefit a large number

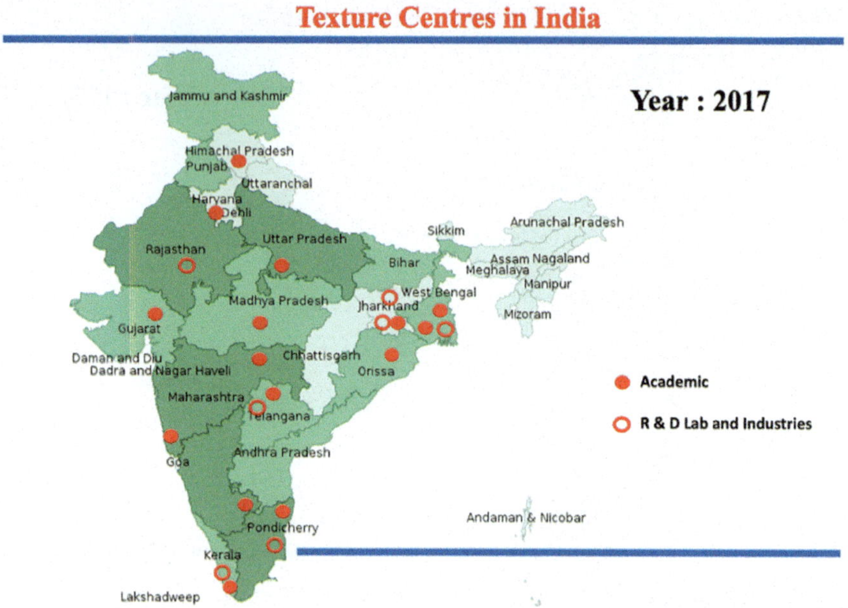

Fig. 32.9 Texture centres in India in 2017

of materials scientists and engineers, who are engaged in the task of enhancing the properties of conventional materials and also in developing newer and better materials possessing unique and novel properties. The 'Property Tetrahedron' (Fig. 32.10) beautifully illustrates this concept.

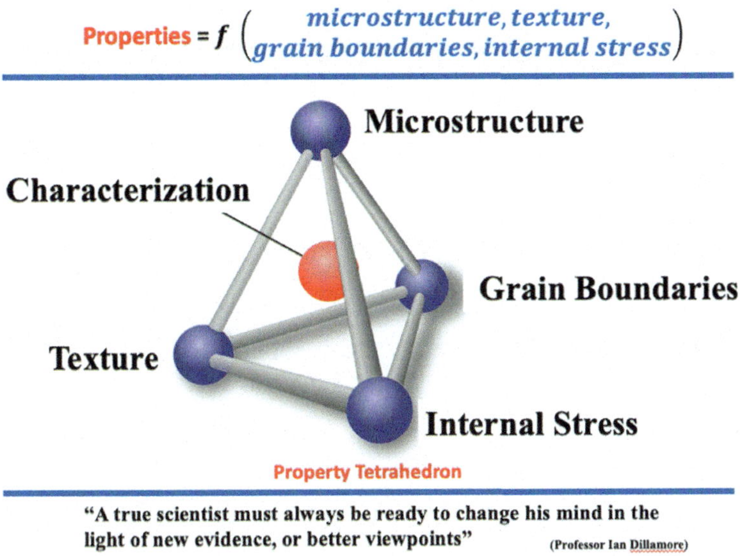

Fig. 32.10 The 'property tetrahedron'

Acknowledgements The author gratefully acknowledges the help rendered by Professor Satyam Suwas and Dr. Nitish Bibhansu in the preparation of the manuscript.

Reference

Suwas, S., Ray, R.K.: Crystallographic Texture of Materials. Springer, London (2014)

Correction to: Nuclear Fuel Complex—A Five Encade Success Story of Indian Nuclear Power Program

Komal Kapoor and Dinesh Srivastava

Correction to: Chapter 17 in: R. Divakar et al. (eds.), *Indian Metallurgy*, Indian Institute of Metals Series, https://doi.org/10.1007/978-981-99-5060-7_17

In the original version of the book, the chapter author name "Komal Kapoor" was updated as corresponding author correctly in Chapter 17. The book and the chapter have been updated with the change.

The updated version of this chapter can be found at
https://doi.org/10.1007/978-981-99-5060-7_17

R. Divakar et al. (eds.), *Indian Metallurgy*, Indian Institute of Metals Series,
https://doi.org/10.1007/978-981-99-5060-7_33

The manufacturer's authorised representative in the EU is Springer Nature Customer Service Centre GmbH, Europaplatz 3, 69115 Heidelberg, Germany. If you have any concerns regarding our products, please contact ProductSafety@springernature.com

Printed and bound by CPI Group (UK) Ltd, Croydon, CR0 4YY
01/10/2026
02225872-0016